T0300011

Computational
Systems Biology of
Cancer

CHAPMAN & HALL/CRC
Mathematical and Computational Biology Series

Aims and scope:

This series aims to capture new developments and summarize what is known over the entire spectrum of mathematical and computational biology and medicine. It seeks to encourage the integration of mathematical, statistical, and computational methods into biology by publishing a broad range of textbooks, reference works, and handbooks. The titles included in the series are meant to appeal to students, researchers, and professionals in the mathematical, statistical and computational sciences, fundamental biology and bioengineering, as well as interdisciplinary researchers involved in the field. The inclusion of concrete examples and applications, and programming techniques and examples, is highly encouraged.

Series Editors

N. F. Britton
Department of Mathematical Sciences
University of Bath

Xihong Lin
Department of Biostatistics
Harvard University

Hershel M. Safer
School of Computer Science
Tel Aviv University

Maria Victoria Schneider
European Bioinformatics Institute

Mona Singh
Department of Computer Science
Princeton University

Anna Tramontano
Department of Biochemical Sciences
University of Rome La Sapienza

Proposals for the series should be submitted to one of the series editors above or directly to:
CRC Press, Taylor & Francis Group
4th, Floor, Albert House
1-4 Singer Street
London EC2A 4BQ
UK

Published Titles

Algorithms in Bioinformatics: A Practical Introduction
Wing-Kin Sung

Bioinformatics: A Practical Approach
Shui Qing Ye

Biological Computation
Ehud Lamm and Ron Unger

Biological Sequence Analysis Using the SeqAn C++ Library
Andreas Gogol-Döring and Knut Reinert

Cancer Modelling and Simulation
Luigi Preziosi

Cancer Systems Biology
Edwin Wang

Cell Mechanics: From Single Scale-Based Models to Multiscale Modeling
Arnaud Chauvière, Luigi Preziosi, and Claude Verdier

Clustering in Bioinformatics and Drug Discovery
John D. MacCuish and Norah E. MacCuish

Combinatorial Pattern Matching Algorithms in Computational Biology Using Perl and R
Gabriel Valiente

Computational Biology: A Statistical Mechanics Perspective
Ralf Blossey

Computational Hydrodynamics of Capsules and Biological Cells
C. Pozrikidis

Computational Neuroscience: A Comprehensive Approach
Jianfeng Feng

Computational Systems Biology of Cancer
Emmanuel Barillot, Laurence Calzone, Philippe Hupe, Jean-Philippe Vert, and Andrei Zinovyev

Data Analysis Tools for DNA Microarrays
Sorin Draghici

Differential Equations and Mathematical Biology, Second Edition
D.S. Jones, M.J. Plank, and B.D. Sleeman

Dynamics of Biological Systems
Michael Small

Engineering Genetic Circuits
Chris J. Myers

Exactly Solvable Models of Biological Invasion
Sergei V. Petrovskii and Bai-Lian Li

Gene Expression Studies Using Affymetrix Microarrays
Hinrich Göhlmann and Willem Talloen

Genome Annotation
Jung Soh, Paul M.K. Gordon, and Christoph W. Sensen

Glycome Informatics: Methods and Applications
Kiyoko F. Aoki-Kinoshita

Handbook of Hidden Markov Models in Bioinformatics
Martin Gollery

Introduction to Bioinformatics
Anna Tramontano

Introduction to Bio-Ontologies
Peter N. Robinson and Sebastian Bauer

Introduction to Computational Proteomics
Golan Yona

Introduction to Proteins: Structure, Function, and Motion
Amit Kessel and Nir Ben-Tal

An Introduction to Systems Biology: Design Principles of Biological Circuits
Uri Alon

Kinetic Modelling in Systems Biology
Oleg Demin and Igor Goryanin

Knowledge Discovery in Proteomics
Igor Jurisica and Dennis Wigle

Meta-analysis and Combining Information in Genetics and Genomics
Rudy Guerra and Darlene R. Goldstein

Methods in Medical Informatics: Fundamentals of Healthcare Programming in Perl, Python, and Ruby
Jules J. Berman

Published Titles (continued)

Chapman & Hall/CRC Mathematical and Computational Biology Series

Computational Systems Biology of Cancer

Emmanuel Barillot   Laurence Calzone

Philippe Hupé   Jean-Philippe Vert

Andrei Zinovyev

CRC Press
Taylor & Francis Group
Boca Raton   London   New York

CRC Press is an imprint of the
Taylor & Francis Group, an **informa** business

A CHAPMAN & HALL BOOK

CRC Press
Taylor & Francis Group
6000 Broken Sound Parkway NW, Suite 300
Boca Raton, FL 33487-2742

© 2013 by Taylor & Francis Group, LLC
CRC Press is an imprint of Taylor & Francis Group, an Informa business

No claim to original U.S. Government works

Version Date: 20120615

International Standard Book Number: 978-1-4398-3144-1 (Hardback)

Library of Congress Cataloging-in-Publication Data

Computational systems biology of cancer / authors, Emmanuel Barillot ... [et al.].
    p. ; cm. -- (Chapman & Hall/CRC mathematical and computational biology series)
    Includes bibliographical references and index.
    ISBN 978-1-4398-3144-1 (hardcover : alk. paper)
    I. Barillot, Emmanuel. II. Series: Chapman & Hall/CRC mathematical and
computational biology series (Unnumbered)
    [DNLM: 1. Computational Biology--methods. 2. Neoplasms--therapy. QZ 26.5]

362.19699'400285--dc23   2012021572

Visit the Taylor & Francis Web site at
http://www.taylorandfrancis.com

and the CRC Press Web site at
http://www.crcpress.com

Contents

List of Figures

List of Tables

List of Acronyms

aCGH	array Comparative Genomic Hybridisation
ANOVA	Analysis of Variance
APC	Anaphase Promoting Complex
API	Application Programming Interface
ATP	Adenosine-5'-Triphosphate
BAC	Bacterial Artificial Chromosome
BAF	B Allele Frequency
BER	Base Excision Repair
BioPAX	Biological Pathway Exchange
caBIG	cancer Biomedical Informatics Grid
CCA	Canonical Correlation Analysis
CDK	Cyclin Dependent Kinase
CGH	Comparative Genomic Hybridisation
ChIA	Chromatin Interaction Analysis
ChIP	Chromatin Immunoprecipitation
CIN	Chromosomal Instability
CML	Chronic Myeloid Leukaemia
CNV	Copy Number Variant
Co-IP	Co-Immunoprecipitation
ComBat	Combating Batch Effects When Combining Batches of Gene Expression Microarray Data
CORG	Condition-Responsive Genes
COSMIC	Catalogue Of Somatic Mutations In Cancer
CSC	Cancer Stem Cell
CTC	Circulating Tumour Cell
CTS	Circadian Timing System
DDBJ	DNA Data Bank of Japan
DIP	Database of Interacting Proteins
DISC	Death-Inducing Signalling Complex
DLDA	Diagonal Linear Discriminant Analysis
DNA	DeoxyriboNucleic Acid
DWD	Distance-Weighted Discrimination
EBI	European Bioinformatics Institute
ECM	ExtraCellular Matrix
ELISA	Enzyme-Linked ImmunoSorbent Assay

EM	Expectation Maximisation
EMBL	European Molecular Biology Laboratory
EMT	Epithelial-to-Mesenchymal Transition
ENCODE	ENCyclopedia Of DNA Elements
ES	Enrichment Score
ESI	Electrospray Ionisation
ESS	Error Sum of Squares
FACS	Fluorescence-Activated Cell Sorting
FBA	Flux Balance Analysis
FC	Fold Change
FDA	Food and Drug Administration
FDR	False Discovery Rate
FFL	Feed Forward Loop
FGED	Functional GEnomics Data
5C	Carbon-Copy Chromosome Conformation Capture
4C	Circularised Chromosome Conformation Capture
FREEC	control-FREE Copy number caller
FWER	Family-Wise Error Rate
GEO	Gene Expression Omnibus
GNEA	Gene Network Enrichment Analysis
GO	Gene Ontology
GPCR	G-Protein Coupled Receptor
GSEA	Gene Set Enrichment Analysis
GWAS	Genome-Wide Association Studies
HCS	High-Content Screening
HMM	Hidden Markov Model
HOT	Highly-Optimised Tolerance
HPC	High-Computing Performance
HPRD	Human Protein Reference Database
HR	Homologous Recombination
HSR	Homogeneously Stained Region
HTS	High-Throughput Screening
HUGO	HUman Genome Organisation
ICA	Independent Component Analysis
ICAT	Isotope-Coded Affinity Tag
ICGC	International Cancer Genome Consortium
IP	ImmunoPrecipitation
ITALICS	ITerative and Alternative normaLIsation and Copy number calling for affymetrix Snp arrays
iTRAQ	isobaric Tag for Relative and Absolute Quantitation
KEGG	Kyoto Encyclopaedia of Genes and Genomes
KiSAO	Kinetic Simulation Algorithm Ontology

kNN	k-Nearest Neighbour
LC	Liquid Chromatography
LCM	Laser-Capture Microdissection
LDA	Linear Discriminant Analysis
LIMS	Laboratory Information Management System
LLE	Locally Linear Embedding
LOH	Loss of Heterozygosity
LR	Logistic Regression
LRR	Log Reference Ratio
MALDI	Matrix-Assisted Laser Desorption/Ionisation
MANOR	MicroArray NORmalisation
MAPK	Mitogen-Activated Protein Kinase
MAPKK	Mitogen-Activated Protein Kinase Kinase
MAPKKK	Mitogen-Activated Protein Kinase Kinase Kinase
MAQC	MicroArray Quality Control
MDE	Matrix Degrading Enzyme
MFA	Multiple Factor Analysis
MIAME	Minimum Information About a Microarray Experiment
MIAME	Minimum Information About a Proteomics Experiment
MIASE	Minimum Information About a Simulation Experiment
MIN	Microsatellite Instability
MINT	Molecular INTeraction database
MIRIAM	Minimum Information Required in the Annotation of Models
miRNA	microRNA
MKL	Multiple Kernel Learning
MMP	Matrix MetalloProteinase
MMR	Mismatch Repair
MPI	Message Passing Interface
MPPI	Mammalian Protein-Protein Interaction database
MRI	Magnetic Resonance Imaging
mRNA	messenger RNA
MS	Mass Spectrometry
MS/MS	Mass Spectrometry/Mass Spectrometry (tandem Mass Spectrometry)
M2H	Mammalian Two-Hybrids
NB	Naive Bayes
NC	Nearest Centroid
NCA	Network Component Analysis
NCBI	National Center for Biotechnology Information
NCI	National Cancer Institute
ncRNA	noncoding RNA

NGS	Next Generation Sequencing
NHGRI	National Human Genome Research Institute
NIH	National Institutes of Health
NMF	Nonnegative Matrix Factorization
OBO	Open Biological and biomedical Ontologies
ODE	Ordinary Differential Equation
OMIM	Online Mendelian Inheritance in Man
OpenMP	Open Multi-Processing
OWL	Web Ontology Language
PA	Process Algebra
PCA	Principal Component Analysis
PCR	Polymerase Chain Reaction
PDE	Partial Differential Equation
PDGFR	Platelet-Derived Growth Factor Receptor
PET	Paired-End Tag
PLDE	Piecewise-Linear Differential Equation
POI	Protein of Interest
PPI	Protein-Protein Interaction
PSI-MI	Proteomics Standards Initiative Molecular Interaction
PTM	Post-Translational Modification
QDA	Quadratic Discriminant Analysis
RDF	Resource Description Framework
RF	Random Forest
RNA	Ribonucleic Acid
ROS	Reactive Oxygen Species
RPPA	Reverse-Phase Protein Array
RTK	Receptor Tyrosine Kinase
SAM	Significance Analysis of Microarray
SBGN	Systems Biology Graphical Notation
SBML	Systems Biology Markup Language
SBO	Systems Biology Ontology
SDE	Stochastic Differential Equation
SED-ML	Simulation Experiment Description Markup Language
SGA	Synthetic Genetic Array
SILAC	Stable Isotope Labelling with Amino acids in Cell culture
siRNA	small interfering RNA
SME	Stochastic Master Equation
SMRT	Single Molecule Real Time
SNP	Single Nucleotide Polymorphism
SOC	Self-Organising Criticality

SOLiD	Sequencing by Oligonucleotide Ligation and Detection
SOM	Self-Organising Map
SQL	Structured Query Language
SSS	Structural Stability Score
SVM	Support Vector Machine
TAP	Tandem Affinity Purification
TCA	Tricarboxylic Acid Cycle
TCGA	The Cancer Genome Atlas
TEDDY	TErminology for the Description of Dynamics
3C	Chromosome Conformation Capture
TV	Total Variation
UAS	Upstream specific Activation Sequence
UCSC	University of California Santa Cruz
UniProtKB	Universal Protein Resource Knowledgebase
VRML	Virtual Reality Markup Language
XML	eXtensible Markup Language
Y2H	Yeast Two-Hybrid
Y3H	Yeast Three-Hybrid
ZMW	Zero-Mode Waveguide

Preface

Cancer is a complex and heterogeneous disease that exhibits high levels of robustness against various therapeutic interventions. It is a constellation of diverse and evolving disorders that are manifested by the uncontrolled proliferation of cells that may eventually lead to fatal dysfunction of the host system. Although some of the cancer subtypes can be cured by early diagnosis and specific treatment, no effective treatment is yet established for a significant portion of cancer subtypes. In industrial countries where the average life expectancy is high, cancer is one of the major causes of death. Any contribution to an in-depth understanding of cancer shall eventually lead to better care and treatment for patients. Due to the complex, heterogeneous, and evolving nature of cancer, it is essential for a system-oriented view to be adopted for an in-depth understanding.

The question is how to achieve an in-depth yet realistic understanding of cancer dynamics. Although large-scale experiments are now being deployed, there are practical limitations of how much they do to convey the reality of cancer pathology and progression within the patient's body. Computational approaches with system-oriented thinking may complement the limitations of an experimental approach. Computational studies not only provide us with new insights from large-scale experimental data, but also enable us to perceive what are the conceivable characteristics of cancer under certain assumptions. It is an engine of thoughts and proving grounds of various hypotheses on how cancer may behave as well as how molecular mechanisms work within anomalous conditions. It is not just computing that helps us fight against cancer, but a computational approach has to be combined with a proper theoretical framework that enables us to perceive "cancer" as complex dynamical and evolvable systems that entail a robust yet fragile nature. This recognition shifts our attention from the magic bullet approach of anti-cancer drugs to a more systematic control of cancer as complex dynamical phenomena. This leads to the view that a complex system has to be controlled by complex interventions. To understand such a system and design complex interventions, it is essential that we combine experimental and computational approaches. Thus, computational systems biology of cancer is an essential discipline for cancer biology and is expected to have major impacts for clinical decision-making.

This is the first book specifically focused on computational systems biology of cancer with a coherent and proper vision on how to tackle this formidable challenge. I would like to congratulate the authors for their vision and dedication.

Hiroaki Kitano
President, The Systems Biology Institute
President and Chief Operating Officer, Sony Computer Science Laboratories, Inc.
Professor, Okinawa Institute of Science and Technology

Acknowledgements

This book would not have been possible without the help of many persons through their careful reading of the text and suggestions, the supply of materials and support. We are very grateful to all of them.

Our colleagues: Valentina Boeva, Valérie Borde, Jacques Camonis, Pierre Chiche, Fanny Coffin, Thierry Dubois, David Gentien, Pierre Gestraud, Alexander Gorban, Luca Grieco, Alexandre Hamburger, Anne-Claire Haury, Toby Dylan Hocking, Laurent Jacob, Hiroaki Kitano, Leanne de Koning, Inna Kuperstein, Séverine Lair, Christian Lajaunie, Philippe La Rosa, Patricia Legoix-Né, Alban Lermine, Damarys Loew, Georges Lucotte, Jonas Mandel, Matahi Moarii, Fantine Mordelet, Elphège Nora, Edouard Pauwels, Stuart Pook, Tatiana Popova, Patrick Poullet, François Radvanyi, Fabien Reyal, Daniel Rovera, Xavier Sastre-Garau, Ron Shamir, Terry Speed, Gautier Stoll, Véronique Stoven, Denis Thieffry, Sylvie Troncale, Anne Vincent-Salomon, Paola Vera-Licona, Yoshihiro Yamanishi and Bruno Zeitouni.

Authors of inspiring books: Alberts et al. (2002), Alon (2007a), Kitano (2001), Komarova (2005), Mukherjee (2010), Palsson (2006), Wagner (2005), Wang (2010) and Weinberg (2007).

Authors of images: Contributors from www.openclipart.org and commons.wikimedia.org.

CRC Press: Rachel Holt, Shashi Kumar, Sunil Nair and Karen Simon.

Our families and friends.

Chapter 1

Introduction: Why systems biology of cancer?

Cancer is probably as old as life, but remains a widespread and devastating disease. Generations of scientists and physicians have dedicated their life to improve patient care and eradicate it. They have contributed to monumental accomplishments in the fight against cancer. Cancer nevertheless remains the second-leading cause of death and disability in the world, behind only heart disease. Further progress is urgently needed to beat it.

Cancer genesis and development are intimately related to the dysfunction of genes. The sequencing of the human genome and subsequent genomic revolution have drastically impacted cancer research, allowing the dissection of cancer at the molecular level. In parallel, the last decade has witnessed the emergence of systems biology, a new field of research aiming at capturing the complexity of biological phenomena involving complex interactions with mathematical and computational tools.

In this book, we hope to enlighten the reader on the computational systems biology approaches applied to cancer research. These approaches offer new promising insights to defeat cancer. This first chapter sets the general context of the book by giving definitions of cancer and systems biology. We introduce the scientific and technological aspects of cancer research including their link with clinics. The chapter also describes the spirit of the book and provides reading guidelines.

1.1 Cancer is a major health issue

1.1.1 Bit of history

Cancer is probably as old as life. Evidence of **metastatic*** cancer was reported in *Edmontosaurus* fossils (Cretaceous) and **neoplasms*** were reported in a Neanderthal skull (35 000 BC), Egyptian and Incan mummies (David and Zimmerman, 2010). The oldest description of cancer in humans was found in Egyptian papyri written between 3000–1500 BC. Among them, the *Georg Ebers papyrus*, the *Edwin Smith papyrus* (circa 1600 BC) and the

Kahun papyri (circa 1825 BC) contain details of conditions which are consistent with modern descriptions of cancer. In Greece, Hippocrates de Cos (460–370 BC), the father of medicine, described cancer in detail in the *Corpus Hippocraticum*[1] and used the Greek terms *carcinos* and *carcinoma* to refer to chronic ulcers or growths which seemed to be malignant tumours and *scirrhus* to refer to a type of cancer with a hard consistency. In Greek, *carcinos* means *crayfish, canker, cancer, tumour*, while *skirros* means *solid tumour* as a noun and *hard, hardened* as an adjective. Celsus (28 BC–50 AD), a Roman doctor, translated the Greek word *carcinos* into the word *cancer*, a Latin word meaning *crab, crayfish, dunce* and *cancer, canker*. It was inspired by some cancerous lesions whose form recalls a crab. Galien (131–201) used the Greek term *oncos*, meaning *mass*, to refer to a growth or a tumour which looked malignant. In art history, testimonies of this disease can also be found. Rubens and Rembrandt were major baroque painters who practised realism (they painted whatever their eyes captured). This has allowed physicians to discover alterations which suggest tumour in the breast of the models they painted. One of the most famous paintings which depicts breast tumour is the oil-on-canvas piece by Rembrandt, *Bathsheba at Her Bath* (see **Figure 1.1**): an Italian surgeon first suggested that Rembrandt might have depicted a breast tumour in his painting, accurately showing the clinical signs (the dark shadow on her left breast) of the fatal disease from which his model, Hendrickje Stoffels, was suffering (Vaidya, 2007). Cancer is not a modern disease but has passed through the ages and likely from the origin of life.

1.1.2   Definition of cancer

Cancer is a pathology that can affect most of the tissues in the human body. It is generally defined by uncontrolled cell growth and, in the case of solid tumour, invasion of underlying tissues. As an additional feature, many cancers, but not all, can also experience migration of tumour cells from the primary site (*i.e.* the **primary tumour***) to a distant one where they settle, a phenomenon called **metastasis*** which is responsible for most cases (around 90%) of lethal issue. This first definition of cancer covers so-called malignant tumours. To be complete, we should also include in that definition **leukaemia***, which gives rise not to tumours but to circulating tumour cells in blood, originating from hematopoietic tissues. A second and broader definition also includes benign tumours, that show uncontrolled cell growth but no invasion, and in most cases do not threaten the patient life. The first definition is often used in clinical **oncology***, whereas in cancer biology the second is usually preferred. This second definition will be used in this book. More details about cancers, their classification and biology will be given in **Chapter 2**.

---

[1]The authorship of the text is unproven.

© 2012 Philippe Hupé

FIGURE 1.1 Bathsheba at Her Bath. Rembrandt Harmenszoon van Rijn (1606-1669), 1654, oil on canvas, 142 x 142 cm, Musée du Louvre, Paris, France.

1.1.3 Few facts about a killer

Although an old disease, cancer still remains a major health issue. Cancer is one of the major killers worldwide carrying responsibility for 7.6 million deaths in 2008 and 12.7 million new cases (source: IARC, GLOBOCAN database, http://globocan.iarc.fr). It accounts for one out of every eight deaths annually, killing more people than AIDS, malaria and tuberculosis combined. In many occidental countries, it is now the leading cause of death, before cardiovascular diseases. Some cancers occur with a high prevalence like breast cancers which affect 1 woman out of 9 during a lifespan. In developing countries the incidence rate is lower, but survival rates are much worse mainly because of late detection and lower quality or inexistent healthcare. Additionally in China, India and many developing countries, cancer deaths are increasing, largely because of smoking and diet habits. It is anticipated that the global yearly number of deaths should reach 17 millions in 2030.

1.1.4 Progress in cancer treatment is real, but insufficient

As a consequence of the importance of cancer in public health, considerable efforts are invested each year to extend and deepen our understanding of **tumour progression***, to develop new therapeutic molecules and to ameliorate cancer medical care. Patient treatments have improved substantially during the last decades and many new therapeutic molecules have been developed. Overall, we now manage to cure more than one cancer case out of two. Incidence and mortality rates are also globally decreasing, and the 2003–2007 period has shown a decline of 1% per year in incidence and 1.6% in mortality by cancer in the USA.

Nevertheless, this encouraging picture requires some deeper examination. First, decrease is largely due to prevention, at least for some cancers. Epidemiological studies have shown the role of environmental factors in the risk of developing cancer (like tobacco exposure in lung cancer), and identifying this cause enabled prevention programs that bear fruits a few years if not decades later. Second, this global reduction in mortality by cancer masks marked differences across geographical areas because of different life habits or risk exposure (*e.g.* oral and lung cancers are progressing in India and China). Third, mortality and therapeutic success vary largely across cancer types, patient age and sex. Altogether this situation calls for improvements in cancer prevention, but also of course in cancer treatment, and in particular the development of new drugs. Early detection of tumours is also a crucial factor impacting the chances of successful treatment. But developing new drugs is now more and more difficult, and more and more expensive. The reasons why the pipeline for drug discovery is not so fruitful are multiple, but in particular many drugs in development show limited efficacy, and present **off-target effects**[*] and toxicity to other cells. In the context of cancer, it is also often the case that a large proportion of the patients do not respond to treatment, or may **relapse**[*] after initial response. Therefore, the one-treatment-fits-all strategy is most of the time unsuccessful. Finding specific targets, and predicting which patient will benefit from a given treatment would therefore be of utmost value.

1.1.5   Progress in cancer drug development needs a qualitative evolution

Historically, the first therapeutic molecules for cancer treatment were targeting fast replicating cells with no specificity to a particular target gene or protein, exerting their action on DNA like nitrogen mustards. The discovery of the so-called **oncogenes**[*] and **tumour suppressor genes**[*] (genes involved in tumour progression, see **Chapter 2**) in the 1980s, followed by the identification of their **signalling pathways**[*] revived an old theory of *magic bullets* (see Strebhardt and Ullrich, 2008, for an historical perspective). According to this paradigm proposed by Paul Ehrlich in 1900 it should be possible to identify specific receptors associated to cancer cells, which can then be targeted specifically with a drug binding to the receptor. As a consequence cancer cells can be killed. The first implementations of this concept were the development of trastuzumab for treating breast tumours overexpressing the *HER2* gene, and of imatinib for treating Chronic Myeloid Leukaemia (CML) in the late 1990s. In both cases, a humanised monoclonal antibody was designed for targeting a specific oncogene (*HER2* in the case of trastuzumab, and the *BCR-ABL1* fusion protein in the case of imatinib), with substantial clinical success. **Amplification*** of *HER2* in the genome of breast tumours, and **translocation*** of chromosomes 9 and 22 in CML cells (*i.e.* the Philadelphia chromosome) brought immediate attention to the oncogene driving these tumours and designated them as natural targets. But other efficient intervention points might be more difficult to identify and require adequate modelling approaches, which are presented in **Chapter 7** and **Chapter 8**.

In addition, in many cases, patients treated with these novel drugs ultimately develop resistance and need **second-line treatment***. This resistance can be explained by the multidimensional nature of cancer, which involves many molecular players (genes, proteins, small molecules, *etc.*) interacting in interconnected pathways that can be thought of as a large and complex network. Targeting only one player is likely to let many bypass roads open in the regulation network and feedback loops which will compensate for the effects of the therapeutic molecule. This advocates the use of combination of specific drugs (or an appropriate sequence in time), or of multi-target inhibitors (*e.g.* muti-kinase inhibitors). Even for one single patient, the magic bullet approach often requires several bullets. But here the mere intuition is of course not sufficient to figure out which points in the network should be targeted. The behaviour of complex networks with many crosstalks and feedback loops is not linear and escapes immediate understanding. Again adequate modelling is essential for rationalising the search of targets (see **Chapters 7** and **8**) and for designing new drugs (a domain of research not covered by this book).

In this book we present how computational systems biology can help in these three essential compartments of the fight against cancer: categorising tumours, finding new targets and designing improved and tailored therapeutical strategies. The question of prevention can also be approached by systems biology, but is out of the scope of this book.

1.2 From genome to genes to network

1.2.1 Accumulation of alterations

Cancer onset is characterised by an accumulation of genetic and epigenetic alterations that can be caused by different stresses, like tobacco, chemical agents, radiations and viruses. These alterations typically modify the structure of DNA and **chromatin***, and consequently the gene products or the regulation of gene expression. They can take different forms (reviewed in **Chapter 2**), like chromosome rearrangements and translocations (formation of chimeric chromosomes), small or large deletions or amplifications of DNA regions (whose copy number will be 0 or 1, *i.e.* a deletion, instead of the normal 2, or 3 or more, up to several dozens, *i.e.* an amplification), variation in ploidy (chromosome number), or point (single nucleotide) **mutations***. In 1971, a model of **retinoblastoma*** occurrence was proposed by Knudson (1971), based on mortality statistics, and became the canonical Knudson two-hit model. In this model, occurrence of retinoblastoma was shown to be in agreement with the progressive accumulation of two mutations, each affecting one allele of the *RB* gene. This model is also applicable to other cancers with a larger number of stages such as colon, stomach and pancreas cancers (Nordling, 1953; Armitage and Doll, 1954), but not to all of them.

Among all alterations that affect a tumour genome, some are causative of the disease (*i.e.* the *driver genes*), whereas others may not play any particular role (*i.e. the passenger genes* which are neighbouring driver genes and are altered concomitantly). Causality can be assessed in different ways. A classical way is to assess it at the genotype level in model animals (mainly mice) where fully controlled comparative experiments can be carried out. The recent use of **antagonist drugs*** or on the contrary **mimetic drugs*** to palliate the deficiency of a protein is another element of proof of causality in the context of human cells.

It is commonly accepted now that cancer development proceeds in a multistep way controlled by Darwinian principles: random genetic changes create a diversity of cells with differences in terms of proliferation and survival rate; both processes are under tight genetic control; time selects the cells with higher propensity to multiply.

1.2.2   Cancer is a gene disease

"The revolution in cancer research can be summed up in a single sentence: cancer is, in essence, a genetic disease." This statement is quoted from a *Nature Medicine* review by Vogelstein and Kinzler (2004) who are pioneers in cancer molecular biology research.

The genetic nature of cancer may seem obvious today but it took long to be established. The first oncogenes (as *BCR-ABL1*, *SRC*) were discovered in the 1970s, and tumour suppressor genes in the following decade (*RB*, *TP53*, *etc.*). In the last two decades, biologists have shown that these genes were arranged into signalling pathways. These pathways have been characterised progressively. At first glance, they appeared to have more or less a linear structure, transducing a signal from a sensor protein (*e.g.* membrane receptor) to an effector protein through a cascade of transducer proteins. Many screening assays for finding new cancer genes were conducted, and as a result many cancer-related genes are identified today: the Catalogue of Somatic Mutations in Cancer (COSMIC) database (v59, June 2012), contains 487 genes implicated in cancer, *i.e.* the Cancer Gene Census (Futreal et al., 2004).

Evidence of the genetic nature of cancer came also from familial studies. Only about 10% of cancers fall in the category of familial syndrome, where the inheritance of some gene allele is associated with an increased risk of developing cancer. A compelling evidence is the *RB* gene (where carriers of mutated alleles have a 90% risk of developing a retinoblastoma, to be compared with the frequency of the pathology in the general population: 1/15000 to 1/20000), and the *BRCA1* and *BRCA2* genes (where carriers of deleterious alleles have 10 to 20-fold increased risk of breast cancer).

Though there are cases of genetic inheritance, we should emphasise that cancer, contrary to most well-known genetic diseases like dystrophies, results mainly from alterations acquired over a lifetime and are not transmitted from parents to offspring.

1.2.3 Cancer is a network disease

At the end of the 20th century, the number of genes whose involvement in cancer was established had increased significantly. It became therefore apparent that they could be organised in a limited number of biological functions, named hallmarks of cancer as described in two landmark papers by Hanahan and Weinberg (2000, 2011). The six, then eight hallmarks are related to cell survival, proliferation and metastatic dissemination and are presented in detail in **Chapter 2**. Originally, it seemed that each function or sub-function had its own, almost independent pathway. As more and more genes controlling these functions were identified and arranged into signalling pathways, it appeared that these pathways are interconnected and present crossroads, talking to each other at different levels. In other words, the hallmarks of cancer should be understood as different ways of looking at a unique process, tumour progression, and are by no means independent. Of course these interconnections obliterate the linearity of the pathways and turn them into a web of interactions between genes, RNAs, proteins and other molecules. Today, it is well established that if cancer is a disease of genes, it is first of all the consequence of a deregulation of the gene networks that control cell growth and dissemination. As a result, methods for modelling gene networks are central to any modern approach of the molecular biology of cancer (see **Chapters 7** and **8**).

1.3 Cancer research as a big science

1.3.1 Cancer research is technology-driven

The origin of the incredibly swift evolution of biological research during the last two decades lies probably in the revolution brought by biotechnologies. Miniaturisation and automation have expanded much faster than anyone anticipated the experimental possibilities for investigating life at the molecular and cellular levels. We can explore many molecular facets of the biological systems that were previously out of reach. In addition we are also capable of building very rapidly a (almost) complete description of one particular level of the system (*e.g.* its genome, or its transcriptome; see **Chapter 3** for details about the technologies used). The first step of this revolution has been the sequencing of the human genome, which was completed in 2001 by contributions from many laboratories in a joint 13 year effort which cost almost 2.7 billion dollars. Though at that time the human genome sequence was in fact only completed at 90%, its availability has opened the road to other large-scale biotechnological approaches (not to mention its crucial interest for biological research), as exemplified in the two following subsections.

1.3.2   Microarray era

The advent of **microarrays*** around 1995 enabled, for the first time, to interrogate expression of many genes simultaneously. After some versions of microarrays addressing a few thousands genes, it has rapidly become the rule to assess the expression of virtually all genes on a microarray. The microarray technology has blossomed in different flavours: genome microarrays (array Comparative Genomic Hybridisation (aCGH) or Single Nucleotide Polymorphism (SNP) arrays) which investigate genome alterations like gain, deletion and point mutations (for SNP arrays); transcriptome arrays for quantifying RNA expression at the transcript or at the exon level, or microRNA (miRNA) expression; proteome arrays which interrogate protein expression and activities; Chromatin Immunoprecipitation (ChIP) arrays for localising on the genome protein-DNA interactions or investigating nucleosome modifications. These technologies and others will be introduced in more detail in **Chapter 3**.

1.3.3   Next-generation sequencing era

Since 2004, a new technology has dramatically accelerated our possibilities in terms of molecular investigation of the cell. The so-called next-generation sequencing technologies (**NGS***) enable parallel sequencing of hundreds of millions of short sequences in parallel (up to one billion per run as of 2012). Besides covering the application of genome, transcriptome and ChIP arrays, this technique also offers new possibilities like the investigation of short mutations, genome rearrangement, fusion transcripts and RNA trans-editing or trans-interactions in the genome. A detailed introduction to these techniques is given in **Chapter 3**. One microarray typically provides today a few hundreds of thousands measurements in a few hours (it grew from a few thousands to 1000 times more within five years). In comparison a NGS experiment outputs up to one billion sequence reads in less than a week, and generates a few hundreds of gigabytes of data (sequence calls and qualities), not to mention the terabytes[2] of raw data (images) and analysis results (alignment and in-

---

[2]mega (10^6), giga (10^9), tera (10^{12}), peta (10^{15}), *etc.* are prefixes to multiply any unit. To illustrate the idea, one terabyte of mp3 music corresponds to two years of listening.

terpretation). Even more striking, the acceleration of the trend is not over, and single molecule techniques and laser-free sequencing chips are promising yet higher throughputs and lower costs of sequencing devices and consumables (see **Chapter 3**). In a near future, one can anticipate that genome sequencing will be routinely used in medical practice, pushing biology in the petabyte era and accentuating the demand for high performance storage and computing, bioinformatics and computational systems biology algorithms and tools (see **Chapter 4** for informatics and bioinformatics aspects).

1.3.4 Cancer research and international data provider consortia

Cancer research has inevitably been impacted by the revolution of biotechnologies. Two major efforts illustrate this phenomenon. In 2006, the National Cancer Institute (NCI) and the National Human Genome Research Institute (NHGRI), two institutes from the National Institutes of Health (NIH) in the USA, launched The Cancer Genome Atlas (TCGA), a program aiming at deciphering the genomic and epigenomic modifications of tumours in more than 20 different cancers. In 2008, the International Cancer Genome Consortium (ICGC) was set up as a joint action of USA, Canada, Europe (France, Germany, Great Britain, Italy, Spain), China, India, Japan and Australia with the goal of establishing the molecular profiles of 25,000 tumours from 50 different cancers. In both efforts, samples are characterised with the relevant clinical features, constitutional genome should also be sequenced, and tumour profiling will cover genome, transcriptome and epigenetic aspects (DNA methylation). Both consortia have agreed to make their data available to the scientific community as rapidly as possible.

1.4 Cancer is a heterogeneous disease

1.4.1 Cancer heterogeneity

Most tissues of the human body can be affected by cancer. Nevertheless the word cancer designates in fact a myriad of different diseases. Cancer from distinct localisations are of course very different, but even for a given organ the heterogeneity is often the rule, and this at several levels.

The first level of heterogeneity is morphological: for example **pathologists***, the physicians who examine tumour samples under the microscope before any treatment is defined, classify breast tumours in eighteen different types (DeVita et al., 2008).

The second level is the clinical heterogeneity: within a tumour type, clinical characteristics, and in particular disease evolution and therapeutic answer, vary largely among patients. There are of course marked differences between tumour types, but when examining the clinical case at the individual level it is often very difficult if not impossible to conclude and give a **prognosis***.

The third level of heterogeneity is molecular: the many portraits of cancer that have been drawn during the last decade with molecular profiling, using microarrays and now NGS, have shown that there are no two tumours identical from a genetic and epigenetic perspective. Of course, tumours from the same type share common traits, like recurrent mutations, deletions, amplifications, rearrangements, and resembling gene expression, but they also display many genetic differences. These molecular differences also concern the tumour microenvironment, which plays a crucial role in tumour progression, and the host of the tumour, the patient, with his personal genetic makeup.

A fourth level of heterogeneity concerns the tumour itself. Although it is often considered as the clonal expansion of the best fit cells, many tumours present internal heterogeneity: several subclones coexist in a dynamic equilibrium, each of them constitutes a specific disease, and drugs active on some of them might be inefficient on others, thus explaining resistance of the tumour to treatment.

In conclusion, if normal tissues are all alike, every tumour tissue is abnormal in its own way.

1.4.2   Sorting out tumour heterogeneity: Classifying tumours

It is tempting to speculate that the molecular heterogeneity can explain the diversity of morphological and clinical phenotypes. Having drafted the molecular profile of tumours at the genetic and epigenetic levels should give us the basic elements that cause or accompany tumour progression, and should provide the necessary information for defining tumour subtypes in a rational way. What was so far mainly defined by morphological observation could also be approached, hopefully better, with full molecular characterisation. This information could maybe not replace, but certainly complements the microscope. This question is much more than the intellectual exercise of ordering our observations. Conceptually, it provides a map of the pathology, and facilitates reasoning and drawing hypotheses that will lead to understanding the nature of the pathology and the biological principles that govern it. In practice, it has of course huge implications for the therapeutic treatment of patients and the atlas of tumour types constitutes a daily tool for the **oncologist***. **Chapter 5** examines how such classification of tumours is carried out.

1.5 Cancer requires personalised medicine

1.5.1 Definition of personalised medicine

Personalised medicine can be defined as a model of medical practice where healthcare of a patient is tailored to his/her personal characteristics. Stated this way, one can argue that traditional medical practice is also considering the patient characteristics, but the difference resides in the philosophy of the approach; personalised medicine is opposed to traditional medicine which is mainly based on reference treatments. The reference treatments have been established on large series of patients and once validated are considered as the universal solution for treating any new patient affected by the disease. Personalised medicine acknowledges that the patient pathology is unique and this unicity is driving the choice of the treatment. There exists no one-fits-all treatment for cancer. Personalised medicine is often (but not always) based on the genetic background of the patient. In the context of cancer, it may be based on this patient's constitutional genetic background, or also on the tumour's genetic and epigenetic landscape, or on the life environment. Personalised medicine may concern preventative as well as curative medicine.

The goal of preventative medicine is to prevent occurrence of diseases or to carry out surveillance to detect the disease at the earliest stage. Surveillance can be adapted as a function of the patient's personal features, like lifestyle (exposure to risk factors) or family history for genetic diseases. This is the case, for example, in breast cancer where *BRCA1* and *BRCA2* genes are used for assessing individual risk and adapting surveillance.

Curative medicine applies when the disease has broken out and been diagnosed. It aims at avoiding treatments that are not beneficial to the patient and optimising strategy in terms of therapeutic response and patient comfort. In particular, avoiding or limiting side effects that one patient may be at risk of, because of his genetic background, is one major goal of cancer curative personalised medicine.

1.5.2 Choosing the adequate treatment: Prediction and prognosis

Many cancer drugs show only limited efficacy in fighting the tumour. In many cases, the therapeutic molecule is not adapted to a significant proportion of patients but oncologists have no means of distinguishing in advance responders from nonresponders. If initial treatment fails, a second-line treatment with another molecule has to be tested. Predicting which therapeutical strategy is the most likely to be effective is still difficult if not impossible today. This situation entails a loss of time in providing the adequate treatment, patient suffering, possibly death from adverse effects of inefficient molecules, and economical waste. The conclusion about cancer treatment is very clear: one cannot expect that a given therapeutic treatment would fit all patients affected by a particular type of cancer. Again, the reason for this clinical heterogeneity is more or less known. To a large extent, it resides in the underlying biological heterogeneity of the cancer, no two tumours being ever identical. Of course, this heterogeneity is, for a large part, genetically determined: the modalities of tumour progression are for a large part rooted in the molecular profile of the tumour. Designing a procedure to classify new tumour cases per type is in general addressed using cases with known outcome, building a prediction rule, estimating its performance and applying it to new tumours. Of course, there are many obstacles to this simple scheme: methodological (*e.g.* how to build a robust procedure, that will perform efficiently on new cases?), technical (how to standardise sample preparation from bedside to analysis?) and also biological (one given tumour is often not genetically homogeneous, but may contain different subclones whose mixture may blur the overall measurements used for prognosis).

Nevertheless, relating the tumour genetic characteristics to the resulting clinical phenotype provides a way for predicting the response to a certain therapeutic molecule and proposing the best treatment to a patient. This idea has been made achievable since the availability of the microarray technology. It is the rationale behind the numerous attempts of deriving from the molecular profiling of tumours (mainly at the transcriptome, but also at the genome and epigenome level) decision rules to help the clinical oncologist in the design of the optimal and individualised treatment. These so-called gene signatures are now proposed as clinical tools for some types of breast cancer, for example Agendia's 70-gene Agilent-based MammaPrint®, *i.e.* the *Amsterdam Signature* (van't Veer et al., 2002; van de Vijver et al., 2002), Veridex's 76-gene signature, *i.e.* the *Rotterdam Signature* (Wang et al., 2005; Foekens et al., 2006), Genomic Health's 21-gene RT-PCR-based Oncotype DX™(Cobleigh et al., 2005; Hornberger et al., 2005) and a 41-gene expression set (Ahr et al., 2001, 2002). It is not known yet whether these tests are announcing the rise of a new strategy for personalised medicine. It is possible that the limited information present in the gene expression profiles they are based on will simply be insufficient. These questions are the subject of **Chapter 6**.

1.5.3   Designing a personalised treatment

The advent of the high-throughput technologies is changing drastically the landscape of cancer medicine. Microarrays are now in clinical use for cancer prognosis. NGS, proteome and metabolite profiling are also opening new possibilities for clinical applications. These technologies are reviewed in **Chapter 3**.

NGS now enables one to sequence the genome of tumours and patients for a moderate cost (a few thousands of euros in 2012), and to use this information to inform decisions about patient care. Many applications are envisaged;

some of them are already being experienced in clinical trials. For example, the choice of a therapeutic molecule like a protein kinase inhibitor can be fully rationalised from the mutation profile of the tumour, following these steps: listing all mutations that affect a tumour, selecting those that may have deleterious consequences and prescribing therapeutic molecules on this basis. This approach has already been tested and encouraging results have been obtained in terms of survival gain (Hoff et al., 2010). It can be anticipated in the near future that patients will be more systematically included in clinical trials proposing the adequate molecule among the expanding list of protein kinase antagonists or other targeted molecules available. Other applications of NGS are the detection of fusion transcripts, genome rearrangements, methylation profiles within the tumour, or the search for Circulating Tumour Cells (CTC) in blood or lymph, thus facilitating prognosis and therapeutic choices without biopsies.

Proposing therapeutic molecules targeting the defective gene products of a patient is undoubtedly a promising approach, which has already borne some fruits. But these molecular targeted therapeutics also have their limitations and failures (Gonzalez-Angulo et al., 2010). The robustness of cancer, whether it is based on crosstalks, compensatory pathways, feedback mechanisms (either present in the normal cell, or acquired by the tumour cell), or tumour heterogeneity often lead to lack of response of patients. Off-target effects and high toxicities of some treatments are another cause for failure. Another fundamental aspect concerns the representativity of the molecular data on which to base a strategy: if it comes from a biopsy of the primary tumour, it may be inadequate for treating metastasis, or new mutations may have appeared that compromises the strategy.

Modelling the functioning of signalling networks and their specificities in a given patient is therefore the next step to achieve fully informed and rationalised personalised treatment of tumours. This should aim at predicting the impact on tumour cell proliferation, migration and death, identifying reasons for relapsing and proposing concomitant complementary treatment or second line strategies, and anticipating adverse effects of drugs on normal cells.

1.6 What is systems biology?

Defining a new field of science is always a difficult task. Young disciplines have fuzzy borders, rapidly evolving topics. Existing scientific communities join the new field, exert various influences and shape it in sometimes unexpected ways. Systems biology does not escape this rule: it is now used in so many contexts, that it seems difficult to give it one simple definition. The attempt below summarises the main features that are generally associated with systems biology approaches. It then retains two main families of definitions that have emerged and survived the first ten years of the discipline.

1.6.1   Operational definition for systems biology

Scientific fields can be defined by their scope and by their method. As to systems biology, the scope is biology, as a whole. We should acknowledge that many definitions of systems biology attach the discipline to molecular biology, though some others have wider definitions, or have even rooted the approach in early attempts to describe biological phenomenon at a higher level, like the movement of the heart and the blood in animals (Auffray and Noble, 2009). But even when attached to molecular biology, systems biology aims at explaining higher level features, like for example cell fate, organ function or individual phenotype. Here comes one essential feature of the discipline: it achieves modelling of a biological phenomenon on multiple layers of description to explain high-level properties. This often means modelling of biological networks like signalling pathways or metabolic pathways. These networks are typically composed of dozens to thousands of nodes (genes, proteins, small molecules, *etc.*) and interactions. Some of these nodes might play a particular role: for example **hubs*** are directly connected to many other nodes and have a coordination mission to answer to stimuli; another example is **routers***, which are bottlenecks controlling many roads from one part of the network to another. Nevertheless the understanding of the network properties cannot be reduced to the study of individual elements, but requires a global approach. Only then will it be possible to predict the behaviour of the network (Will it drive the cell to proliferation? To death? And how?). In other words, the reductionist gene-per-gene approach that prevailed in the last decades of biological research is no longer fruitful in the context of biological networks. These features of networks that are not accessible at the gene level, but only at the global level, are often referred to as **emergent properties***. One typical example is also the robustness of the normal or tumour cells (the fact that the cells maintain their functions despite a fluctuating environment), which is discussed in **Chapter 9**.

Whatever definition of systems biology we retain, one constant aspect of systems biology is the interdisciplinary nature of the approach, which gathers mathematicians, biologists and clinicians about a biomedical question. This is even a crucial feature of any systems biology group to be built on such a polymorphism of skills, and to achieve coordinated efforts both at the experimental and theoretical levels. This combination is rooted in the remarkable and unparalleled efficacy of mathematics in bringing the necessary support to other sciences, and in particular physics and chemistry, in explaining the universe, and the fecundity of the use of mathematical laws in achieving concrete realisations that impact every day life or overcome limits we thought absolute.

It has been objected that mathematical modelling in biology can only bring little understanding because biology is in essence too complex. One could easily revert the argument: it is because biology is complex that mere intuition does not suffice and the abstraction and rigour of mathematical modelling are required. Complexity also means that the reductionist approach is not fruitful enough to understand the object of study, and another approach should be adopted. Systems biology can precisely be defined as an attempt to decipher the complexity of the cell and higher level biological systems.

The huge volume of data to analyse is then another reason why modelling should complement and in many cases replace intuition. The expression *systems biology* is born concomitantly with the deciphering of the human genome sequence, and a large part (but not all) of the discipline incorporates high-throughput molecular and cellular biology data at some level in its approach. Indeed high-throughput technologies enable the complete description of biological systems at some level (*e.g.* genome or transcriptome). These data inventories, and how complete they could be, are not yet biological knowledge, and are not enough *per se* to explain and interpret biological systems. Mathematical modelling is then required to extract coherent and useful knowledge out of these heaps of observations and overcome a purely phenomenological representation. The purpose here is three-fold: organising these facts into categories, explaining the functioning of the system, and predicting new behaviours of the systems. Thus the systems biology approach is defined as the classical scientific method in biology (Ayala, 1968).

But one should also be aware of all that mathematical modelling can offer: first it provides a formal language to describe knowledge in an unambiguous way; second it enables organisation, classification of the large sets of data that biology is now producing, achieving a first step toward understanding; third it provides a framework for reasoning about the biological system, and eventually it enables drawing of hypotheses and designing experiments for validating them. After iterative cycles of experimental validations and model improvements the model may become predictive and also may even identify explanations of the system behaviour. These aspects are described in detail in **Chapter 7** and **Chapter 8**. Finally, mathematical tools can also introduce new concepts or give formal substance to concepts that intuition had proposed. In the context of cancer systems biology, robustness is one of the main concepts that illustrate this function of mathematics. **Chapter 9** describes in detail this subject.

1.6.2 Systems biology: Is it data-driven or model-driven?

Several definitions of systems biology have been proposed, which can be broken down into two main categories: data-driven approaches, and model-driven approaches.

The past of computational molecular biology research has been dominated by data-mining approaches, which aim at finding regularities in large datasets like genome sequences. A typical example could be deciphering the structure of genes and of all elements that constitute a genome (introns, exons, promoters, enhancers, insulators and other regulatory sequences). These approaches make use of sophisticated statistical approaches like Hidden Markov Model (HMM) and Support Vector Machine (SVM). The models they produced are heuristics, make in general little assumptions about the knowledge of internal mechanisms of the system, and are well suited precisely in these situations of uncertainty.

The second type of approaches is based on models which simulate the dynamics of the biological system, which is typically described as a network of interacting nodes (*e.g.* genes, proteins, small molecules, complexes). Nodes and edges in this network reflect our knowledge of the biological mechanisms that locally control the functioning of parts of the systems. The global modelling of the network enables to simulate the behaviour of the system. Note that there are many different types of networks, depending on the system modelled and the level of precision required or achievable.

This separation of systems biology approaches in two categories, one statistical and one dynamical, is of course a simplification of reality, and ignores intermediary approaches, like prior knowledge integration into statistical learning, as presented in **Chapter 6**. These two parts can also be presented from a historical perspective. Systems biology as a scientific discipline, was proposed in the beginning of the millennium and defined independently by two schools:

1. On the one hand, Leroy Hood and co-workers defined systems biology as a pragmatic approach which consists of four steps (Ideker et al., 2001): first, large-scale data are collected to describe all the components of the system (*e.g.* at DNA level, RNA level, *etc.*); second, the components of the system are systematically perturbed (by genetic means, or drugs, or control of the environment) and monitored, if possible on a global scale (typically all genes are assayed); third, a model is built and iteratively refined so that its predictions fit experimental observations; finally specific perturbations are designed and performed to test the model and distinguish between competing hypotheses.

2. On the other hand, another school which can be represented by Hiroaki Kitano (Kitano, 2002b) defines systems biology as a science which studies dynamical behaviours of biological systems, by focusing on the interactions of their components. The idea is that these behaviours, and in particular biological functions, are intrinsic properties of the systems that emerge from the interactions between components and cannot be revealed by the study of individual components. The robustness of biological systems, the fact that biological systems maintain their states and functions despite external perturbations, is a typical example of such a property.

The first approach can be envisaged as a bottom-up, data-driven approach of pragmatic biologists where models are built from data. Mathematical tools are heuristics which integrate observations. The risk is to stay at the phenomenological level, and achieve little prediction beyond the observed cases.

The second one is rather a top-down, model-driven approach of mathematicians which captures knowledge in a model and then challenges it with reality. Here the trap could be to build elegant mathematical abstractions without biological impact.

1.6.3   Systems biology: Yet another definition

To the authors of this book it seems that computational systems biology should be defined as an attempt to reconcile the two schools introduced above, which can be historically related to the long-standing split between experimental biologists and theoretical biologists. The availability of close-to-exhaustive descriptions of biological systems should make this reconciliation possible in a near future.

Finally scientific disciplines can also be defined from the ultimate goal they pursue, even or in particular if this goal seems out of reach in a foreseeable future: systems biology aims at constructing a virtual model of any biological organism, or even ecosystem, on which *in silico* experiments could be conducted, thus accelerating hypotheses testing and avoiding all limitations of *in vivo* and *in vitro* experiments. One appealing application in medicine is of course a virtual patient, on which innovative therapeutic strategies could be designed and tested *in silico* before experimental validation.

A first move in that direction was done in 2008 by Japanese and British researchers during a joint meeting held under the auspices of two major funding agencies (JST and BBSRC). A declaration, the *Tokyo declaration*, was issued. It states that progress in the field of systems biology is now such that it constitutes a new paradigm in biology. This paradigm shift is needed to make sense out of all the results of molecular biology, which though impressive fall short in extending our understanding of life, and converting this knowledge to practical realisations in health, environment or agriculture domains. Establishing a road map of systems biology should therefore be a priority. They propose an effort to *"generating a comprehensive molecules-based computational representation of human physiology,"* with a scaling approach on pathway, organs, and then organisms, starting with animal models and aiming at building a *silicon human*: *"Recent advances in systems biology indicate that the time is now ripe to initiate a grand challenge project to create over the next thirty years a comprehensive, molecules-based, multiscale, computational model of the human (the virtual human), capable of simulating and predicting, with a reasonable degree of accuracy, the consequences of most of the perturbations that are relevant to healthcare."*

Another example is the virtual physiological human program, a European initiative which funded fifteen projects in diverse domains of human health (http://www.vph-noe.eu/): *"The concept of a virtual physical human is a sophisticated computer modelling tool, which compares observations of an individual patient and relates them to a vast dataset of observations of others with similar symptoms and known conditions. By processing all this information, the model can simulate the likely reaction of the individual patient to possible treatments or interventions. Such tools will not only improve the quality of treatment offered to patients who are already ill or injured, but could also be used in preventive medicine, to predict occurrence or worsening of specific diseases in people at risk, for example through family history."*

1.6.4   Systems biology of cancer

How can we define systems biology of cancer? Several authors have given definitions, in general emphasising a data-driven approach, or a model-driven one (Kitano, 2003, 2004b; Khalil and Hill, 2005; Hornberg et al., 2006; Gonzalez-Angulo et al., 2010; Kreeger and Lauffenburger, 2010; Sonnenschein and Soto, 2011). In this book both cancer and systems biology have been defined above, and application of systems biology approach in cancer research is very natural, since cancer is a disease of interactions: interactions within signalling pathways, and between signalling pathways (Hanahan and Weinberg, 2011), interactions between cells, and with the microenvironment (Sonnenschein and Soto, 2011). But are there some other features specific to cancer systems biology?

The first feature is undoubtedly the objective of cancer systems biology: finding intervention points and therapeutic strategies for curing the disease. Also we already underlined the relative failure of the reductionist approach in providing full understanding of cancer biology and defeating the pathology. Both from a cognitive and from an applicative point of view systems biology is required.

A concept from systems biology that has shown its strong pertinence in studying both biological and clinical aspects of cancer is the notion of robustness (Kitano, 2003, 2004b). Any definition of cancer systems biology should mention this contribution.

Another characteristic of cancer systems biology, and the reason why cancer research requires systems biology more than any other biological system, is linked to the very nature of the pathology. No two tumours are identical. The genetic and epigenetic profile of any tumour has its own specificities which makes any extrapolation (*e.g.* assuming that treatment should be equally efficient) from one situation to another a risky bet. The rewiring of the biological networks in pathological conditions, though not yet precisely documented, is probably specific to any tumour. In these conditions, building a model of the patient, considering his genetic makeup, would solely allow us to simulate the action of potential drugs, thus assessing efficiency and anticipating possible negative effects in the specific context of the patient body and environment. This could save a lot of time and effort in the development of drugs, in clinical trials (*e.g.* by selecting *in silico* the patient to include in a trial or to exclude) and in the design of innovative, multidrug therapeutic strategies.

Cancer systems biology has then a flavour of cancer systems medicine. One of the early visions of this idea was developed at the Institute of Systems

Biology in Seattle by Leroy Hood, and termed *P4 medicine* (P4 stands for predictive, preventive, personalised and participatory). The ingredients of the approach are the same as other visions: high-throughput technologies for characterising samples, computational and mathematical tools for making sense out of the data, modelling of biological regulatory networks and identification of therapeutic intervention points. As mentioned earlier, they develop a data-driven, pragmatic approach based on heuristics. Another characteristic of this proposition lies in the fundamental change in the practice of medicine that they anticipate. The practice of medicine is mainly reactive today, in the sense that the physician mainly reacts to the diseased state of the patient and little is done to prevent occurrence of disease. In comparison the systems medicine will become *predictive*, based on the genetic makeup of an individual and his environment; this will allow *preventive* actions, like adapting lifestyle, taking preventive drugs, to avoid diseases for which the individual will be at high risk. This idea was of course older, and the term predictive medicine was first introduced by Jean Dausset (Nobel prize in medicine, 1980). There are two consequences to this new practice: first the medicine has become *personalised*, tailored to the unique genetic feature of the individual; second it will become *participatory*, as it opens a incredibly wide variety of options about personal healthcare, and requires in-depth exchanges between the individual and his physicians.

Finally the most famous vision of cancer systems biology came from Hanahan and Weinberg (2000). In a landmark paper in 2000, they recapitulated a quarter century of molecular oncology research, and anticipated a major change in our paradigm of cancer research. They made the following prophecy:

"Two decades from now, having fully charted the wiring diagrams of every cellular signalling pathway, it will be possible to lay out the complete integrated circuit of the cell upon its current outline. We will then be able to apply the tools of mathematical modelling to explain how specific genetic lesions serve to reprogram this integrated circuit in each of the constituent cell types so as to manifest cancer. With holistic clarity of mechanism, cancer prognosis and treatment will become a rational science, unrecognisable by current practitioners. It will be possible to understand with precision how and why treatment regimens and specific antitumour drugs succeed or fail. We envision anticancer drugs targeted to each of the hallmark capabilities of cancer; some, used in appropriate combinations and in concert with sophisticated technologies to detect and identify all stages of disease progression, will be able to prevent incipient cancers from developing, while others will cure preexisting cancers, elusive goals at present. One day, we imagine that cancer biology and treatment - at present, a patchwork quilt of cell biology, genetics, histopathology, biochemistry, immunology, and pharmacology - will become a science with a conceptual structure and logical coherence that rivals that of chemistry or physics."

It is stunning to realise that they described quite precisely the coming rise of cancer systems biology, which at that time had not yet hatched.

1.7   About this book

1.7.1   What does this book try to achieve?

Authors of this book are members of a cancer computational systems biology laboratory in Institut Curie, Paris, a centre for cancer research and cure. This book arose from our experience during one decade as computational biologists working in close collaboration with experimental biologists and clinicians. Our daily practice led us to observe and take part in many aspects of cancer research, that this book tries to summarise. We have experienced the exciting progression of cancer research toward a big science, and would like to share this experience with the reader. Over the years, we have developed a conviction that the future of cancer research and the development of new therapeutic strategies rely on our ability to capture biological and clinical questions and convert them into mathematical models, integrating both our knowledge of tumour progression mechanisms and the tsunami of information brought by high-throughput technologies like microarrays and NGS. There are of course many excellent books presenting the computational aspects of cancer research. This book combines the following:

- Offers a comprehensive overview of concepts and methods in computational systems biology of cancer
- Dissects the computational and design principles behind some of the existing tools
- Provides listings of additional available bioinformatics resources relevant for a computational systems biology approach to cancer
- Illustrates with biological applications
- Presents dynamic modelling of cancer related networks and data mining approaches
- Deals in-depth with clinical aspects and biological questions

Finally it is a self-consistent monograph written by colleagues from different fields of computational biology working door-to-door for several years on common projects.

1.7.2   Who should read this book?

The main audience of this book is the increasing number of graduate students, engineers and researchers involved in cancer bioinformatics and systems biology research, both in academia and in the pharmaceutical and biotechnological industries. Our goal is to provide a textbook and a guide for students and bioinformatics professionals from both computational and life sciences backgrounds. Moreover, we expect to provide an entry point to the field of cancer research for students and researchers in mathematics and computer science, and conversely an entry point to the use of computational and modelling approaches for students and researchers in biology and oncology.

This book and accompanying web site can also be used as a central resource for teaching systems biology and bioinformatics. It comes with a summary of main messages for each chapter, a series of exercises with data and algorithms, example of applications, and it offers links to other materials for deepening the subject if needed.

To ensure such a wide audience, we have tried to make the book as self-contained as possible, and assume very limited knowledge in biology, mathematics or computer science from the reader. The readers with good knowledge of the biology of cancer may skip **Chapter 2**, those familiar with high-throughput technologies could skip **Chapter 3**. Some parts of **Chapter 5** to **Chapter 11** introduce some mathematical aspects and might be out of reach to the reader with no background in mathematics, but skipping the most technical parts does not prevent understanding of the rest of the book.

1.7.3 How is this book organised?

This book presents concepts, algorithmic methods, bioinformatics tools and biological applications, with accompanying tables and illustrations. It introduces theoretical elements, and illustrates them with applications on real data.

Each chapter contains:

- An overview of the problem

- A presentation of the main concepts (which are also highlighted in boxes) and state-of-the-art methods; some chapters end with a list of key concepts

- A description of existing tools

- An application to concrete cases

- A listing and brief description of publicly available resources (data and software)

- References to further reading and to more advanced material

- A set of knowledge-testing exercises for several chapters

This chapter, **Chapter 1**, presents the subject and the rationale of the book. It defines systems biology in general and its application to cancer study. A short historical perspective is given.

Chapter 2 introduces the *basic principles of the molecular biology of cancer*. It assumes that the reader is familiar with the biology of the cell which is nevertheless described in the appendices. This chapter describes the series of events that transform a normal cell into a cancer cell during the tumour progression. The traits common in all cancers and referred to as *hallmarks* are presented in turn.

Chapter 3 presents the main *Experimental high-throughput technologies for investigating cancer*: microarrays, NGS, mass spectrometry and cellular phenotyping. These *omics* technologies appeared during the last two decades and have enabled an unprecedented high-throughput description of normal and tumour cells, thus opening the way to the systems biology of cancer.

Chapter 4 provides an overview of the *bioinformatics tools and standards for systems biology* which are necessary to any systems biology project, as enabling resources. This includes experimental design, data normalisation and quality control.This chapter also introduces a series of resources that are used in computational systems biology: raw data repositories, pathway and network databases, and standards for describing data and models.

Chapter 5 *Exploring the diversity of cancers*, examines how exploratory analysis of large amounts of cancer omics data can unveil the heterogeneity of cancers at the molecular level, and shed light on biological processes underpinning this diversity.

Chapter 6, *Prognosis and prediction: Towards individualised treatments*, focuses on the problem of predicting the evolution of cancer (prognosis) and response to therapies (prediction). It deals in particular with gene signatures, the statistics behind the construction of biomarkers, the supervised classification setting, the methods for feature selection, the problems of validation, and the possibility to include prior knowledge in predictive models.

Chapter 7, *Mathematical modelling applied to cancer cell biology*, exposes our motivation, goals and methods of modelling biological systems using formal mathematical tools. Two examples of mathematical models of cell cycle using both chemical kinetics and logical formalisms are presented. Motifs of feedback loops are also studied.

Chapter 8, *Mathematical modelling of cancer processes*, presents a review of the current status of knowledge around some of these hallmarks with a mathematical perspective and focus on some specific aspects of each hallmark and characteristics as an example.

Chapter 9, *Cancer robustness*, introduces the notion of robustness, one typical property whose study requires a computational systems biology approach. It provides a general review of relevant existing ideas in this rapidly evolving field. It describes the general mechanisms leading to robustness in biological systems in general and robustness of cancer in particular, and the use of the notion of robustness in cancer treatment strategies. Then in **Chapter 10**, the mathematical principles of biological and cancer robustness will be reviewed.

In **Chapter 11**, *Finding targets*, presents mathematical techniques for finding perturbations that should be applied to the cancer cell in order to disrupt some of its tumorigenic properties and achieve either reversal of the tumorigenic phenotype or pushing the cancer cell further on the way to cell death. Data-driven approaches based on application of statistical techniques are exposed. Techniques based on network analysis are then presented.

Chapter 12, the conclusion, tries to examine what the perspectives and challenges ahead in computational systems biology of cancer are, and also mention connected fields of research that we intentionally left out of the scope of this book.

In the **Appendices**, a reminder of the *basic principles of molecular biology of the cell* is provided and serves as a prerequisite to read **Chapter 2**. Moreover, the tools, software, databases and important genes mentioned in this book are compiled.

This book also contain a glossary; words which are defined therein appear as follows in the text: **tumour progression***.

The reader will find accompanying materials available on the website http://www.cancer-systems-biology.net:

- Most of the figures in the book are distributed under the Creative Commons license CC-BY-SA. Anybody is free to copy, distribute, transmit and adapt the figure under the following conditions: she/he must attribute the figure citing the present book, and if she/he alters, transforms the figures, she/he may distribute the resulting work only under the license CC-BY-SA.

- A tutorial provides an overview of an analysis scenario which can be performed on high-throughput data. As an illustration, the characterisation of gene expression patterns and DNA copy number alteration is presented on breast cancer data. Scripts and data are provided such that the reader can reproduce the analysis on her/his computer.

Chapter 2

Basic principles of the molecular biology of cancer

The human body consists of billions of different cells and thousands of cell types organised in many tissues. These cells complete their specific function according to a program using an information flow proposed by Crick (1970) in the *central dogma of molecular biology* (see **Appendices**). Cancer cells arise from these normal cells. Cancers are classified into four large groups according to the normal cell from which they originate. The majority of cancers (80%) arise from ephitelial normal cells and forms the first group called **carcinomas*** (*e.g.* breast, ovary, cervix, prostate, lung, pancreas, colon, *etc.*). All the other cancers arise from nonepithelial normal cells. The second group contains **sarcomas*** which derive from connective or supportive tissue (*e.g.* bone, cartilage, fat, muscle, blood vessels) and soft tissues. The third group arises from hematopoetic tissues (*i.e.* blood-forming cells) and includes **lymphomas*** and **leukaemias***. The last group consists of tumours arising from the central and peripheral nervous system and includes **glioblastomas***, **neuroblastomas***, **schwannomas*** and **medulloblastomas***. All cancers form solid tumours except in the case of leukaemias which generate circulating tumoral cells in the blood. Cancer is a heterogeneous disease in terms of morphological, clinical and molecular characteristics (see **Section 1.4**).

Despite this heterogeneity, cancers share many characteristics. They are described throughout this chapter which presents the series of events that transform a normal cell into a cancer cell during **tumour progression***. The molecular alterations which affect gene regulation and signal transduction mechanisms are detailed. The traits common in all cancers referred to as *hallmarks*, are enumerated in turn with a particular focus on chromosome aberrations. A brief overview of basic principles that are essential to understand the biology of cancer is proposed here. Comprehensive and excellent books such as Weinberg (2007) and Alberts et al. (2002) as well as the seminal reviews by Hanahan and Weinberg (2000, 2011) deserve to be studied for more details. This chapter assumes that the reader is familiar with the *basic principles of molecular biology of the cell* which are nevertheless reviewed in the **Appendices**.

2.1   Progressive accumulation of mutations

During a lifetime, cells die and have to be replaced to maintain the integrity of the organism. Therefore, cells reproduce during the cell cycle (see **Box 2.1**). In a normal human body, about 10^{16} cell divisions take place in the course of a lifetime. During cell division, there are fundamental limitations on the accuracy of DNA replication and repair so that **mutations*** occur spontaneously with a rate around 10^{-7} mutations per gene per cell division (Komarova, 2005). Main alterations are point mutations (*i.e.* modification of one nucleotide), insertion, deletion, or **amplification*** of DNA sequences. Mutation in the gene sequence can alter the structure and function of the gene product. However, in some case the mutation is *neutral*, (*i.e.* it results in a different but chemically similar amino acid) or even *silent* (*i.e.* it does not change the amino acid). In the context of cancer, a mutation is referred to as a genetic change and corresponds to a modification of the DNA sequence followed by modifications of gene products and related cellular mechanism. Mutations can also occur due to exposure to mutagenic agents such as chemical agents (*e.g.* tobacco smoke), physical agents (*e.g.* UV light) or biological agents (*e.g.* viruses). As a result, in a lifetime, each gene may have undergone a mutation on about 10^9 separate occasions. Among the resulting mutant cells, there will be many having perturbed mechanisms which cause disturbances in gene regulation. As a consequence, the harmonious behaviour of the cell with respect to its neighbours will be affected too. These changes are transmitted to the cell offspring. Inheritable nongenetic changes can also occur during the cell cycle. They correspond to epigenetic characteristic modifications (see **Section 2.4**) and are called **epimutations***. Thus, the number of mutations, including genetic mutations and epimutations is likely to be greater than 10^9 per gene. According to the aforementioned figures, the question about cancer does not seem to be *"why it occurs?"* but *"why it occurs so infrequently?"*. Clearly, if a single mutation is enough to convert a typical healthy cell into a cancer cell which proliferates without restraint, humans and animals would not be viable organisms. Hopefully, protective mechanisms such as apoptosis can eradicate abnormal cells from the organisms. Therefore, many mutations are needed to transform a normal cell into a cancer cell which can bypass these control mechanisms.

Why are so many mutations needed? One explanation is that cellular processes are controlled in complex and interconnected ways: cells use redundant regulatory mechanisms to help them to maintain tight and precise control over their behaviour and integrity. Thus, many different regulatory systems have to be disrupted before a cell can throw off its normal restraints and behave defiantly as a malignant cancer cell. In addition, tumour cells may meet new barriers to further expand at each stage of the evolutionary process and therefore need to acquire additional mutations. Therefore cancer is a multistep process which requires an accumulation of several mutations (Vogelstein and Kinzler, 1993). This explains why its prevalence increases with the age of individuals. However, some cancers, such as paediatric cancers, involve less complex mechanisms of **tumorigenesis***, when only one driving mutation is able to trigger tumor growth.

❏ **BOX 2.1: Cell cycle**

The cell cycle can be divided in four successive phases: G1, S, G2 and M. During G1, the cell grows and prepares itself for S phase, during which DNA replication occurs. From one double-stranded DNA molecule (chromosome), two identical double-stranded DNA molecules (called sister chromatids) are formed and held together by cohesion proteins. The G2 phase is the temporal gap between the end of replication and the beginning of mitosis. During the M phase, replicated DNA molecules are segregated to daughter cells. The sister chromatids separate so that the daughter cells receive one copy of each chromosome. The cell cycle is the progression through these four phases.

In essence, cancer is a genetic disease but is caused by **somatic mutations*** while other genetic diseases are caused by germ-line mutations. However, familial forms of cancer exist. In these cases, mutations which have been inherited from parents can be already present in all the cells of the body and transmitted from one generation to another. In breast cancer, we consider that about 15% of cancer cases can be attributed to inherited predisposition due to the presence of gene mutations. The well-known examples in breast cancer are the mutations of *BRCA1* and *BRCA2* genes which are involved in DNA repair during the cell cycle. These two genes account for about 16% of the familial risk of breast cancer. Since a mutation is constitutively present in the cells, the normal function of *BRCA* genes relies only on the remaining wild type allele. Therefore, a patient who carries *BRCA1* or *BRCA2* mutations has a 10- to 20-fold higher risk of developing breast cancer (Stratton and Rahman, 2008). Whatever the cancer, the identification of new susceptibility alleles has direct application in the implementation of cancer prevention strategies.

2.2 Cancer-critical genes

Cancer is a disease of genes caused by the accumulation of several mutations. The most important genes whose alterations are causal in tumour progression are named cancer-critical genes (Alberts et al., 2002) and most of the known ones are reported in Vogelstein and Kinzler (2004), Bunz (2008) and in the Catalogue of Somatic Mutations in Cancer (COSMIC) (Forbes et al., 2008,

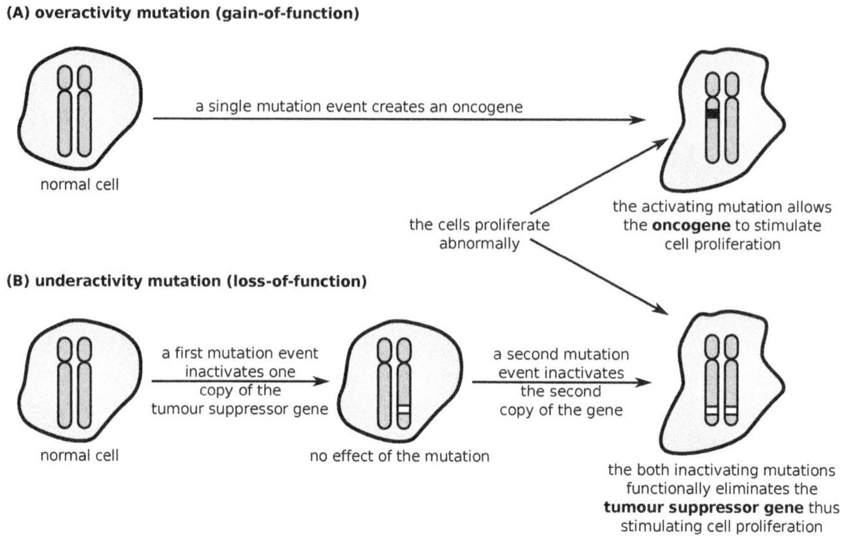

(A) overactivity mutation (gain-of-function)

a single mutation event creates an oncogene

normal cell

the activating mutation allows the **oncogene** to stimulate cell proliferation

the cells proliferate abnormally

(B) underactivity mutation (loss-of-function)

a first mutation event inactivates one copy of the tumour suppressor gene

normal cell

no effect of the mutation

a second mutation event inactivates the second copy of the gene

the both inactivating mutations functionally eliminates the **tumour suppressor gene** thus stimulating cell proliferation

FIGURE 2.1 Oncogene and tumour suppressor gene. (A) Oncogenes are dominant: a gain-of-function in one copy of a gene can induce cancer. (B) Tumour suppressor genes are generally recessive: both alleles of the gene must be lost to induce cancer. Exceptions to this rule exist and are detailed in **Figure 2.3**. Activating and inactivating mutations are represented by black and white boxes respectively. Image and legend adapted from Alberts et al. (2002).

2011). They are grouped into two categories, according to whether cancer risk is increased due to a high or little activity of the gene product.

2.2.1   Oncogenes

The first category contains genes for which a gain-of-function mutation increases cancer risk. They are called **proto-oncogenes** and their mutant and overactive forms are called **oncogenes** (see **Figure 2.1A**). Oncogenes encode proteins which control cell proliferation, apoptosis, or both. They can be activated by structural alterations resulting from mutation (see **Figure 2.2A**), amplification (see **Figure 2.2B**) and chromosome rearrangements causing a juxtaposition to enhancer elements (see **Figure 2.2C1**) or a fusion gene (see **Figure 2.2C2**) (Croce, 2008; Alberts et al., 2002). Mutations and **translocations**[*] can occur as initiating events or during tumour progression, whereas amplification usually occurs during tumour progression. The products of oncogenes can be classified in six broad groups: transcription factors (*e.g.* amplification of *MYCN* in neuroblastoma), chromatin remodelers (*e.g.* fusion gene with *MLL* in leukaemia), growth factors (*e.g.* amplification of *FGF4* in

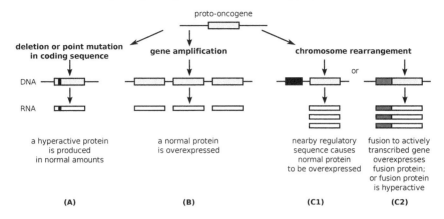

FIGURE 2.2 From proto-oncogene to oncogene. Three mechanisms can transform a proto-oncogene into an oncogene: (A) mutations, (B) gene amplification and (C) chromosome rearrangements. Image adapted from Alberts et al. (2002).

Kaposi's[1] sarcoma), growth factor receptors (*e.g.* mutation of *FGFR3* in bladder cancer, see **Section 5.5.5**), signal transducers (*e.g.* mutation of *KRAS* in colon cancer), and apoptosis regulators (*e.g.* fusion gene with *BCL2* in lymphoma).

The example of the *MYC* oncogene family. In neuroblastoma an amplification of *MYCN* which overproduces this protein is frequently observed (see **Figure 2.2B**). Other mechanisms involving translocation and the fusion gene also play an important role (Mitelman et al., 2007). For example, if a proto-oncogene is translocated near a DNA domain which normally regulates the constitutional expression of a gene at a high-level, then the proto-oncogene will be highly transcribed and becomes an oncogene (see **Figure 2.2C1**). This occurs in Burkitt's lymphoma where *MYC* is juxtaposed with regulatory elements of the immunoglobulin heavy chain *IGH* gene. The *MYC* gene is constitutively activated because its expression is driven by immunoglobulin regulatory elements. In this case the translocation leads to an upregulation of the proto-oncogene.

The example of the *EWS/FLI1* oncogene. Translocation can also create a new gene called a chimeric gene or the fusion gene (see **Figure 2.2C2**) by fusion of two existing genes. This phenomenon occurs in Ewing's sarcoma[2] where a translocation between chromosomes 11 and 22 merges a portion of the *EWS* gene and the *FLI1* gene into the chimeric oncogenic transcription factor gene *EWS/FLI1*. In these tumours, the fusion gene often involves *EWS* mainly with *FLI1* and less frequently with other genes.

---

[1]Kaposi's sarcoma is a tumour caused by a human herpesvirus.
[2]Ewing's sarcoma is a paediatric cancer in which tumours are found in the bone or in soft tissue.

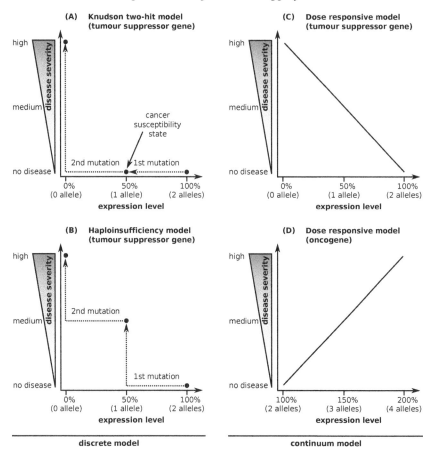

FIGURE 2.3 Models of tumour suppression. For different models, the severity of the disease is explained as a function of the expression level of the gene. (A) In the Knudson two-hit model, two mutations are required to inactivate the tumour suppressor gene and promote tumorigenesis. This is the case in retinoblastoma with the *RB* gene. (B) In the case of haploinsufficiency, a monoallelic inactivation of the tumour suppressor gene can form cancer. A biallelic inactivation enhances tumour progression and metastasis. This mechanism occurs for *TP53*. (C) In the continuum model for the tumour suppressor gene, decreasing slightly the expression level or the protein activity increases the severity of the disease. (D) The continuum model can be applied to oncogenes too; increasing the expression level of the oncogene increases the severity of the disease. Image adapted from Berger et al. (2011)

2.2.2   Tumour suppressor genes

The second category contains genes for which a loss-of-function mutation creates the danger. They are called **tumour suppressor genes** and have cancer-preventive effects which usually require the presence of only a single functional gene. To give rise to cancer these genes have generally to undergo biallellic inactivation in tumours; this is known as the Knudson two-hit model (see **Figure 2.1B**, **Figure 2.3A** and Knudson, 1971). Inheritance of a single mutant allele increases tumour susceptibility, because only one additional mutation is required for complete loss of gene function. This is why some tumour suppressor genes have been identified in familial forms of cancer such as *RB* in **retinoblastoma*** (Knudson, 1971). *RB* encodes a protein involved in the control of the cell cycle and its activity is deregulated in virtually all types of cancers (Weinberg, 2007), including both familial and **sporadic*** forms (osteo**sarcomas***, breast **carcinomas***, small cell lung carcinomas, bladder carcinomas, **melanomas***, *etc.*). As for oncogenes, tumour suppressor genes are involved in many functions (Sherr, 2004). Among tumour suppressor genes let us mention *TP53* which encodes a transcription factor involved in genome integrity maintenance, *ATM* which encodes a protein kinase involved in DNA damage signal transduction (see **Figure A.6**), *BRCA1* and *BRCA2* which encode proteins involved in DNA repair during the cell cycle.

The example of the *RB* tumour suppressor gene. If a mutation inherited from one parent or arisen spontaneously in one allele during replication occurs in *RB*, the protein coded by the allele is nonfunctional. Then, a second event occurs. For example, the duplication of the chromosome carrying the mutation is followed by the loss of the chromosome carrying the functional gene (*i.e.* the Loss of Heterozygosity (LOH) event depicted in **Figure 2.13B**). A a result, *RB* is present in two copies of nonfunctional forms and will no longer be able to control the cell cycle. This is the Knudson two-hit model. From the different aberration configurations depicted in **Figure 2.13** we can imagine other combinations of alterations which also lead to the same effect.

The Knudson two-hit model implies the inactivation of both alleles to initiate tumorigenesis (see **Figure 2.1B** and **Figure 2.3A**). However, there is evidence that not only biallelic inactivation but also monoallelic inactivation of tumour suppressor genes can cause cancer. The main mechanisms are called **negative dominance*** and **haploinsufficiency*** (Berger and Pandolfi, 2011; Berger et al., 2011). For example, in the case of haploinsufficiency (see **Figure 2.3B**), a single-copy loss of *TP53* is sufficient for the development of cancer, but the complete loss of *TP53* further enhances and promotes tumour progression and metastasis (Berger and Pandolfi, 2011). The Knudson two-hit model and the haploinsufficiency model are *discrete models* in the sense that the successive mutations fully inactivate each allele in a step-wise fashion. Berger and Pandolfi (2011) proposed a *continuum model* which assumes that even subtle decreases in gene expression level or protein activity of the tumour suppressor gene can be relevant to cancer formation. The severity of cancer is related to a continuum of decreasing tumour suppressor gene expression level, rather than to discrete changes in DNA copy number (see **Figure 2.3C**). The continuum model holds for the case of an oncogene too. Increasing its expression level will generally be positively correlated with the severity of the disease (see **Figure 2.3D**). This continuum model theory is still under debate.

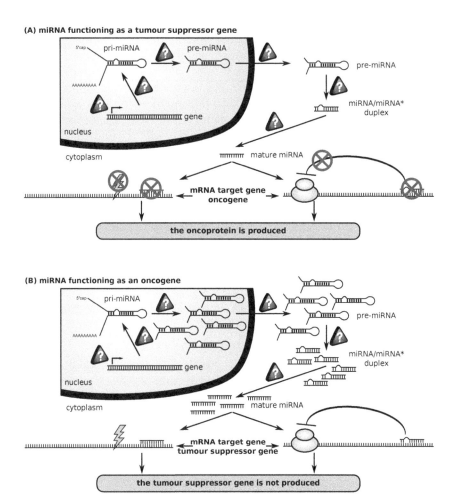

FIGURE 2.4 Role of miRNA in a cancer cell. Any defects in microRNA (miRNA) biogenesis (indicated by question marks) can lead to inappropriate levels of the target proteins. (A) The reduction or deletion of a miRNA which acts as a tumour suppressor leads to an unexpected high level of the target oncoprotein. (B) The amplification or overexpression of a miRNA which acts as an oncogene leads to the elimination of the target tumour suppressor protein. Image and legend adapted from Esquela-Kerscher and Slack (2006).

2.2.3 Non-protein-coding cancer-critical genes

Besides protein-coding genes, noncoding RNA (ncRNA) and especially microRNAs (miRNA) can act as either an oncogene or a tumour suppressor gene (Esquela-Kerscher and Slack, 2006; Fabbri et al., 2008). A defect in miRNA gene regulation leading to a loss or an amplification of miRNAs has been reported in a variety of cancers (Calin and Croce, 2006). Typically, an underexpression of a miRNA which targets an oncogene or an overexpression of a miRNA which targets a tumour suppressor gene will have an impact on cancer development (see **Figure 2.4**).

2.3 Evolution of tumour cell populations

2.3.1 Clonal origin

The clonal evolution of tumour cell populations was proposed by Nowell (1976) and claims that tumour cells originate from a single cell which has acquired an accumulation of mutations as described previously. The model argues that tumour development proceeds via a process similar to Darwinian evolution, in which a series of random genetic changes, each conferring one or another type of growth advantage, are selected and lead to the progressive transformation of normal cells into cancer cells (Hanahan and Weinberg, 2000). From an initial population of slightly abnormal cells, descendants of a single mutant ancestor evolve from bad to worse through successive cycles of mutation and natural selection. At each stage, one cell acquires an additional mutation which gives it a selective advantage over its neighbours, making it better able to thrive in its environment which, inside a tumour, may be harsh, with low levels of oxygen and scarce nutrients. The progeny of this well-adapted cell will continue to divide, eventually becoming the dominant clone in the developing cancer. Thus, tumours grow in fits and starts, as additional advantageous mutations occur and the cells bearing them proliferate. Tumour progression usually takes many years.

2.3.2 Stemness of cancer cells

Normal stem cells have the capacity to self-renew, meaning undergo divisions which allow their number to remain constant and give rise to a variety of differentiated cells. By analogy to stem cells, **Cancer Stem Cells (CSC)**[*] have the same properties. In the CSC model, the tumour is viewed as an abnormal organ with stem cells driving the growth (see **Figure 2.5**). This model implies that, in a cancer with a defined set of genetic alterations, a mixture of cell types with a different malignant potential are present. In a

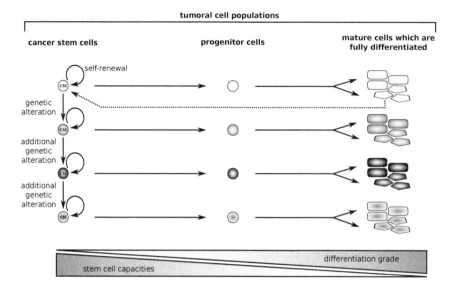

FIGURE 2.5 Hierarchical organisation of a tumour according to the CSC model. CSCs acquire additional genetic alterations during tumour progression which can be beneficial for the tumour. These additional alterations form new subpopulations within the tumour. Each CSC has the ability to self-renew and to produce more differentiated cells. In this model, selection pressure is predicted to act on the CSC level. The dotted arrow indicates the possibility for differentiated cells to be converted back to stem cells due to the accumulation of mutation. Image and legend adapted from Vermeulen et al. (2008).

tumour both differentiated cells which have lost the capacity to propagate a tumour, and cells which retain a clonogenic capacity exist. The proposed hierarchical organisation of a tumour could be easily integrated into the classical clonal selection proposed in Nowell's theory. As explained before, this theory views a tumour as a clonally-derived cell population, which acquires new potentially advantageous mutations and gives rise to new more rapidly proliferating clones. When one integrates the CSC theory into this model, the selection pressure is predicted to act at the level of the CSC compartment. This does not mean, however, that certain features present only in more differentiated cells in the tumour could not be subject to selection, especially if they increase the growth rate of the CSCs from which they are derived (Vermeulen et al., 2008).

The CSC model is still under debate and the question is whether the cell of origin of the CSC has to be a stem cell or whether the accumulation of mutations converts differentiated cells back into stem cells. In this theory, the only cells capable of initiating and driving tumour growth are CSCs and it is logical to assume that a **metastasis*** arises from CSCs.

2.4 Alterations of gene regulation and signal transduction mechanisms

In normal cells, the gene regulation and signal transduction held in the central dogma of molecular biology involve different mechanisms (see **Section A.1.2**). Here we give some examples of the alterations of these mechanisms occuring in cancer cells due to the progressive accumulation of mutations.

2.4.1 Modification of transcription factor activity

In cancer, many transcription factors are involved in the tumour progression mechanism. The most famous one is incontestably the tumour suppressor gene *TP53*, also known as *the guardian of the genome*; it plays a key role in preserving the integrity of the genome (see **Section A.1.2**). *TP53* is directly involved in many cancers due to the presence of mutations in the gene which encodes this protein (30 to 50% of common human cancers have a *TP53* mutation — figures from Weinberg, 2007). As a consequence, mutated *TP53* loses a part of its transcription factor functions since it can no longer bind to all its target genes which can also no longer be transcribed into messenger RNA (mRNA) (Vogelstein and Kinzler, 2004). Therefore, when mutated, *TP53* can no longer play its guardian's role efficiently.

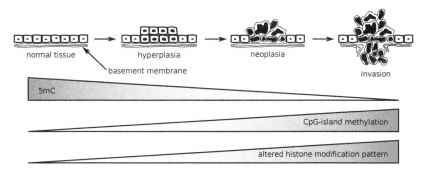

FIGURE 2.6 Epigenetic alterations during tumour progression in carcinomas. During the tumour progression normal cells undergo transformation and progress through different stages including **hyperplasia**[*], **neoplasia**[*] and invasion. In this process, there is a progressive loss of total DNA methylation content, an increased frequency of hypermethylated CpG islands, and an increased histone-modification imbalance. 5mC stands for 5-methylcytosine which is the result of the DNA methylation. Image adapted from Esteller (2008).

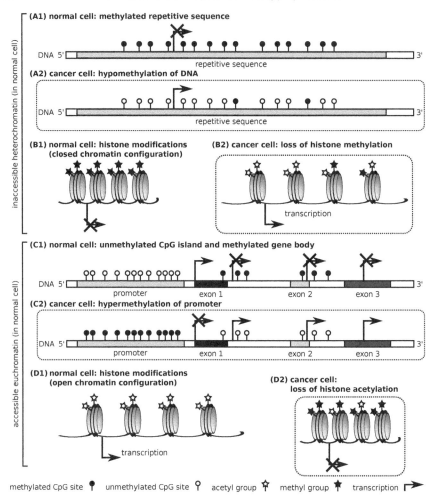

FIGURE 2.7 Epigenetics patterns in a normal and cancer cells. The figure describes the DNA methylation and histone modification patterns observed in cancer cells (dotted boxes) with respect to their normal-tissue counterparts. (upper part) Repetitive sequences are silenced by DNA methylation (A1) and histone methylation (B1). These modifications prevent chromosomal instability, translocations and gene disruption through the reactivation of endoparasitic sequences. In cancer cells, there is a loss of DNA methylation (A2) and a loss of histone methylation (B2). (bottom part) For transcriptionally active genes their CpG islands in the promoter region are unmethylated (C1). However, methylation in the gene body facilitates transcription and prevents spurious transcription start sites. The histone acetylation (D1) allows the acessibility of transcription factors and polymerase in order to initiate tanscription. In cancer cells, modifications of methylation and acetylation marks in histones impact the gene regulation (B2 and D2). Image and legend adapted from Portela and Esteller (2010); Rodríguez-Paredes and Esteller (2011); Gibney and Nolan (2010); Taby and Issa (2010); Schones and Zhao (2008).

2.4.2 Epigenetic modifications

In normal cells, epigenetic patterns serve as mechanisms to regulate gene transcription and are also involved in the genomic stability (see **Appendices**). In cancer, these patterns are altered according to three types of modifications which occur during tumour progression (see **Figure 2.6** and **Figure 2.7**).

First, **hypomethylation** of DNA in tumours (see **Figure 2.7A2**) as compared to the level of DNA methylation in their normal-tissue counterparts (see **Figure 2.7A1**) was one of the first epigenetic alteration found in human cancers. Cancer cells are characterised by a massive global loss of DNA methylation with about 20–60% less overall 5-methyl-cytosine (Portela and Esteller, 2010, for a review, see). The loss of methylation is mainly due to hypomethylation of repetitive DNA sequences and extensive hypomethylated genomic regions in gene-poor areas. During the development of the disease, the degree of hypomethylation of genomic DNA increases as the lesion progresses from a benign proliferation of cells to an invasive cancer. This hypomethylation increases chromosome instability leading to deletion, translocations and chromosome rearrangements (see **Section 2.8**). This was observed by Shann et al. (2008) in breast cancer cell lines. They have also shown that genes with intragenic hypomethylation had a low level of expression. The loss of methyl group from DNA can also cause loss of **genomic imprinting**[*] and can contribute to gene activation in some types of cancers. This is the case for *IGF2*, which increases the risk factor for colorectal cancer (Esteller, 2008).

Then, **hypermethylation** of promoter CpG island is a key process in tumour progression and leads to the transcriptional silencing of tumour suppressor genes (see **Figure 2.7C2**). In a normal cell, the promoter is generally not methylated to allow the transcription (see **Figure 2.7C1**). Within the promoter CpG islands, the hypermethylation of specific locations (*i.e.* the core regions) have a crucial role in gene silencing (van Vlodrop et al., 2011). The profiles of hypermethylation of the CpG islands in tumour suppressor genes are specific to the type of cancer. Hypermethylation can be one of the lesions in the Knudson two-hit model.

Finally, **global alterations of histone modification patterns** have the potential to affect the structure and the integrity of the genome, and to disrupt normal patterns of gene expression. According to its transcriptional state, the human genome can be divided into actively transcribed euchromatin and transcriptionally inactive heterochromatin. In normal cells, heterochromatin is characterised by low levels of acetylation and high levels of certain types of methylation (see **Figure 2.7B1**) while euchromatin is characterised by high levels of acetylation (see **Figure 2.7D1**). In cancer cells, methylation and acetylation marks are altered and impact on the gene transcription (see **Figure 2.7B2** and **Figure 2.7D2**). As alterations in DNA methylation, these modifications may be causal factors in cancer. In bladder cancer, Stransky et al. (2006) have shown that certain types of histone methylation can occur in large genomic regions and lead to the loss of expression of neighbouring genes inside this region. This phenomenon is known as long-range epigenetic silencing. In colorectal cancer, Frigola et al. (2006) have found that this long-range epigenetic silencing at the level of histone methylation could also be associated with DNA methylation.

In breast cancer, the distribution of aberrantly methylated regions across the genome was also found to be nonrandom and tended to concentrate in relatively small genomic regions spanning up to several hundred kilobases (Novak et al., 2006, 2008). All the aforementioned epimutations can be one of the lesions of the Knudson two-hit model as they can silence one allele (or even both) of tumour suppressor genes. The understanding of all these epigenetic changes and their contribution to tumour progression is very important for further progress in the field of diagnosis, **prognosis**[*] and therapy as reported by Jovanovic et al. (2010).

2.4.3 Modification of the post-transcriptional regulations

After transcription, a RNA transcript can be processed into different mRNAs to increase the diversity of proteins in a process called alternative splicing (see **Figure A.4**). In cancer cells, there are aberrant alternative splice forms which are not found in normal cells (Venables, 2004; Srebrow and Kornblihtt, 2006; Kim et al., 2008). They provide the cell with new functions. Moreover, as mentioned in **Section 2.2**, miRNAs can act either as oncogenes or tumour suppressors. Indeed, any alteration which affects its biogenesis and its ability to bind target genes will prevent its important role in protein-coding gene regulation (Calin and Croce, 2006).

2.4.4 Disruption of signal transduction

Disruptions in **signalling pathways**[*] caused by any alteration which modifies the signal transmission are involved in cancer development. For example, an inactivation of the tumour suppressor kinase gene *ATM* will prevent TP53 from being activated and will impact the integrity of the cell in case of DNA damage. Another example is the overexpression of the *HER2* oncogene (also known as *ERBB2*). *HER2* is membrane surface-bound receptor tyrosine kinase and is normally involved in the signal transduction pathways leading to cell growth and differentiation. In about 20% of breast cancers, the *HER2* kinase is overexpressed[3] due to an amplification (see **Section 2.8**) and induces an extensive activation of the signal transduction cascade, which causes very aggressive cancers with high metastatic risk.

In the general case, the alteration of the signal transduction can be either an absence of the signal amplification while the signal should be amplified or an overamplification of the signal while it should not. Protein kinases are key

---

[3]A normal cell has 20,000 HER2 receptors while there are about 1.5 million in a HER2 positive cancer cell.

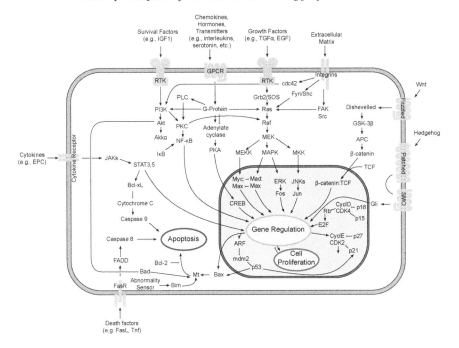

FIGURE 2.8 Signalling pathways in the cell. Many different and inter-connected signalling pathways allow the cell to integrate environmental stim-uli. This complex network mediates commands such that the cell can live, re-produce or die. Image adapted from Hanahan and Weinberg (2000), source: http://commons.wikimedia.org/wiki/File:Signal_transduction_pathways.svg.

players which control the signal transduction cascade and are often altered in cancers. Therefore, they represent therapeutic targets for drugs. For example, drugs such as imatinib and gefitinib are small-molecule kinase inhibitors, and trastuzumab[4] is an antibody which targets the HER2 kinase receptor.

2.5 Cancer is a network disease

In the cell, many different signalling pathways and gene regulatory net-works regulate specific cellular processes (see **Figure 2.8** and **Box 4.3**). Al-though each pathway and network are involved in a specific function they

---

[4]Trastuzumab is an antibody which interferes with the kinase activity of HER2 and recruits immune effector cells which are responsible for antibody-dependent cytotoxicity. Trastuzumab can also induce complement-dependent cytolysis and enhance phagocytosis by Fc-receptors bearing antigen-presenting cells (Hudis, 2007).

are not independent but intertwined with complex crosstalks. As discussed previously, cancer is a gene disease but requires a multistep accumulation of mutations and defects during the lifetime. This can be easily understood as the huge and complex interconnections between regulators aims at making the cell robust with respect to perturbations (see **Chapter 9**). Therefore tumour progression cannot only be viewed as a gene disease but a network disease too. As a result the systems biology approach offers a natural framework for modelling tumour progression.

2.6   Tumour microenvironment

Solid tumours consist of a mixture of cancerous cells and noncancerous cells. The latter include fibroblasts, immune cells, pericytes and endothelial cells, and are referred to as stromal cells. This makes a tumour a very complex system, as many different cell types are present simultaneously. Moreover, as each cell releases many different biomolecules, the medium in which the tumour lives is very complex too. Clearly, the biology of tumours can no longer be understood simply by enumerating the traits of the cancer cells but instead must take into account the contributions of the *tumour microenvironment* to tumour progression. A major component of the microenvironment is the **Extracellular Matrix (ECM)***. In cancer the ECM is commonly deregulated, becomes disorganised and affects tumour progression (Lu et al., 2012). Understanding the role of tumour microenvironment is a huge challenge and will require sophisticated integrative approaches. Mathematical models integrating the role of the tumour microenvironment will be reviewed in **Section 8.1.3**.

The microenvironment not only plays a role in tumorigenesis but is also involved in drug response and the resistance of tumours to treatment. Indeed, for an anticancer drug to kill cancer cells, it must be distributed throughout the tumour vasculature, cross blood vessel walls, and diffuse in the tumour tissue (Trédan et al., 2007). Moreover, the efficiency of the drug transport, absorption and metabolism depends on the genetic background of the patient, as Single Nucleotide Polymorphism (SNP) can affect pharmacokinetics and pharmacodynamics (Wiechec and Hansen, 2009). Taking into account the genetic variability at the level of drug response will be an essential component of personalised treatment of cancer.

2.7 Hallmarks of cancer

The progressive accumulation of mutations in cancer-critical genes alters gene regulation and signal transduction mechanisms. Although a large variety of mutations can occur, all cancers have traits in common. Indeed, to lead successfully to a tumour, a cell must acquire a whole range of aberrant properties and subversive new skills as it evolves. Different cancers require different combinations of properties. Nevertheless, we can draw up a short list of the key typical behaviours of cancer cells.

2.7.1 Hallmark capabilities

Hanahan and Weinberg (2000) suggest that the vast catalog of cancer cell genotypes is a manifestation of six *acquired capabilities* which alter cell physiology and dictate malignant growth (see **Figure 2.11**). They are the following:

sustaining proliferative signalling: cancer cells can become independent of exogenous growth signals to proliferate.

evading growth suppressors: cancer cells are insensitive to signals which block cell proliferation.

activating invasion and metastasis: a tumour can invade adjacent tissues. It can spawn pioneer cells out of the **primary tumour*** site and become metastatic.

enabling replicative immortality: cancer cells become immortal because they are able to pass through an unlimited number of successive cell cycle divisions[5].

inducing angiogenesis: a tumour can induce the formation of new blood vessels such that it can be supplied with nutrients and evacuate wastes.

resisting cell death: the cell death machinery is inefficient.

These six capabilities' hallmarks can be merged into the three main following properties which allow the cancer cells to proliferate, survive and disseminate.

The first property is the **defective control of the cell cycle** (*i.e. sustaining proliferative signalling* and *evading growth suppressors* hallmarks). The cell cycle normally ensures that the number of cells within the organism remains constant so that when a cell dies a new one is born. Therefore, the cell cycle must be precisely controlled. In cancer, this control is inefficient and

---

[5]Normal cells can divide between 40 and 60 times at most which is called the Hayflick limit.

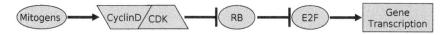

FIGURE 2.9 Simplified representation of the mechanism regulating RB activity. External signals such as mitogens trigger the activation of the Cyclin D/CDK protein complex which in turn starts phosphorylating and inhibiting RB, thereby freeing E2F which can then activate the transcription of cell cycle genes. More realistic and comprehensive representation is shown in **Figure 4.5**.

the cells are continuously reproducing. Many mechanisms are involved in the cell cycle. We will only mention here the key role of RB, which controls the initiation of the cell cycle. When RB is active, it functions as a cell cycle brake by inhibiting transcription. This inhibition is done by sequestering a family of transcription factors, the E2F, responsible for the transcription of genes involved in cell cycle regulation, DNA replication and apoptosis. The brake or checkpoint is lifted when the cell receives the appropriate external signals, such as growth factors. The cell can then enter the replicative phase. RB is active when hypophosphorylated and loses its grip on the *E2F* family members when phosphorylated, first by G1 cyclin-dependent kinases, sensors of growth (*CDK4/CCND1* or CyclinD1), and then by other cyclin-dependent complexes later in the cycle (*CDK2/CCNE* (CyclinE) and *CDK2/CCNA* (CyclinA), not shown) (see **Figure 2.9**). The deficiency of the RB pathway is observed in cancers when (1) the external signals are constant (*e.g.* when cell growth factors are always present or when their receptors are mutated and as a result the cell believes that the signal is always on) (see **Section 8.1.1** for a review of mathematical models), (2) the cell is insensitive to anti-growth factors (*e.g.* TGFβ) (see **Section 8.1.2** for a review of mathematical models), or (3) when processes that control the passage through the restriction point (see **Box 2.2**) are perturbed by mutations or cellular dysfunctions (Sherr, 1996) (see **Chapter 7**). More precisely, in many cancers, alterations of RB activity itself arise from the deletion of the gene, promoter hypermethylation, the presence of viral oncoproteins, LOH or as a result of a deregulation of kinases or kinase inhibitors.

The second property is the **defective control of cell death** (*i.e. enabling replicative immortality* and *resisting cell death* hallmarks). The cell death process is the process of cell destruction. First, this mechanism occurs when cells have reached a limited number of cell cycles; this process is called *senescence* and is caused by the telomere shortening which occurs during each cell cycle. Second, it happens when the genome integrity is compromised during the cell cycle and the cell commits suicide; this process is called *apoptosis*. In the apoptosis process, the machinery can be broadly divided into two classes of components: sensors and effectors. The sensors (*i.e.* the proteins involved in the signal transduction) are responsible for monitoring the extracellular and intracellular environment for conditions of normality or abnormality leading

---

❑ **BOX 2.2: Restriction point**

The restriction point is an irreversible transition occurring in the late G1 phase after which the cell commits to DNA replication and cell cycle progression. In the absence of growth factors, the cell enters a G0 (or quiescent) state. In the presence of growth factors, growth is stimulated and the cell advances through the G1 phase. During the time preceding the restriction point transition, the cell is responsive to both positive and negative external factors (Pardee, 1974). However, when it passes through the restriction point in late G1, the cell commits to cell cycle events and proceeds to DNA synthesis. After that point, even if the growth factors are removed, and in the absence of other cell dysfunctions, the cell completes its cycle before stopping at the following round.

---

to life or death of a cell (see **Section A.1.2**). These signals regulate the second class of components, which function as effectors of apoptotic death. One of the major players among effectors is TP53 which can trigger apoptosis in case of DNA damage during replication. Defective apoptosis can be due to a mutation of both TP53 and the sensors like ATM (see **Section 8.1.5** for a review of mathematical models).

The last property is the **invasiveness and metastatic potentials** (*i.e. activating invasion and metastasis* and *inducing angiogenesis* hallmarks). The uncontrolled proliferation of cells leads to damage of the organ in which the cancer cell originates. In addition, if adjacent tissues are invaded, it may also damage neighbouring organs (see **Section 8.1.3** for a review of mathematical models). Moreover, the tumour has the ability to induce the formation of new blood vessels. This process is called *angiogenesis* (see **Section 8.1.4** for a review of mathematical models). Therefore the tumour has the possibility to spawn cancer cells in the blood. As a result, Circulating Tumour Cells (CTC) are generally observed in the blood of patients with advanced primary carcinomas. These cells constitute a reservoir to seed metastases able to colonise distant tissues. By spreading throughout the body, a cancer becomes almost impossible to eradicate surgically or by localised irradiation, and thus can become deadly. Metastases are the cause of 90% of human cancer deaths (Hanahan and Weinberg, 2000). Chaffer and Weinberg (2011) proposed that the complex metastasis process can be divided into two major phases. The first phase consists of the translocation of a cancer cell from the primary tumour to the distant tissue and the second phase consists of the colonisation. The concept of **Epithelial-to-Mesenchymal Transition (EMT)*** used in the field of developmental biology to explain early embryogenesis morphology and the reverse process (MET) may provide the basis to decipher the mechanism of the mestastatic process (Chaffer and Weinberg, 2011; Thompson and Haviv, 2011).

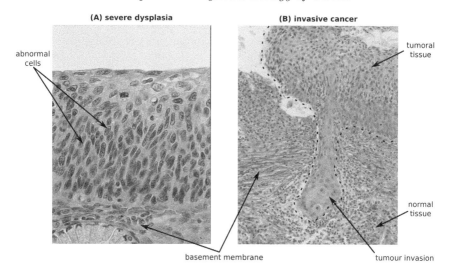

FIGURE 2.10 Tumour progression and invasion in cervix cancer. (A) In the severe dysplasia state, cells with abnormal morphology are present. (B) In invasive cancer, the cancer cells infiltrate adjacent tissues as a close analogy with a water drop. The dashed line separates the cancer cells (upper part) from the adjacent tissue (lower part). Image courtesy of Dr. Xavier Sastre-Garau, Institut Curie. © 2012 Institut Curie. (See colour insert.)

Besides developmental biology, fluid mechanics theory provides an interesting framework to model the tumoral invasion (Guiot et al., 2007b,a; Shieh and Swartz, 2011; Shieh et al., 2011; Wirtz et al., 2011). Indeed, the tumoral tissue and the normal tissue on which it lies can be regarded as two different fluids with their own rheological properties such as viscosity and elasticity. As long as these properties remain stable during tumour progression, both fluids are one above the other and the tumour remains *in situ*. However, biophysical and biochemical modifications occur during tumour progression. They can affect the rheological properties within the tumoral microenvironment and the surrounding tissues. As a result, the equilibrium state can be broken and both fluids can mix together thus allowing the tumour to invade adjacent tissues, as a close analogy with a water drop (see **Figure 2.10**).

2.7.2   Emerging hallmarks

Hanahan and Weinberg (2011) proposed two other hallmarks called *emerging hallmarks* which are also involved in tumorigenesis:

reprogramming energy metabolism: cancer cells have developed adjustments of energy metabolism in order to fuel cell growth and division and therefore sustain uncontrolled cell proliferation.

evading immune destruction: tumours manage to avoid detection by the immune system or are capable of limiting the extent of immunological killing, thereby evading destruction.

Regarding the energy metabolism, **Section 8.1.8** will review the mathematical models which have been proposed.

2.7.3 Enabling characteristics

Distinct mechanisms allow the normal cells to acquire the capabilities at different stages during the tumour progression. Hanahan and Weinberg (2011) proposed that this acquisition is made possible by the two following *enabling characteristics*:

genome instability and mutation: the genome instability generates mutations (see **Section 2.1**) including chromosome aberrations (see **Section 2.8**), thus providing the cancer cells with selective advantage. This enabling characteristic is causally associated with the acquisition of hallmark capabilities.

tumour-promoting inflammation: immune cells are normally expected to eradicate abnormal cells in the organism. However, the inflammatory response has a paradoxical effect of enhancing tumorigenesis and progression. Indeed, inflammation can contribute to multiple hallmark capabilities by supplying bioactive molecules useful to the tumour microenvironment (*e.g.* growth factor for the tumour cells and blood vessels, proteases which facilitate invasion).

Sections **8.1.6** and **8.1.7** will review the mathematical models which have been proposed.

In total, Hanahan and Weinberg (2000, 2011) proposed eight different hallmarks and two enabling characteristics which are common to almost every cancer. **Chapter 8** will describe different mathematical models based on systems biology approaches addressing these hallmarks.

2.8 Chromosome aberrations in cancer

The *genome instability and mutation* hallmark described in **Figure 2.11** leads to chromosome aberrations. They are detailed in this section as they have largely been studied in particular using high-throughput technologies described in **Chapter 3**.

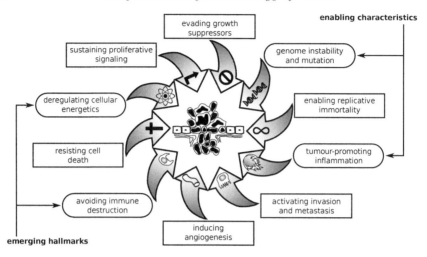

FIGURE 2.11 Hallmarks of cancer. Most if not all cancers have acquired the same set of six functional capabilities during their development, albeit through various mechanistic strategies. This acquisition has been made possible due to enabling characteristics. Moreover, tumours show common emerging hallmarks. Image adapted from Hanahan and Weinberg (2000, 2011).

2.8.1 Abnormal karyotype in cancer

The normal configuration of chromosomes is often termed the **euploid** karyotype state. Euploidy implies that each of the chromosomes is present in normally structured pairs. Deviation from the euploid karyotype is **aneuploidy** and is observed in many cancers (see **Figure 2.12**). Often, this aneuploidy is merely a consequence of the general chaos which reigns within a cancer cell due to the progressive accumulation of mutations. Indeed, once a critical number of mutations is reached, the cell cannot correctly perform the duplication and the segregation of the chromosomes because of defects in DNA repair and cell cycle checkpoints leading to a genome instability (Aguilera and Gómez-González, 2008). This instability occurs at both nucleotidic (*i.e.* imperfect copy of the DNA sequence) and/or chromosomal levels (*i.e.* improper number of chromosomes). As a result, the daughter cells will have pertubated functionalities. In 1914, Theodor Boveri proposed the hypothesis that cancer cells derive from cells with an irreparable defect within the chromosomes. This hypothesis of a chromosomal or genetic cause of cancer was only reconsidered in recent decades in the light of new findings on genomic rearrangements and cancer genetics (Satzinger, 2008). Rearrangements occur due to DNA breaks and fusions, and lead to an abnormal karyotype of the daughter cells. In neuroblastoma, such breaks have been shown to occur preferentially within early replicating regions during the S phase (Janoueix-Lerosey

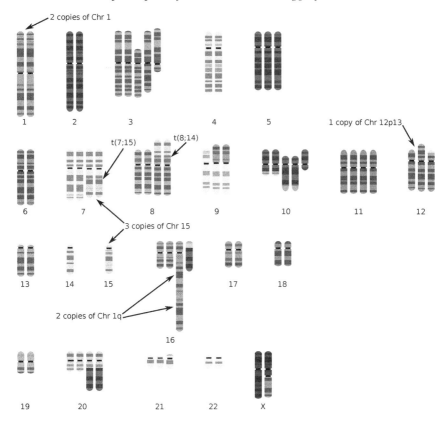

FIGURE 2.12 Karyotype of the T47D breast cancer cell line. The karyotype shows that the cell line has undergone various chromosome aberrations due to genome instability. The letter t indicates a translocation involving the two chromosomes in brackets. Only some alterations are indicated on the figure. **Figure 3.8** and **Figure 3.15** show application of high-throughput technologies in order to characterise the chromosome aberrations within this cell line. Data from Roschke et al. (2003), source: http://www.ncbi.nlm.nih.gov/sky. (See colour insert.)

et al., 2005). Typical chromosomal aberrations which produce an abnormal karyotype are illustrated in **Figure 2.13** and explained below:

polyploidy: each chromosome is present in p copies where $p > 2$ represents the ploidy ($p = 2$ is normal ploidy).

aneuploidy: extra or missing copies of some chromosomes.

translocation: chromosome abnormality caused by exchange of parts between homologous or nonhomologous chromosomes. The translocation can be reciprocal (or balanced, *i.e.* exchange of two extremities of chro-

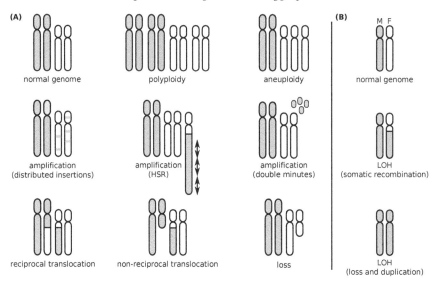

FIGURE 2.13 Schematic illustration of chromosomal aberrations. (A) Common aberrations include gain, loss and translocations. Amplifications may be visible as double minutes, chromosomes with Homogeneously Stained Region (HSR) or the amplified DNA may be distributed at multiple sites. Two different chromosomes are represented in grey and white. (B) LOH without DNA copy number change leave the chromosome apparently intact while either the whole chromosome or part of the chromosome is lost for one parental origin. The grey chromosome comes from the mother (M) and the white chromosome from the father (F). Image adapted from Albertson et al. (2003).

mosomes) or nonreciprocal (or imbalanced, *i.e.* one extremity is gained and/or lost during the exchange).

amplification: one part of a chromosome (as a gene or a group of a few contiguous genes) is present in a high number of copies (from 4 to more than 50 copies). These numerous copies are either incorporated into chromosomes in nearly contiguous Homogeneously Stained Regions (HSR), interspersed in the genome or form acentric fragments (double minute).

The following terms can be also associated to the copy number for a given region of the chromosome: deletion (no copy is present anymore), monosomy (one copy is present), trisomy (three copies are present), tetrasomy (four copies are present), *etc.* These rearrangements can be either complete (*i.e.* the whole chromosome is concerned) or partial (*i.e.* only a region of the chromosome is concerned).

Some chromosome aberrations do not produce an abnormal karyotype and chromosomes appear to be present with the expected number of two copies (see **Figure 2.13B**). In a normal genome, chromosomes are heterozygous:

there is one copy from the father and one copy from the mother. In cancer, some chromosomes come from the same parental origin. In this case, the two copies of the chromosome are the same and therefore the chromosomes are homozygote. This phenomenon is termed LOH without any DNA copy number change since one parental chromosome (or only a portion) has been lost in a first event and the missing chromosome (or only the missing portion) has been duplicated from the remaining parental chromosome. When the LOH rearrangement concerns the whole chromosome this duplication is called uniparental disomy (or isodisomy). When it concerns only a portion of the chromosome it is called somatic recombination or partial isodisomy.

Both mutations and chromosome aberrations accumulate during tumour progression lasting several years. However, Stephens et al. (2011) demonstrated that a single cellular catastrophe can lead to tens to hundreds of genomic rearrangements at once. During this catastrophic event named *chromothripsis*, DNA is fragmented, its subsequent repair is not precise and leads to chromosomal aberrations, as well as the loss of chromosomal regions (see **Figure 2.14**). This phenomenon can occur in at least 2%–3% of all cancers.

Different types of chromosome aberrations occur in cancer. Some of them modify the copy number of entire or small portions of chromosomes and are called DNA copy number alterations. Others do not modify the number of chromosomes. Many high-throughput technologies including **microarrays**[*] and **NGS**[*] have been used to characterise these aberrations genome-wide and they are presented in **Chapter 3**.

2.8.2   Impact of chromosome aberrations on cancer-critical genes

Why are these chromosomal aberrations so important in cancer? In many cases they cause tumour progression because they can directly affect the critical genes involved in cancer. More generally, oncogenes are expected to be found in gain or amplification regions, or in fusion genes, while tumour suppressor genes are expected to be found in loss regions or LOH regions without DNA copy number change. Therefore, the characterisation of chromosome aberrations should help to find new candidates for cancer-critical genes.

2.9   Conclusion

In this chapter, we have shown that a multistep accumulation of events transform a normal cell into a cancer cell, causing defects in the gene regulation and signal transduction mechanisms. Many alterations arise during tumour progression including point mutations, epigenetics modifications, chromosome

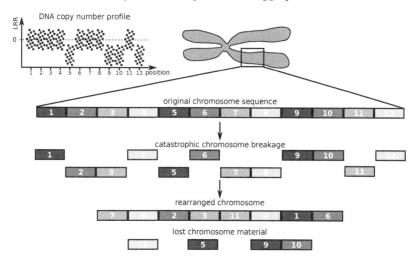

FIGURE 2.14 Chromothripsis. During a catastrophic event, the DNA is fragmented. Some of the fragments are then randomly ordered and merged while other fragments are lost. The plot in the upper left corner represents the expected DNA copy number profile (DNA copy number profile is explained in **Figure 3.8**). Image adapted from Tubio and Estivill (2011).

aberrations, *etc.* Whatever their nature, these alterations can be classified into *drivers* or *passengers* according to their causality on cancer development (Stratton et al., 2009). Driver alteration is causally involved in the development of cancer. Such an alteration has been positively selected according to Darwinian selection because it conferred a growth advantage on cancer cells at some step during the tumour progression. The passenger alteration has not been selected nor conferred growth advantage and has therefore not contributed to tumour progression. Passenger alterations have merely occurred during the growth of the cancer. Among the vast variety of alterations which are observed in cancer a small minority are drivers. Therefore, it is essential we can distinguish the drivers from the passengers on the road to cancer development. Genome-wide characterisation of molecular alterations occuring in cancer, in particular using high-throughput technologies (see **Chapter 3**), combined with computational systems biology approaches (see **Chapter 4** to **Chapter 11**) offer insights into the discovery of drivers and the understanding of their causality in cancer-related cellular processes. These approaches will not only allow the identification of new **biomarkers*** in order to help clinicians in their prognostic and predictive decisions but also the discovery of new drug targets and the better understanding of tumour progression mechanisms.

✎ **Exercises**

- Using the possible chromosome aberrations listed in **Figure 2.13**, enumerate those which can be observed in the karyotype of the T47D breast cancer cell line provided in **Figure 2.12**.

- In a cell, an abnormal karyotype is observed showing an Homogeneously Stained Region (HSR) (see **Figure 2.13**). Further investigations indicate that the miRNA miR-21 is present in the amplified region. The tumour suppressor gene *PTEN* is targeted by the miR-21. What could be the effect of this amplification on *PTEN* expression and its impact for the cell?

⇨ **Key notes of Chapter 2**

- Cancer is caused by an accumulation of mutations during the lifetime of the organism.

- Cancer is a genetic disease due to the deregulation of gene expression involving tumour suppressor genes and oncogenes.

- Cancer is a network disease.

- A tumour is a complex and heterogeneous system which encompasses different cancer cell types and normal cell types.

- The tumour microenvironment plays an important role in tumour progression.

- The tumours have common characteristics called hallmarks.

- Abnormal karyotypes are very frequently observed in cancer cells.

Chapter 3

Experimental high-throughput technologies for cancer research

Chapter 2 described a series of dysregulations that occurs at different molecular levels when a normal cell becomes a cancer cell. The sequential accumulation of **mutations*** and events occurring during **tumour progression*** can disrupt the normal behaviour of the cell (see **Figure 3.1**) at the level of:

1. DNA, including:
   - Mutations of the DNA sequence
   - Changes in DNA copy number
   - Loss of Heterozygosity (LOH)
   - **Translocations***
2. Noncoding RNA expression including microRNA (miRNA).
3. Messenger RNA (mRNA) expression, including:
   - Modifications in alternative splicing
4. Protein and particularly:
   - Their quantity
   - Their modification including phosphorylation of protein kinases which play a key role in signal transduction
5. Epigenetic characteristics, including:
   - DNA Methylation
   - Histone modifications (methylation, acetylation, *etc.*)
6. Interactions between the different molecules, such as:
   - Interactions between transcription factors and DNA
   - Interactions between proteins
7. As a consequence, these alterations lead to change in the phenotypic characteristics of the cell.
8. Interactions of cancer cells with their environment:
   - Blood supply
   - Immune response
   - Interaction with the **Extracellular Matrix (ECM)***

Understanding tumour progression and improving the classification of tumours require to unravel the alterations which have occured at these different molecular levels. Current biotechnologies allow us to accurately characterise the molecular profiles of each tumour sample and the information retrieval must be as exhaustive as possible. For instance, we aim at determining the DNA copy number of as many loci as possible over each chromosome, quantifying the messenger RNA (mRNA) expression of all known genes, detecting what alternative splice forms are present, *etc.* This exhaustive search might be reachable for some molecular profiles but in some cases, especially for proteins, it is intractable for both complexity and technological reasons (see **Section 3.4**). Since the quantification of the molecular profiles is supposed to be as exhaustive as possible, the techniques which allow their measurement are often referred to as *genome-wide* techniques. More generally, the name of the technology which studies a particular type of molecular profile is the concatenation of the molecular entities or the biological functions under investigation with the *-omics* suffix. For example, as illustrated in **Figure 3.1**, *genomics* investigates the DNA alterations (mutation, copy number, *etc.*), *miRNomics* the microRNA (miRNA) expression, *transcriptomics* the mRNA expression, *spliceosomics* the different alternative splice forms, *proteomics* the different proteins, *kinomics* the phosphorylation state of protein kinases, *epigenomics* the epigenetic modifications, *interactomics* the interactions between different molecular entities and *phenomics* the observable traits of the cells. The *-omics* suffix comes from the Greek stem *omes*, which stands for *all, every, whole* or *complete*, reminding us of the fact that these techniques aim at achieving an exhaustive search. These techniques are also called high-throughput technologies because they produce a huge amount of information within a short time. It is important to pinpoint that besides high-throughput technologies, there exist other approaches or technologies but they will not be addressed in this book. In this chapter, technical details will be given regarding *omics* technologies.

3.1   Microarrays

3.1.1   General principles and microarray design

The better understanding of biological molecular processes combined with the improvement in DNA technologies have allowed researchers to use *in vitro* some chemical reactions which happen *in vivo*. The discovery of **restriction enzymes*** and **reverse transcriptase*** in 1970, DNA sequencing in 1977, **Polymerase Chain Reaction (PCR)*** in 1985 (for a history of DNA technologies, see the milestones timeline in Nature Publishing Group, 2007) have been main revolutions in biotechnology. The improvements in chemistry,

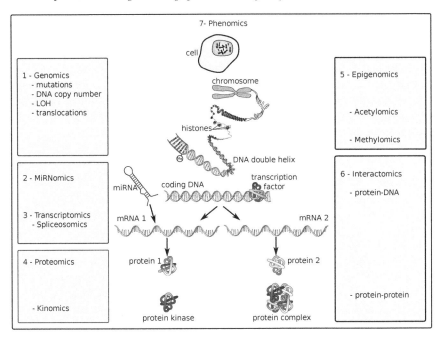

FIGURE 3.1 Omics technologies in oncology. The main omics technologies used in cancer research are listed in this illustration. Note that others omics approaches exist.

physics, optics, robotics, software engineering and molecular biology have allowed the development of new tools for genome-wide quantification; the *microarray* technology also called *biochip* or *chip* has provided miniaturised sensor tools such that it is possible to quantify the mRNA expression of the whole genome on a slide glass smaller than two square-centimetres. Microarrays appeared in 1995 and can be considered as one of the major biotechnological revolutions of the last 15 years. Originally, microarrays emerged in the field of transcriptomics and they have been widely transposed for all the omics approaches mentioned in **Figure 3.1**. As a result, a large variety of microarray techniques have been developed for various applications as reported in Hoheisel (2006) and also in the *Chipping Forecast* supplement series published by *Nature Genetics* in 1999, 2002 and 2005. However, all microarray technologies rely on similar characteristics presented below.

The basic principles in microarray technology are the following: *probes* (DNA, RNA or protein) are tethered to a solid support (*i.e.* the chip) such as glass, plastic or silicon (Southern et al., 1999). They act as a specific reporter either to quantify the DNA copy number at a known locus on the genome, the expression of a known gene or the amount of a protein. Probes are supposed to be chosen specifically in order to report the quantification of their expected target. In the case of DNA or RNA probes, the specificity is guaranteed by the choice of a unique base-paired complementarity between the probe sequence and the target sequence, and by the choice of an appropriate antibody in case of proteins. Probes are deposited on a microscopic area of the chip called spot or feature. Then, either DNA, RNA or proteins are extracted from a tumour sample and hybridised on the chip. If present within the sample, a given DNA sequence, RNA sequence or protein will be hybridised on its matching probes. In a microarray, thousands or even millions of such spots are present which makes it a very powerful tool for genome-wide screening.

❏ BOX 3.1: Fluorochrome

A fluorochrome is a molecule which has the following interesting property: when excited by a light with the appropriate wavelength, it has the ability to fluoresce at another wavelength (*n.b.* the absorption-emission spectrum is known for the fluorochrome). For example, the cyanine 5 (Cy5) fluorochrome is excited at 650 nm and fluoresces maximally at 670 nm. Once the targets hybridise on their matching probes, a scanner allows the Cy5 to be excited using a laser at 650 nm wavelength and the measurement of the emitted fluorescence at 670 nm. Cy5 and Cy3 emit maximally in infra-red and orange and absorb maximally in red and green respectively.

Another characteristic of the microarray is the use of fluorescent markers called fluorochromes (see **Box 3.1**) in order to measure DNA, RNA or protein quantities. Indeed, a measurement strategy is required to quantify what is the amount of each target attached to their respective probe since the direct quantification of the target is not possible on the microarray. That is the reason why a fluorochrome is used to overcome this limitation. During the preparation of the sample, specific chemical reactions allow the fluorochrome to be incorporated in the nucleotide sequence or in the protein. The signal intensity of the fluorescence light is quantified and directly related to the amount of target which attached to the probe. Different fluorochromes emitting at different wavelengths (or colours) can be used simultaneously in some microarray platforms. Such microarrays offer the possibility to label and analyse two different samples at the same time. For example, a common reference can be used across different experiments. These microarrays are referred to as two-colour or two-channel microarrays. For proteomic microarrays, the reporter antibody is generally coupled with a fluorochrome. Since protein studies require techniques that take into account the chemical properties of the protein, **Section 3.4** will be specifically devoted to proteomics.

In genomics, transcriptomics and miRNomics studies, the Affymetrix GeneChip® has been widely used and measures about 6.5 million features in a single experiment (see **Figure 3.2A** and Dalma-Weiszhausz et al. (2006) for a review on this technology). The number of features (on a given surface)

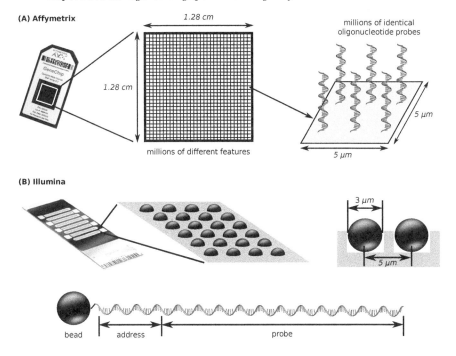

FIGURE 3.2 Affymetrix GeneChip® and Illumina BeadChip designs. (A) The chip consists of a 1.28×1.28 square-centimetre modified quartz wafer. This surface carries about 6.5 millions of 5 μm × 5 μm features. Each feature, in turn, is composed of millions of identical **oligonucleotide** probes. The oligonucleotide is a 25-base long single-strand sequence and acts as a specific reporter of a known locus on the genome. Image adapted from Dalma-Weiszhausz et al. (2006). (B) Silica beads, each 3 μm in diameter, self-assemble randomly into micro-wells with 5 μm centre-to-centre spacing. Each probe is represented with an average of 30-fold beads on each array. Each bead contains a probe sequence of interest and an address sequence which allows its identification according to a decoding system described in Gunderson et al. (2004). Both the address and the probe represent a specific oligonucleotide sequence for each bead. Each bead is covered with hundreds of thousands copies of that specific oligonucleotide sequence. Image adapted from Fan et al. (2006) and http://www.illumina.com.

is continuously increasing due to improvements in the fabrication process allowing the reduction of feature size. Other microarrays provided by Agilent, Nimblegen or Illumina companies have been widely used as well. For example, the Illumina company proposed a chip named BeadChip. While the principles of probes and target sequence still hold, probes are not deposited on the microarray slide surface anymore; probe sequences are attached to silica beads (see **Figure 3.2B**). These beads self-assemble in the micro-wells that cover the chip (Fan et al., 2006). Since the self-assembly of the beads in micro-wells

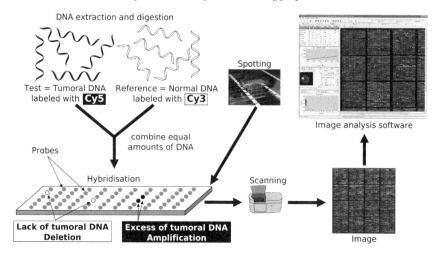

FIGURE 3.3 Array-CGH protocol. The protocol includes the extraction and labelling of the DNA, the hybridisation on the chip, the scanning and image analysis to quantify the signal. (See colour insert.)

is a random process, each bead contains a probe sequence for the target of interest and an address sequence which permits its identification according to a decoding system described in Gunderson et al. (2004). In addition to commercial platforms, many in-house microarrays have been produced. All these microarray technologies have been widely applied in **oncology*** as reported by Cowell and Hawthorn (2007).

3.1.2   DNA copy number study based on microarray experiment

The study of genome-wide characterisation of DNA copy number changes was originally performed using the Comparative Genomic Hybridisation (CGH) technique developed in the early 1990s. In the first version of this technique, total genomic DNA is isolated from a tumour and normal control cells, labelled with different fluorochromes and hybridised to normal metaphase chromosomes (Kallioniemi et al., 1992). This technique is therefore termed chromosomal CGH. Differences in the tumour fluorescence with respect to the normal fluorescence along the metaphase chromosomes are then quantified and reflect changes in the DNA copy number in the tumour genome.

Subsequently, array Comparative Genomic Hybridisation (aCGH) was established (Solinas-Toldo et al., 1997; Pinkel et al., 1998). In this technique, microarrays with genomic sequences inserted into **Bacterial Artificial Chromosome (BAC)*** replace the metaphase chromosomes as hybridisation reporters. aCGH solved many of the technical difficulties and problems caused by working with cytogenetic chromosome preparations. The main advantage of aCGH is the ability to perform copy number analyses with much higher resolution than was ever possible using chromosomal CGH. aCGH has already been widely used in oncology for many purposes such as global analysis of copy number aberrations, identification of putative target genes, tumour classification or assessment of clinical significance of copy number changes (Kallioniemi, 2008). A typical aCGH microarray experiment works as follows (see **Figure 3.3** and Pinkel and Albertson, 2005):

- Total genomic DNA is isolated from a tumour sample (*i.e.* the test DNA) and from a normal sample (*i.e.* the reference DNA). Genomic DNA is then generally digested with a restriction enzyme and the DNA fragments are differentially labelled: the tumoral DNA is labelled with a red fluorochrome (*e.g.* Cy5) and the normal DNA with a green fluorochrome (*e.g.* Cy3).

- Equal amounts of tumoral and normal DNA are combined.

- The mixture of both tumoral and normal DNA fragments is hybridised on the chip. Within each spot, there is a competitive hybridisation between the tumoral DNA target sequences and the normal DNA target sequences.

- A scanning step quantifies the signal intensity for the red and green channels. An image file is created in which each pixel is given a red and green intensity.

- Image analysis software accurately reconstructs the signal intensity for each spot from the image.

Once this protocol has been performed, how do we expect the signal to vary with respect to the DNA copy number of each sample? For each spot, a competitive hybridisation takes place between the tumoral and normal DNA. The relative hybridisation intensity of the test signal over the reference signal at a given location is (ideally) proportional to the relative DNA copy number of those sequences in the test and reference genomes. If the tumoral DNA copy number is greater than the normal DNA copy number, then the signal will be shifted towards red. On the contrary, if the tumoral DNA copy number is lower than the normal DNA copy number, then the signal will be shifted towards green. Therefore, the DNA copy number of the tumoral DNA is directly proportional to the red/green ratio and its theoretical value is given in **Figure 3.4**. For statistical reasons, we generally do not use the ratio of red/green but the $\log_2$ of this ratio, therefore named $\log_2$-ratio[1]. In practice, due to technical variability, there is a fluctuation of the signal around its expected value and statistical methods are necessary to retrieve the true signal.

---

[1]The log transformation allows the distribution of the values to be closer to the normality, which is generally preferred in statistics.

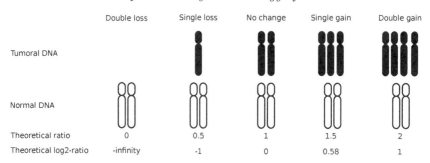

FIGURE 3.4 Theoretical array-CGH quantification. The theoretical ratios and log$_2$-ratios are given for different DNA copy number alterations occurring in the tumoral DNA.

Moreover, the quantified signal is generally less than expected for three reasons. First, the quantification made with the technology is not perfect and the signal is often less than proportional with respect to the true DNA copy number (Pinkel et al., 1998; Pollack et al., 1999). Then, the tumoral DNA generally contains contamination by normal cells coming from adjacent normal tissue; they can represent a significant proportion within the sample and reduce the signal from the cancer cells. Finally, the tumour might be heterogeneous since it can derive from different clonal populations (see **Figure 2.5**) which share different patterns of DNA copy number alterations.

The aCGH technology relies on the assumption that the reference DNA is diploid. In practice, this is not the case since DNA copy number variations exist even in normal individuals: some parts of the DNA sequence can be present in many copies inside the genome. Such a part of the genome is called Copy Number Variant (CNV) (Iafrate et al., 2004; Freeman et al., 2006; Redon et al., 2006) and the Database of Genomic Variants provides a catalog of such variations. For instance, Perry et al. (2007) found that the copy number of the salivary amylase gene (*AMY1*) is correlated positively with salivary amylase protein levels, and that individuals from populations with high-starch diets have on average more *AMY1* copies than those with traditionally low-starch diets. This was the first example of positive natural selection on a copy number variable gene in the human genome. Ideally, to avoid the identification of DNA copy number changes due to CNVs between the test DNA and the reference DNA the two DNAs used in the aCGH protocol should come from the same patient (in this case the DNAs are referred to paired). However, normal DNA from the patient is not always available and a normal reference DNA from a normal standard individual is generally used. Importantly, CNV may have some impact on cancer risk and using paired DNAs can be a drawback as it prevents the identification of CNV for the patient under study.

The typical graphical representation of a DNA copy number molecular profile is depicted in **Figure 3.5**: the x-axis represents the probe location

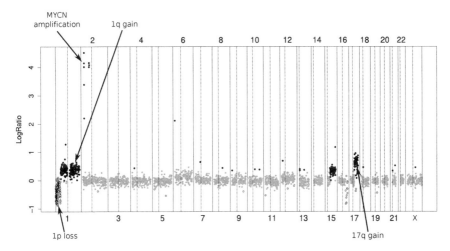

FIGURE 3.5 aCGH profile of the IMR32 neuroblastoma cell line. The $\log_2$-ratios for each probe ordered along the genome from chromosome 1 to 22 and X are represented. Vertical black lines indicate separation between chromosomes. Vertical dashed black lines indicate centromere position. The imbalanced translocation 1p-17q and the 1q gain are identified by aCGH. An alteration of small size like *MYCN* amplification can be detected thanks to the high resolution of the aCGH technology. Data from Janoueix-Lerosey et al. (2005). (See colour insert.)

ordered along the genome from chromosome 1 to 22 and X; the y-axis represents the $\log_2$-ratio value of the DNA copy number. In the profile of the IMR32 **neuroblastoma**[*] cell line, a loss of chromosome 1p[2] and a gain of chromosome 1q and 17q due to an imbalanced translocation clearly appear. Alterations of small size like that of *MYCN* **amplification**[*] on chromosome 2 can be detected thanks to the high resolution of the aCGH technology with respect to chromosomal CGH.

The recent advances in microarray technologies has shifted from BAC aCGH to **oligonucleotide**[*] aCGH allowing an increase in the number of loci per chip (Davies et al., 2005; Ylstra et al., 2006). BAC arrays are mainly in-house microarrays while oligonucleotide microarrays are from commercial companies. Among the widely used commercial technologies, let us mention Agilent Human Genome CGH Microarray, Nimblegen Human Whole Genome Tiling arrays, Illumina BeadChip and Affymetrix GeneChip® (note that for Affymetrix and Illumina technologies, no normal DNA is needed in the protocol and they are one-colour microarrays in contrast to the other technologies that use both normal and tumoral DNA and are two-colour microarrays). At the early period of BAC array, the number of loci investigated was around 1,000-

---

[2]p and q define the short and long chromosome arm respectively.

2,000 and never exceeded 32,000 loci[3] (Ishkanian et al., 2004). The use of oligonucleotide arrays has allowed a huge increase in the number of loci investigated on a single chip. At the time of the writing of the present chapter, the number of loci quantified in the human genome with a single oligonucleotide array ranges from 1 million to 2.5 millions allowing a maximum theoretical resolution of 1.2Kb. This number is very likely to increase. Although the most recent chips cover more exhaustively the genome, their exact resolution does not only depend on the number of loci but also on their sensitivity. Coe et al. (2007) has proposed a definition of resolution for aCGH technology, termed *functional resolution*, which incorporates the uniformity of loci spacing on the genome, as well as the sensitivity of each platform to single-copy alteration detection. From their study, the current commercial platforms allow a single-copy detection of the order of 35-55Kb while it was 10Mb for chromosomal CGH and 1Mb for BAC aCGH (at the time of the study by Coe et al. (2007), the highest number of loci in a single chip was offered by Nimblegen Human CNV arrays and allowed the quantification of 385,000 loci over the whole human genome). The oligonucleotide chips that permit to scan the genome for more than 50,000 loci are often termed high-density or high-resolution chips. Haraksingh et al. (2011) compared the performance of these technologies. Besides the huge increase in resolution offered by these oligonucleotide arrays, Illumina and Affymetrix incorporated in their design polymorphic probes to measure Loss of Heterozygosity (LOH) in addition to DNA copy number, as we will see in the next section.

3.1.3 LOH study based on microarray experiment

Though two individuals are genetically very similar, their DNA sequences still differ enough to explain a large part of the variability of phenotypes, including the susceptibility to develop many diseases. This makes determination of polymorphism profiles very helpful in biomedical sciences. In 2002, the International *HapMap Project* started with the goal to determine the common patterns of DNA sequence variation in the human genome and to make this information freely available in the public domain (International HapMap Consortium, 2003). For this purpose, 270 samples of individuals originated from Asia, Africa and Europe were used. More recently, *The 1000 Genomes Project* was initiated in 2008 in order to obtain the most detailed catalogue of human genetic variation (1000 Genomes Project Consortium, 2010). Single Nucleotide Polymorphisms (SNP) are the most important source of genetic variability between individuals (see **Box 3.2**). Therefore, they represent very valuable probes to be considered in a microarray design to study the genomic variation across different individuals or population. Moreover, in cancer studies SNP probes can assess LOH as we will explain in the next paragraph. Both

---

[3]A BAC generally contains a human DNA sequence of 100Kb; 32,000 BACs allow the coverage of the whole human genome.

Affymetrix and Illumina companies offer microarray designs which incorporate one specific probe for each SNP allele. Such microarrays are generally referred to as SNP arrays.

❑ **BOX 3.2: Single Nucleotide Polymorphism (SNP)**

A SNP (SNP, pronounced snip) is a DNA sequence variation occurring when a single nucleotide (A, T, C, or G) in the genome differs at the same genomic position between two individuals (Sachidanandam et al., 2001; Bunz, 2008). Here is an example of SNP (C/G) for which two alleles (arbitrary named A and B) exist:

A allele: gtcacccatccctc $\boxed{\text{c}}$ gtgctggtaatcaga

B allele: gtcacccatccctc $\boxed{\text{t}}$ gtgctggtaatcaga

SNPs occur once every 1,000–2,000 nucleotides and this variation is only called a polymorphism if it occurs in 1% or more of the population. About 10 millions of these variants have been indexed in the dbSNP database hosted in the NCBI (Sherry et al., 2001). About 96% of SNPs occur outside protein-coding regions: some of them can be phenotypically silent while others can have a functional impact (*e.g.* if a SNP is in a regulatory sequence, in an alternative splice site, *etc.*). Other SNPs called nonsynonymous affect protein sequences. Both types of SNPs can serve as landmarks in the search for genes associated with diseases, drug responses and complex phenotypes.

To illustrate how SNP probes can be used in order to evaluate LOH in a tumour, let us consider the normal cell in **Figure 3.6A**. In this case, one chromosome comes from the mother (M) and one chromosome from the father (F). Along the chromosome, there are different SNPs. Without loss of generality, let us assume that four SNPs are present on the chromosome, each one having two alleles (either A or B). For each locus, B Allele Frequency (BAF) can be computed as follows:

$$\text{BAF} = \frac{n_B}{n_B + n_A},$$

where n_A and n_B represent the number of alleles A and B respectively. For the normal cell, in the case of a heterozygous locus (SNP2 and SNP3), the BAF is equal to 0.5 while for a homozygous locus (SNP1 and SNP4) it is equal to either 0 if the allele A is present or 1 if the allele B is present. A SNP is called informative if it is heterozygote. Let us consider a cancer cell which has experienced a loss of the paternal chromosome followed by a duplication of the maternal chromosome (see **Figure 3.6B**). These correspond to the copy neutral LOH case (*i.e.* there are two copies of the chromosome as in a normal cell but the two chromosomes have the same parental origin).

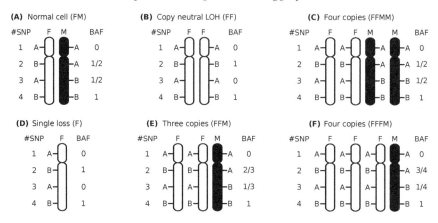

FIGURE 3.6 Illustration of BAF values. The paternal chromosome F chromosome is depicted in white and the maternal M chromosome in black. From the normal cell state (A), five different possible alterations occuring in a cancer call are represented (B to F). The BAF value for each SNP is computed.

In this case, the BAF values for heterozygous loci SNP2 and SNP3 switch from 0.5 to 1 and 0 respectively. In the case of a single loss of the paternal chromosome, the BAF values are the same as in the copy neutral LOH case (see **Figure 3.6D**). In the three copies case in which there is an additional maternal chromosome, the BAF value for heterozygous loci SNP2 and SNP3 switch from 0.5 to 0.66 and 0.33 respectively (see **Figure 3.6E**). In the four copies case, two situations can occur. The first case (FFMM) corresponds to a duplication of both paternal and maternal chromosomes and leads to BAF values like in the normal cell case (see **Figure 3.6C**). In the second case (FFFM) in which there are three copies of the paternal chromosomes, the BAF values for heterozygous loci SNP2 and SNP3 switch from 0.5 to 0.75 and 0.25 respectively (see **Figure 3.6F**). It is important to notice that the BAF values for homozygous loci (SNP1 and SNP4) remain the same whatever the case considered and thus are called noninformative loci. Moreover, the BAF values are symmetric with respect to 0.5 (*e.g.* 0.33 and 0.66 in the three copies case).

The examples given in **Figure 3.6** are not exhaustive and many other situations can be imagined. Importantly, it must be pointed out that the BAF values computed from tumoral samples can differ from the theoretical value due to the contamination by normal cells. Indeed, the theoretical BAF values can be formulated as follows:

$$\text{BAF} = \frac{(1-p).n_B^c + p.n_B^n}{(1-p).(n_B^c + n_A^c) + 2p},$$

where p is the normal DNA proportion due to contamination, n_A^c and n_B^c correspond to the number of A and B alleles in the tumour, n_A^n and

n_B^n correspond to the number of A and B alleles in the normal sample. We expect $n_A^n + n_B^n$ to be equal to 2 since the normal DNA is diploid. In practice, the proportion p is generally unknown. It can be estimated based on the **pathologist***'s expertise from **histological sections*** or using dedicated biostatistics approaches (Popova et al., 2009).

In a microarray experiment, the BAF value will be computed as follows:

$$\text{BAF} = \frac{S_B}{S_A + S_B},$$

where S_A and S_B are the signal intensities quantified on the chip for both A and B alleles using the intensity values from their respective probes.

As we already mentioned in the previous section, the DNA copy number can be assessed using this type of microarray technology. In the case of polymorphic probes, the DNA copy number (CN) can be obtained by summing the number of each allele as follows:

$$\text{CN} = (1 - p).(n_B^c + n_A^c) + p.(n_B^n + n_A^n).$$

The signal obtained from a microarray experiment is an intensity value that needs to be transformed into a more comprehensible value. Even for a one-colour microarray, the signal measurement from a normal DNA sample obtained on another microarray experiment (either a paired normal sample or a pool of normal samples not related to the tumour sample under investigation) is used in order to compute a log-ratio as in aCGH experiment. In the case where no normal reference DNA is available, individuals from the HapMap project can be used since the microarray experiments have been done on different SNP microarray platforms including both Illumina and Affymetrix. The Log Reference Ratio (LRR) with a reference normal sample is computed as follows:

$$\text{LRR} = \log_2\left(\frac{S_A + S_B}{S_A^{ref} + S_B^{ref}}\right),$$

where S_A and S_B are the signal intensities quantified on the chip for both A and B alleles in the tumour sample, and S_A^{ref} and S_B^{ref} are the signal intensities quantified on the chip for both A and B alleles in the reference normal sample.

Both the BAF and LRR values provide complementary information and help to characterise DNA alterations in tumoral samples. For example, both values will discriminate between the normal cell case and the copy neutral LOH case depicted in **Figure 3.6**. Indeed, in the normal cell, while the LRR values are equal to 1, the BAF values are equal to 0.5 in the normal cell case for informative SNPs and either 0 or 1 in the copy neutral LOH case. Similarly, the combination of both values helps to distinguish between the four copies case FFMM and the four copies case FFFM. In these two cases, while the LRR values will be equal to 2, the BAF values will be equal to 0.5 for

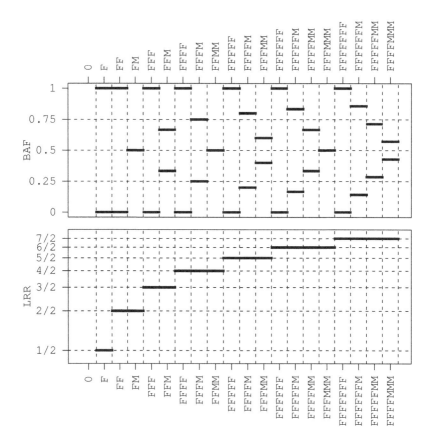

FIGURE 3.7 Theoretical BAF and LRR values. For different possible alterations, both the theoretical BAF and LRR values are represented. F and M stand for paternal and maternal chromosomes respectively.

informative SNPs and either 0.75 or 0.25 in the FFFM case. The **Figure 3.7** illustrates different possibilities for both LRR and BAF values in a situation where there are 0 to 7 copies of the same chromosome. Note that, the 0 copy state is very particular in the sense that neither LRR nor BAF can be computed for mathematical reasons. In practice, the signal intensities S_A and S_B are generally never equal to zero due to background noise and the possible contamination by normal cells. Therefore, the LRR will have a very low value (*e.g.* -2 or even lower) and the BAF will be equal to 0.5 for informative SNPs.

The **Figure 3.8** shows the LRR and BAF profiles (each containing about 50 thousand loci) of a real experiment using Affymetrix GeneChip® SNP microarray technology on the breast cancer cell line T47D (data from Hu et al., 2009). Let us describe a few chromosomal alterations in this tumour cell line:

Chromosome 1 The LRR shows that the 1p chromosome arm is present in 2 copies while the 1q chromosome arm is present in 4 copies. This is confirmed from the karyotype (see **Figure 2.12**) which also provides the additional information that the 2 additional copies of the 1q chromosome arm have been merged and fused to chromosome 16 due to a translocation. The BAF indicates that the 1p chromosome arm comes from the same parental origin as the BAF values are near 0 or 1, while there are 2 copies from the paternal 1q chromosome arm and 2 copies from the maternal 1q chromosome as the BAF values equal 0.5 for informative SNPs.

Chromosome 12 The LRR profile shows a single loss of the p12-pter (*i.e.* a sub-region of chromosome 12 which goes until the terminal extremity of its p arm) of chromosome 12 which is confirmed on both the BAF profile and the karyotype.

Chromosome 15 The LRR profile shows that there are 3 copies of the chromosome 15. The karyotype indicates that two copies of the chromosome 15 have been each fused to two different chromosomes 7 and one chromosome 15 remains alone. The BAF profile indicates that there are two copies from one parent and one copy from the other as the BAF values are either around 0.33 or 0.66 for informative SNPs.

Note that, neither the LRR profile nor the BAF profile can indicate what translocations are involved in the alterations while the karyotype can provide this information. Moreover, balanced alteration will never be observed from the LRR profile as the copy number remains unchanged in that case. We will see that NGS technologies can help to unravel this information (see **Section 3.2**).

DNA copy number alterations can impact the modification of gene expression and the following part describes how to perform microarray studies at the level of RNA.

3.1.4 RNA study based on microarray experiment

As already mentioned, the development of microarray technology was first initiated in the field of transcriptomics and it has been largely addressed in the literature. Initially, the experimental protocol was quite similar to the aCGH protocol described in **Figure 3.3** except that mRNA is used instead of genomic DNA. New technologies developed by the Affymetrix company have appeared, which changed the chip building and the protocol in such a way that the reference sample is not needed anymore. The Affymetrix GeneChip® (see **Figure 3.2**) thus provides an approach to have a semi-quantitative level of mRNA instead of a relative value with respect to a reference. Recently, a new probe design has been proposed in order to identify alternative splice forms from a gene expression microarray experiment. Such microarrays are called exon-arrays. For each gene, different probes targeting the different exons of

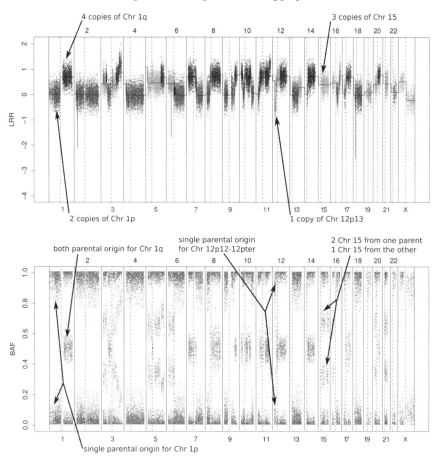

FIGURE 3.8 LRR and BAF profiles for the T47D breast cancer cell line.
For each probe ordered along the genome from chromosome 1 to 22 and X, the LRR
values (top profile) and the BAF values (bottom profile) are represented. Vertical
black lines indicate separations between chromosomes. Vertical dashed black lines
indicate centromere position. In the LRR profile, the piecewise black line corresponds
to the mean copy number for each genomic region; Green = 1 copy, Yellow = 2 copies,
Red = 3 copies and Blue = 4 copies. The Affymetrix Human Mapping 100K Xba chip
for the T47D cell line has been retrieved from the NCBI GEO database. We have
analysed the data using CRMAv2 (Bengtsson et al., 2009) and GLAD (Hupé et al.,
2004) for the LRR profiles, and ACNE (Ortiz-Estevez et al., 2010) for the the BAF
profile. The profiles can be compared with the karyotype of the breast cancer cell
T47D provided in **Figure 2.12** and the results obtained with NGS in **Figure 3.15**.
Data from Hu et al. (2009). (See colour insert.)

the genes are deposited on the chip. However, such a design allows the identification of exons which are differentially expressed between two conditions

for example, but makes it very difficult to pinpoint precisely which isoform is expressed. To overcome this limitation, probes overlapping two consecutive exons have been added (this chip is therefore termed exon-junction microarray). Besides mRNA, chips to study miRNA have also been developed and rely exactly on the same principles.

3.1.5 DNA–protein interaction study

In the cell, interactions between DNA and proteins are essential for many biological processes such as DNA replication, recombination, DNA repair and the regulation of transcription. For example, in response to environmental stresses, transcription factors bind to their DNA-binding site and regulate the transcription of their target genes (see **Figure A.6**). A transcription factor can regulate many different genes which are not alway easy to predict by sequence analysis or *in vitro* studies. Therefore, the identification of all potential target genes of transcription factors is a challenge. For that, a high-resolution genome-wide approach based on a combination of Chromatin Immunoprecipitation (ChIP) and DNA microarrays (chip) can be used. This technology is referred to as ChIP-on-chip. The protocol for a typical ChIP-on-chip experiment is the following (see **Figure 3.9** Bulyk, 2006; Buck and Lieb, 2004):

- Cells are grown in culture under the desired experimental condition.

- In the cell culture, the proteins are cross-linked to DNA, generally using formaldehyde. This step forms reversible bonds between the DNA-associated proteins and the DNA.

- After cross-linking, the cells are lysed and the **chromatin*** is sheared into fragments of 1Kb size or smaller.

- The DNA fragments cross-linked to the Protein of Interest (POI) are enriched by immunoprecipitation using an antibody which recognises specifically the POI.

- The formaldehyde cross-links are then reversed and the DNA is purified.

- A DNA amplification step is generally required since the immunoprecipitation yields low DNA quantity.

- Enriched DNA is then labelled with a fluorescent molecule such as Cy5. This is referred to as the ImmunoPrecipitation (IP) fraction.

- In two-colour microarray platforms, an aliquot of the lysate before immunoprecipitation is kept, from which DNA is purified. This serves as a reference and is similarly amplified and labelled with a different fluorochrome, such as Cy3. This is referred to as the input fraction.

- Both the IP and the input fractions are combined and hybridised to a single DNA microarray in the same way as the aCGH protocol (see **Figure 3.3**). The IP signal and the input signal are quantified for each probe on the microarray using a scanner.

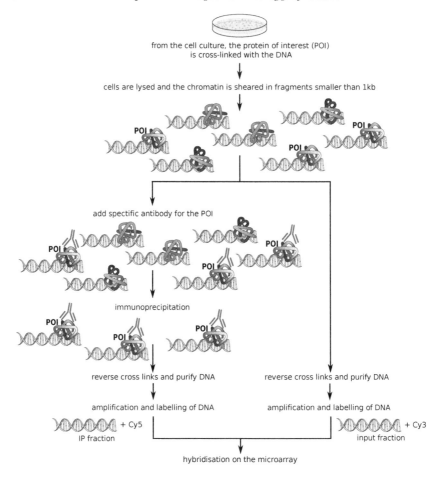

FIGURE 3.9 ChIP-on-chip protocol. From a cell culture, the interactions between the Protein of Interest (POI) and the DNA are isolated by immunoprecipitation. Both the ImmunoPrecipitation (IP) fraction and the input fraction are hybridised on a microarray. Image adapted from Buck and Lieb (2004).

Ideally, to provide a comprehensive and high-resolution survey of DNA-protein interactions, the ChIP-on-chip must contain probes which cover the entire genome for both coding and noncoding regions. For this purpose, oligonucleotide *tiling arrays* are used. In this design, the probes are selected to cover the entire genome or contiguous regions of the genome. The probes are either partially overlapping or contiguous. Since the precise location of the selected probes is known, a genome-wide map of the protein–DNA interactions is built as shown in **Figure 3.10**. For each genomic locus, the $\log_2$-ratio between the ImmunoPrecipitation (IP) signal and the input signal is computed. The regions which are bound by the Protein of Interest (POI) have a

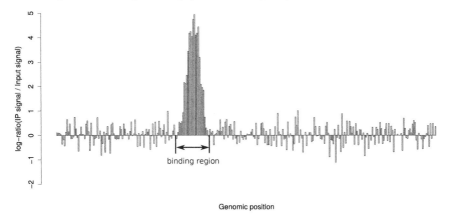

FIGURE 3.10 ChIP-on-chip profile. The $\log_2$-ratios between IP signal and input signal is represented along the genome. The DNA binding region of the POI appears as a peak in the genomic profile.

high $\log_2$-ratio value and appear as a peak in the genomic profile. The resolution of the method depends mainly on two factors: the length of the sheared chromatin and the length and spacing of the probes on the microarray.

Besides the identification of DNA protein binding sites, the ChIP-on-chip approach has been widely used to investigate the chromatin structure such as the nucleosome position map and the location of histone modifications. For the latter, an antibody which is specific for the modification of interest is used, thus allowing the histone code to be deciphered (Schones and Zhao, 2008). Indeed, epigenetic modifications are very important in cancer both at the histone and DNA levels. The next section will present how to investigate the epigenetic modifications at the DNA level.

3.1.6 DNA methylation

DNA methylation is an epigenetic modification which plays an important role in gene regulation and genome stability (see **Section A.1** in the **Appendices** for an introduction to the epigenetics mechanisms). As the genomic sequence remains the same for both methylated and unmethylated states, hybridisation-based microarray experiments cannot be applied as such to query the methylation state of CpG dinucleotides. Therefore, almost all sequence-specific DNA methylation analysis techniques rely on a methylation-dependent treatment of the DNA before amplification and hybridisation on a DNA microarray. Three main approaches are used and reviewed in Schones and Zhao (2008) and Laird (2010).

The first technique is based on a restriction enzyme which can specifically differentiate methylated and unmethylated CpG. The restriction enzyme cleaves unmethylated CpG while methylated CpG remains uncleaved, allowing the distinction between both methylation states after hybridisation on the microarray (Schumacher et al., 2006).

The second technique is based on affinity enrichment using an antibody which specifically recognises methylated cytosine. The protocol is very similar to the protocol described in the ChIP-on-chip experiment. These techniques are known as methylated-DNA IP (MeDIP, mDIP, mCIP).

The third technique is based on a bisulfite conversion (Reinders et al., 2008). The analysis of bisulfite-treated DNA using oligonucleotide arrays utilises the fact that after bisulfite treatment, unmethylated DNA contains uracil (analogous to thymine) in place of cytosine and hybridises poorly to array oligonucleotides that contain guanine. Methylated DNA, however, will hybridise on a probe corresponding to its complementary strand. Therefore, to detect the methylation state at a given genomic locus, two different probes are designed in order to discriminate between the methylation states as shown in **Figure 3.11**.

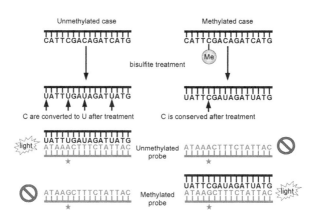

FIGURE 3.11 DNA methylation probe design. In a bisulfite treatment based approach, two different probes are designed for each locus in order to investigate its methylation state. While the probe pairs its target locus for the methylated state, all the G nucleotides are replaced by A in the unmethylated probe. Depending on the methylation state of the locus under investigation, a fluorescent light is emitted or not.

3.2 Emerging sequencing technologies

In 1977, a revolution in the era of genetic engineering was the development of technologies to sequence DNA. The same year, Maxam and Gilbert, and Sanger and Coulson proposed methods to sequence genomes (Gilbert and Sanger were awarded the Nobel prize in 1980 for their contribution to DNA sequencing). Sanger's method has been widely used for almost two decades and has led to a number of monumental accomplishments, including the completion of the first human genome sequence. In 1990, the Human Genome Project was launched as an international consortium aiming at sequencing with the Sanger's method (also called first-generation sequencing) a complete human genome. In 2001, the consortium (International Human Genome Sequencing Consortium, 2001) and Celera Genomics (Venter et al., 2001) each reported draft sequences providing a first overall view of the human genome. In 2003, the sequencing of the human genome was completed (Collins et al., 2003) and the international collaboration worked to convert the first draft into a genome sequence with high accuracy and nearly complete coverage (International Human Genome Sequencing Consortium, 2004). Therefore, the Human Genome Project took 13 years, involved about 3,000 scientists worldwide and cost 2.7 billion dollars to obtain the first human genome sequence (Wadman, 2008). Sanger's method was further improved and used by Levy et al. (2007) to publish the second human genome sequence (J. Craig Venter's genome). It took 4 years to be completed, involved 30 scientists and cost 100 million dollars. In spite of the improvement, Sanger's method was still not suitable in order to sequence genomes either in a reasonable time or at a reasonable price.

To overcome these limitations, a second-generation sequencing (also called next-generation sequencing) appeared in 2004, providing a dramatic increase in the throughput capacity with a lower cost. The main companies offering next-generation sequencing platforms are Illumina (Genome Analyzer^{TM}, HiSeq 2000^{TM} and MiSeq platforms), Life technologies / Applied biosystems (SOLiD^{TM} platform), Life technologies / Ion Torrent (Personal Genome Machine PGM^{TM} and Ion Proton^{TM} platforms) and Roche Applied Science (The 454 Genome Sequencer FLX platform) (Rusk and Kiermer, 2008; Chi, 2008; Niedringhaus et al., 2011; Rothberg et al., 2011). The latter platform was used to sequence James Watson's genome (Wheeler et al., 2008). It took 4.5 months to be completed, involved about 30 scientists and cost less than 1.5 million dollars. High-throughput sequencing technologies are a very competitive field and third-generation sequencing already appeared in 2008 (called next-next generation sequencing). This last generation sequencing is based on single-molecule analysis. The main companies are Helicos BioSciences (HeliScope platform) and Pacific Biosciences (PacBio RS platform) (Blow, 2008; Niedringhaus et al., 2011; Thompson and Milos, 2011). The HeliScope platform was used by Pushkarev et al. (2009) to sequence a human genome in several days, involving three scientists and costing about 50 thousand dollars. Fourth-generation sequencing combined single-molecule and nanopore sequencing technologies, Oxford NANOPORE Technologies (GridION platform) being the main competitor. As the sequencing cost is dramatically decreasing while the throughput capacity is increasing, these new technologies should sequence a human genome in several minutes for 1,000 dollars or even less before 2014 (Netterwald, 2010). **Box 3.3** defines the key concepts used in the sequencing field. Note that we will use the acronym **NGS*** which stands for next-generation sequencing to refer to any high-throughput sequencing techniques from second-generation until the most recent.

3.2.1   General principles of high-throughput sequencing

The different NGS platforms rely on high-level technologies including enzymology, chemistry, high-resolution optics, hardware, and software engineering. Although each platform has its own specificities (see **Table 3.1**), they generally share the following common steps that last between several hours to 10 days depending on the platform (see Mardis, 2008b; Metzker, 2010; Glenn, 2011):

- From the genome under investigation (here genome means any kind of nucleotide sequence of DNA or RNA), small fragments are prepared generally using a random shearing of the sequence. The small fragments obtained are referred to as the template and this step is called the library preparation. Depending on the application and the type of nucleotide sequence under investigation, specific library preparation protocols are used.

- The templates are amplified using PCR in order to provide a sufficient signal (this step is not required for single-molecule sequencing). To start the reaction, **primers*** are used.

- A glass slide or a chip encompasses all the different templates obtained from the genome under investigation allowing thousands to billions of sequencing reactions simultaneously.

Most sequencing machines rely on an optical detection in order to detect each nucleotide being sequenced which implies:

- The use of modified oligonucleotides labelled by fluorochromes. During the sequencing process, light is emitted by the fluorochrome and registered by the optics after laser excitation.

- An image analysis step allowing the quantification of the signal. For each template, a read which corresponds to the sequence of the template (or at least a subpart of it) is obtained.

Nevertheless, we will see that alternative detection methods without optics (*i.e.* laser-free) appeared in the most recent sequencers.

3.2.2 Principles of high-throughput sequencing based on amplification

High-throughput sequencing based on amplification requires a PCR step to amplify the amount of DNA for each template. This way, sufficient material

---

❏ **BOX 3.3: Key concepts in sequencing**

template: it is the true nucleic sequence which has to be read by the sequencer.

read: it corresponds to the sequence of the template which has been read by the sequencer. It consists of tens to hundreds bases depending on the technology used. Current techniques produce several millions up to billions of such reads in a single experiment.

sequencing error: it is a base in the read which does not correspond to the true base in the template.

depth of coverage: it represents the number n of reads which overlap a given position in the genome and is noted nX. The depth of coverage can be summarised by its average over all the positions within the genome. As sequencing errors are made by the sequencer, increasing the depth of coverage will improve the accuracy of the sequence obtained after alignment or assembly.

coverage: it represents the percentage of the genome which has been covered at least by one read.

reference genome: it is the nucleic sequences for each of the chromosomes from a sample which is considered as representative of a given species. A reference genome is obtained using *de novo* sequencing. The Human Genome Project has lead to the first human reference genome.

alignment: it is the process of mapping (*i.e* to obtain the position on the chromosome they belong to) the reads on the reference genome. As sequencing errors generally occur in the read, mismatch is allowed between the read and the reference genome.

sequence assembly: process which consists in merging reads into much longer DNA fragments in order to reconstruct the sequence of the sample under study.

de novo **sequencing:** it is the process of assembling reads together so that they form a new and previously unknown sequence.

run: set of steps which are performed by a sequencer in order to generate the reads.

GC-content it is the percentage of bases on a DNA sequence that is either a G or a C.

Platform	ST	AM	DM	Description
454	S	ePCR	OD	First second-generation sequencer with long reads
Genome Analyzer	S	bPCR	OD	First second-generation sequencer with short reads
SOLiD™	L	ePCR	OD	Second second-generation sequencer with short reads
Ion PGM	S	ePCR	H	First post-fluorescence sequencer
HeliScope	S	N	OD	First single-molecule sequencer
PacBio RS	S	N	OD	First real-time single-molecule sequencer
GridION	S	N	CD	First nanopore real-time single-molecule sequencer

TABLE 3.1 Sequencing platform characteristics. The table indicates the sequencing technique (ST), amplification method (AM), detection method (DM) and the description of different high-throughput sequencers. S = Synthesis; L = Ligation; OD = Optical detection with fluorescence; H = Hydrogen ion; CD = Current Disruption; N = none; ePCR = emulsion PCR; bPCR = bridge PCR.

is provided for a reliable detection of the signal. We propose to take as example the Sequencing by Oligonucleotide Ligation and Detection (SOLiD™) platform. After DNA shearing (see **Figure 3.12A**), the templates are separated into single strands and captured onto beads under conditions which favour one DNA molecule per bead (see **Figure 3.12B**). The templates (with a length ranging from 150 to 180bp) are amplified using emulsion PCR in order to provide a sufficient signal during the sequencing reactions. In an oil phase, aqueous droplets encompass one bead and form micro-reactors for the PCR reaction to take place. The beads contain an adaptor P1 which is ligated to the DNA template in the 5' end and a second P2 adaptor ligated in the 3' end. These two adaptors are known as DNA sequences and are essential to initiate the PCR. The same P1 and P2 adaptors are used whatever the template. The beads are then deposited on a glass slide.

   The main limitation of this sequencing technology is the read size that can be obtained. Indeed, the probability of detecting the right base generally decreases with its position on the read. As a result, reads longer than 75 bases are the maximum so far allowed by the SOLiD™ platform. Longer reads would not be reliable for quality reasons. In order to obtain longer reads, the company progressively increases the number of cycles while preserving the quality of the sequence. However, using a modified enzymology and chemistry, it is possible to perform a second sequencing from the position n' which represents the start of the P2 adaptor (see **Figure 3.12A**). This time, the sequencing moves from the 3' end towards the 5' end. This approach is called Paired-End Tag (PET) sequencing. Sequencing both ends of a given template is useful for

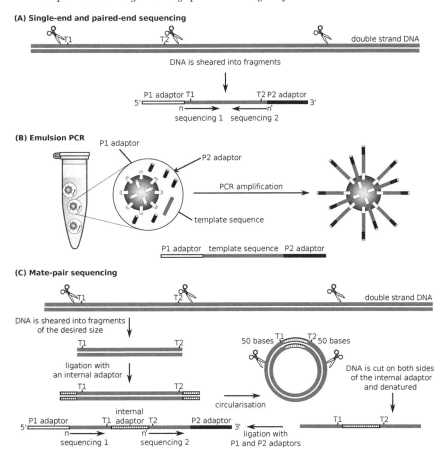

FIGURE 3.12 Library preparation for the SOLiD platform. (A) For both single-end and paired-end sequencing, the DNA is sheared into fragments (the fragment or template with the extremities T1 and T2 ranges from 150 to 180bp). One sequencing step is performed for single-end (from position n) and two for paired-end (from position n and n'). (B) An oil-aqueous emulsion is created to encapsulate 1-μm bead with P1 and P2 adaptors and a single template. After emulsion PCR within the droplet, the bead contains several thousand copies of the initial template sequence. (C) In the mate-pair library, the DNA fragments are first selected according to the desired size (for example, the fragment with the extremities T1 and T2 is chosen to be 3Kb). Then an internal adaptor is ligated at both ends of the fragments, the DNA is circularised and cut at both sides of the internal adaptor such that both resulting templates are 50 bases long. Two sequencing steps are performed. Adapted from Sequencing by Oligonucleotide Ligation and Detection (SOLiD^{TM}) documentation.

understanding the structure of the genome and *de novo* sequencing (Fullwood et al., 2009a). Using PET, the current 5500 XL SOLiD^{TM} is theoretically able to generate in seven days 300Gb of sequences corresponding to 4.8 billions of reads having 75 bases (P1 side) and 35 bases (P2 side) in length. An alternative to PET is mate-pair library sequencing (see **Figure 3.12C**). It enables the generation of DNA fragments with a desired size ranging from 2 to 10Kb. After DNA shearing, a ligation with an internal adaptor enables the circularisation of the DNA fragment. The circularised DNA is cut on both sides of the internal adaptor and ligated to usual P1 and P2 adaptors. The PCR can take place as described previously. Two sequencing steps are performed in the same manner as in PET. The first sequencing starts at position n with a primer that can pair with the P1 adaptor and the second sequencing starts at position n' with a primer that can pair with the internal adaptor. Combining data generated from mate-pair library sequencing with data from PET provides a powerful combination of read lengths for maximal genomic sequencing coverage across the genome. The **Figure 3.13** illustrates how genome rearrangements can be identified using a mate-pair sequencing with 5500 XL SOLiD^{TM}.

As we have previously mentioned, a recent platform called post-fluorescence (*i.e.* laser-free) sequencers does not require fluorochromes any-more thus avoiding scanner, cameras and laser. Among sequencers based on amplification, the Ion Torrent PGM (Rothberg et al., 2011) replaced the optical detection by semiconductor technology (see **Figure 3.14A**). In that case, the well lies under an ion-sensitive layer. As the polymerisation reaction releases a hydrogen ion as a byproduct, the charge from that ion can be detected by the ion-sensitive layer. In the homopolymer region, the intensity of the voltage detected indicates how many nucleotides have been incorporated. All the Ion Torrent system is encapsulated into a chip. The chip version 318 contains 12 million wells which generate roughly 1Gb of sequence in about 2 hours with an average read length of 200 bases.

3.2.3   Principles of single-molecule sequencing

In contrast to second-generation sequencing where the signal is registered for a template population, third-generation sequencing registers the signal from a single template DNA molecule (Metzker, 2010; Efcavitch and Thompson, 2010; Hohlbein et al., 2010). In this case, no PCR amplification is needed anymore as the sequence can be directly obtained from a single-strand DNA template placed into a well. Such a method avoids both the costs and errors related to the PCR step. Single-molecule sequencing can be separated into two main categories.

The first one consists of sequencing the DNA template by a cycling process. For each cycle, only one nucleotide is added, incorporated by a DNA polymerase if its complementary nucleotide is met at the position being se-quenced, and then, the remaining nucleotides are washed out (in the same manner as Ion Torrent PGM does). The HeliScope platform was the first one to propose such an approach (Harris et al., 2008). During each cycle, the fluo-

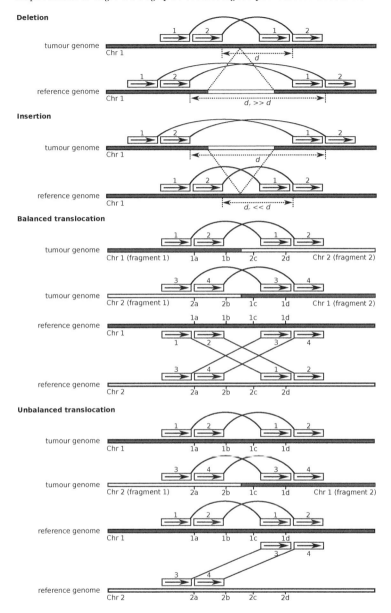

FIGURE 3.13 Identification of genome rearrangements using mate-pair sequencing. After mate-pair sequencing of the tumour genome (with SOLiD™ sequencer), the mate-pair reads are aligned on the reference genome. As the expected distance between mate-pairs is known (for example, $d = 3$Kb), deletion or insertion can be identified based on variation of the observed distance d_r between mate-pairs once aligned on the reference genome. Balanced and unbalanced translocations can be identified too. Tools like SVDetect (Zeitouni et al., 2010) allow the identification of these rearrangements. See **Figure 3.15** for an application of translocation detection.

rescence of the labelled nucleotide is quantified by the platform and indicates whether a given nucleotide has been incorporated or not.

The second category consists of sequencing in real time. Such platforms are referred to as Single Molecule Real Time (SMRT) sequencing. In SMRT, the polymerisation reactions performed by the DNA polymerase can be registered continuously in real time and the process happens as it would *in vivo* during the DNA replication process. Pacific bioscience has developed such a technology, and the principles are the following (Eid et al., 2009). A glass slide contains millions of holes called Zero-Mode Waveguide (ZMW) each with 70 nm in diameter and 100 nm in depth providing a reaction volume of 1.5 zeptoliter (10^{-21} litre). Within each ZMW, a single DNA polymerase molecule is anchored to the bottom glass surface. Nucleotides, each type labelled with a different coloured fluorochrome, are then flooded above an array of ZMWs at the required concentration. At the bottom of each ZMW lies an immobilised DNA polymerase enzyme which can replicate the complementary strand from a single-strand DNA template. The ZMW nanostructure allows the polymerisation reaction to take place. After laser excitation below the glass slide, the emitted fluorescence light indicates what nucleotide has been incorporated as far as the polymerase goes along the DNA sequence (see **Figure 3.14B**). The current technology allows 75,000 ZMWs to sequence in parallel. Interestingly, Flusberg et al. (2010) noticed that the methylation state of a given nucleotide impacts the DNA polymerase kinetics. During the polymerisation process, the fluorescence pulses in SMRT sequencing are characterised not only by their emission spectra but also by their duration and by the interval between successive pulses. Both pulse duration and interval between pulses are affected by the epigenetic modification of DNA, allowing the methylated and unmethylated cytosines to be discriminated. As a result, the principal challenge for single-molecule sequencing based on fluorescence detection was to avoid unwanted background noise created by the labelled nucleotides. The ZMW was especially designed for this. McCarthy (2010) reported that the company claims to obtain reads up to 10,000 bases (10 times longer than Sanger sequencing), with a sequencing speed 10,000–20,000 faster than the current second-generation sequencing technology (1–3 nucleotides per second can be incorporated by the polymerase).

Other strategies based on single-molecule DNA analysis use nanopore structures which consist of an orifice slightly larger than the width of a double-stranded DNA molecule (Stoddart et al., 2010). The nanopores are inserted into a lipid bilayer biomembrane. The base detection is possible through the measurement of conductivity through a membrane via the pore. The chemical differences of each base result in different magnitudes of current disruption which differentiate the four bases. Interestingly, the magnitudes of current disruption also depend on the methylation of the bases which allows the detection of epigenetic modification at the DNA level (Wallace et al., 2010). Oxford NANOPORE Technologies currently uses a bionanopore combined with an exonuclease bound to the inside of a protein nanopore. The exonuclease serves as a DNA binding site and cleaves individual bases from the DNA strand. Each cleaved base passes through the pore and the detected current allows its identification. The exonuclease regulates the motion of the DNA; otherwise it would move too fast for an accurate detection (see **Figure 3.14C**). Assuming a steady 1 ms per base sequencing rate, a single pore would require 69 days to process 6 billion bases. One hundred thousands pores operating perfectly at that rate could theoretically sequence a genome with 30X depth of coverage in 30 minutes. For faster and more accurate sequencing, Oxford NANOPORE Technologies is working toward the development of a strand sequencing technology in which a single-stranded DNA fragment is passed through the pore and the identification of single bases is achieved as they pass through it (see **Figure 3.14D**). A future generation of nanopore technology is solid-state nanopores. These are man-made holes in synthetic materials such as graphene sheet (Garaj et al., 2010). The use of these synthetic nanopores alleviates the difficulties of biomembrane stability and protein nanopore positioning used in both previous methods but requires control of the motion of the DNA strand (Luan et al., 2011). Fourth-generation sequencing allows sequencing in real time, avoids optical detection like Ion Torrent and does not require synchronous reagent wash steps, which makes it a very promising approach.

3.2.4   Targeted sequencing

Although the cost for genome sequencing is dramatically decreasing, it still remains an expensive technology. Therefore, it is not yet feasible to sequence many different whole human genomes. Consequently, different protocols have been developed to narrow the scope of investigation such that genomic regions of interest (*i.e.* the targeted sequences) can be selectively sequenced after enrichment. The first approach consists of the amplification of the regions of interest using PCR followed by NGS. The second approach is based on capture sequencing. In this case, the genomic DNA of interest are first captured using probes tethered to either a microarray or beads in a solution (Mamanova et al., 2010). The main limitation of the method is that probes are designed to target *a priori* known regions. From the sample under investigation, the DNA is extracted, sheared and hybridised on the probes. The targeted fragments attach to their respective probes while the nontargeted fragments are washed away. Subsequently, the targeted DNA fragments can be sequenced by NGS.

Targeted sequencing is generally combined with barcode multiplexing such that different samples can be processed simultaneouly in the same run. A different barcode (*i.e.* a short and known DNA sequence, for example a four base-pair barcode, allows theoretically to sequence 256 samples in the same time) is merged to the templates of each sample. Each sample is uniquely identified by its barcode. The sequences from all the samples are then pooled and sequenced. The barcode is sequenced during the run and allows the assignment of the template to its sample.

A typical application of target-enrichment followed by NGS is the exome

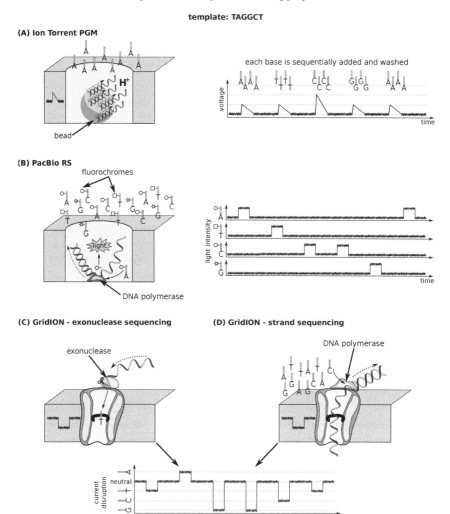

FIGURE 3.14 From second- to fourth-generation sequencing, illustration on TAGGCT template. (A) Second-generation sequencing. In Ion Torrent PGM, each base is sequentially added and washed. The voltage shift due to hydrogen ion emission indicates how many bases have been incorporated. (B) Third-generation sequencing. In PacBio RS, the four labelled nucleotides are flooded above the array. The light intensity for each colour indicates what nucleotide has been incorporated as far as the polymerase goes along the DNA sequence. (C) Fourth-generation sequencing. In GridION with exonuclease sequencing, the exonuclease attached to the nanopore cleaves each base from the template. As the base passes through the pore, it transiently binds to an adaptor molecule which causes a characteristic current disruption. (D) Fourth-generation sequencing. In GridION with strand sequencing, the template is threaded through the nanopore due to the polymerisation reaction. As long as the template moves through the pore, each base causes a characteristic current disruption as in the exonuclease sequencing. (See colour insert.)

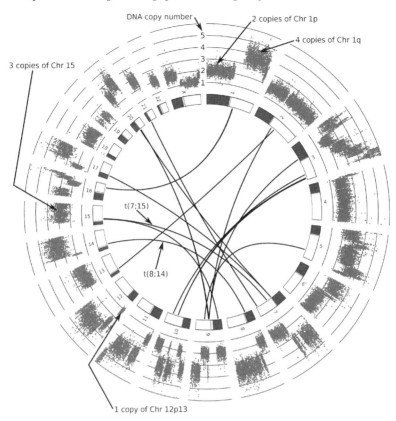

FIGURE 3.15 Characterisation of DNA copy number and identification of inter-chromosomal translocations in T47D using mate-pair sequencing. Chromosomes (black = p arm, grey = centromere, white = q arm) are represented around a circle. Black links in the inner part represent inter-chromosomal translocations. The translocations indicated in **Figure 2.12** are also identified in this experiment. The outer part represents the DNA copy number profile which looks very similar to the profile in **Figure 3.8**. We have analysed the data with the following algorithms: reads have been aligned with the bowtie algorithm (Langmead et al., 2009), DNA copy number estimated with FREEC (Boeva et al., 2011a) and translocations identified with SVDetect (Zeitouni et al., 2010). Circos has been used to plot the results (Krzywinski et al., 2009). **Figure 3.13** explains how translocations are identified. Data from Hillmer et al. (2011).

sequencing (Teer and Mullikin, 2010). This application aims at sequencing all exons corresponding to the protein-coding regions (the human genome contains about 180,000 exons covering 30Mb, *i.e.* about 1% of the human genome sequence). This will allow the identification of mutations which affect gene functions. Importantly, annealing conditions (*i.e.* the conditions in which two DNA strands bind together) allow mismatches in pairing during the capture such that SNPs and mutations can be detected.

3.2.5 Application of high-throughput sequencing in oncology

Why is it important to mention these NGS technologies? Among the different molecular levels, many include nucleotide sequences, either DNA or RNA (see **Figure 3.1**). Microarrays have been considered so far as the favourite tool for genomics, transcriptomics or miRNomics. NGS technologies appear as a very suitable tool to perform comprehensive experiments in tumoral genomes in order for molecular profiling at the level of DNA, mRNA and miRNA. Therefore, this is a cutting-edge technology which is very likely to replace microarray experiments in a near future. Moreover, these new platforms can explore new areas of biological inquiry, including the investigation of ancient genomes, the characterisation of ecological diversity, and the identification of unknown etiologic agents. NGS offers many applications, especially in the field of medical science (Schuster, 2007; Mardis, 2008a) and particularly in oncology in order to (note that points 9 to 13 were not possible with microarrays):

1. Quantify mRNA expression (this is called RNA-seq)
2. Quantify miRNA expression
3. Identify alternative splice forms
4. Quantify DNA copy number (see **Figure 3.13**, **Figure 3.15**)
5. Identify LOH
6. Identify protein-DNA interactions using ChIP-seq, *i.e.*, ChIP followed by sequencing (Farnham, 2009)
7. Map nucleosome position with respect to the DNA sequence
8. Study epigenomic modifications
9. Discover mutations
10. Discover polymorphism
11. Map chromosomal rearrangements (translocation, fusion gene, deletion, amplification, *etc.*) at a resolution of one base (Chen et al., 2008; Campbell et al., 2008) (see **Figure 3.15**)
12. Discover noncoding RNAs (ncRNA)
13. Study the spatial organisation of the chromatin

While most second-generation sequencing relies on alignment on a reference genome, the longer reads obtained by third and fourth generation sequencing will allow assembly (Martin and Wang, 2011) of tumoral genomes. This gives new insights to investigate genome rearrangements in a tumour.

Note that as all the sequencing technologies have different characteristics, some sequencers are more adapted to a given application (Thompson and Milos, 2011). For daily routine diagnosis using targeted sequencing Ion Torrent PGM is well adapted, while other sequencers might be more appropriate for research purposes.

3.2.6   Towards single-cell sequencing

A tumour consists of a complex mixture of cancer cells and normal cells (see **Section 2.6**). Normal cells can represent a significant proportion within the sample and mask the signal from the cancer cells. Moreover, tumours are heterogeneous as they often consist of different clonal subpopulations (see **Section 2.3**). Omics experiments generally require sufficient material extracted from several thousands of cells in order to obtain detectable signals. Therefore, molecular profiling with NGS provides an average overview across the different cell populations from the tumour sample. A deeper characterisation of tumours at the level of subpopulations or even at the level of single cells offers a more accurate picture of its complexity and its heterogeneity. Typically, comparing the tumour heterogeneity with respect to survival on one hand, and to resistance to treatment on the other hand, would be very valuable from a clinical perspective and that of personalised medicine. Indeed, the higher the heterogeneity, the higher the probability to have some cell subpopulation surviving when exposed to treatment.

Single-cell sequencing gives insights to answer the question of whether pre-existing rare cells in the **primary tumour**[*] can escape the treatment or whether resistant cells emerge in response to treatment by acquiring *de novo* mutations (Navin and Hicks, 2011; Navin et al., 2011). Another application of single-cell sequencing is the identification of Circulating Tumour Cells (CTC) and characterisation of their genomic alterations as the presence of these cells can be correlated with patient survival. The feasibility of single-cell sequencing relies on single-cell isolation techniques such as flow cytometry using Fluorescence-Activated Cell Sorting (FACS) or Laser-Capture Microdissection (LCM). Obviously, single-cell sequencing could enter clinical practice, as both cost and time of sequencing are decreasing.

3.3   Chromosome conformation capture

Microarray and NGS technologies help to reconstruct the structure of a tumoral genome as a one-dimensional linear succession of genetic elements (translocations, gain regions, loss regions, SNPs, mutations, *etc.*). However, in the cell nucleus, the genome is organised into a complex tridimensional structure. For example, chromatin loops and bridges bring distant elements of the chromosome into close physical proximity. As a result, these chromosomal interactions between distant genetic elements can contribute to the silencing or activation of genes. Among these elements, there are enhancers and promoters involved in the transcription regulation (see **Section A.1.1**). While it is more likely that a genetic element interacts with its neighbours in the same chromosome, it may be that the transcription of co-regulated genes on different chromosomes occur in the same spatial localisation inside the nucleus. Such models are referred to as transcription factories (Sutherland and Bickmore, 2009; Cook, 2010). Technical advances in detecting these interactions contribute to our understanding of the functional organisation of the genome, as well as its adaptive plasticity in response to environmental changes during development and disease (Göndör and Ohlsson, 2009). As the interaction between different DNA elements mainly occurs via protein complex, all the protocols used to investigate the chromosome conformation use formaldehyde to fix cells. As a result, it cross-links proteins to other proteins and to DNA elements which are in close proximity in the nuclear space such that interacting DNA elements are linked together. Four different approaches are used to unravel DNA-DNA interactions (see **Figure 3.16** and Simonis et al., 2007; Tanizawa and Noma, 2011).

Chromosome Conformation Capture (3C). For two loci of interest chosen in advance, the 3C technology (Dekker et al., 2002) quantifies the frequency of interactions between them (see **Figure 3.16A**). After cross-linking, chromatin is digested with a restriction enzyme. DNA ends are ligated under conditions that favour junctions between cross-linked DNA fragments. Cross-links are then reversed. Finally, a real-time quantitative PCR amplifies and quantifies the ligation product using primers designed to pair the two loci of interest. The main limitation of this technology is that only two loci can be investigated during the experiment.

Circularised Chromosome Conformation Capture (4C). To overcome the limitation of 3C, an alternative protocol based on the 3C approach has been developed to screen physical interactions between chromosomes without a preconceived idea of the interacting partners (see **Figure 3.16B**). The technique is termed 4C (Göndör et al., 2008). A circularisation step permits the identification of interacting sequences using two primers positioned on the sequence of interest but close to the junction between the sequence of interest and the interacting sequence. After PCR, high-throughput single-end sequencing (or microarray) can detect DNA loci which interact with the locus of interest.

Carbon-Copy Chromosome Conformation Capture (5C). Another extension to 3C can investigate all potential interactions within a limited region of interest (see **Figure 3.16C**). Basically, the technique performs many 3C experiments in parallel using a multiplex ligation-mediated amplification (Dostie et al., 2006, 2007). The technique is termed 5C and requires one to design as many primers as the number of loci to investigate within the region of interest. Typically, a thousand primers can be used such that one

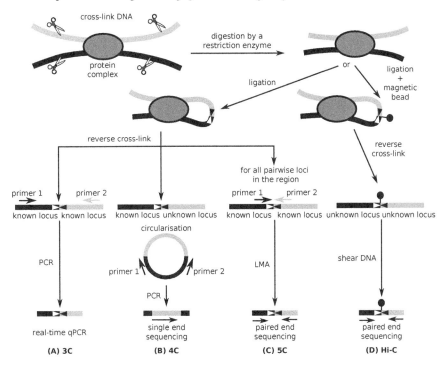

FIGURE 3.16 Protocols for 3C-based approaches. (A) 3C quantifies the interaction between two loci of interest, (B) 4C the interaction between one loci of interest and the whole genome, (C) 5C between loci located inside a region of interest and (D) Hi-C between all possible loci within the genome. Image adapted from Simonis et al. (2007); Tanizawa and Noma (2011).

million interactions can be tested. Both ends of interacting loci are identified with high-throughput PET sequencing. The size of the region which can be studied is limited by the number of primers which can be used simultaneously. Therefore the technology is not suitable for genome-wide scans.

    Hi-C. The detection of chromosomal interactions using 3C and its subsequent adaptations requires the choice of a set of target loci of interest. As it requires primers to be designed by the biologist, which is the most limiting part, it makes genome-wide studies impossible. To overcome this limitation, a protocol called Hi-C has been proposed to perform a genome-wide investigation of the chromosome conformation (see **Figure 3.16D** and Lieberman-Aiden et al., 2009; van Berkum et al., 2010). Hi-C allows the preparation of a genome-wide library of ligation products corresponding to pairs of fragments which were originally in close proximity to each other in the nucleus. After cross-linking and digestion by the restriction enzyme, a reporter is added at the junction of interacting DNA fragments. The reporter is a biotin-streptavidin complex fixed to a magnetic bead. After shearing the library, only the frag-

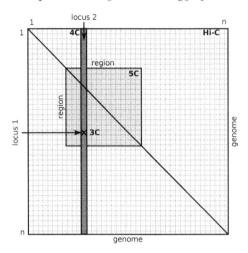

FIGURE 3.17 Schematic representation of the 3C-based approaches. In the case of a Hi-C experiment, the data can be represented as a symmetric matrix in which one row and one column correspond to two different loci such that all rows and columns span the whole genome. For cell of the matrix (*i.e.* a pair of loci), a value ranging from 0 to 1 can be affected and represents the frequency of their interaction. 3C, 4C and 5C correspond respectively to a cell, one row, and a a sub-matrix from the whole matrix obtained with Hi-C.

ment which contains a junction and thus having magnetic beads can be extracted using a magnet. The purified junctions can be subsequently analysed using high-throughput PET sequencing, resulting in a catalog of interacting elements.

Interestingly, 3C-based technologies can be combined with ChIP in order to analyse interactions between specific protein-bound DNA sequences. This approach is termed ChIP-loop (Simonis et al., 2007) or ChIA-PET (Fullwood et al., 2010, 2009b) and works as follows. After cross-linking, the use of a specific antibody allows the enrichment in DNA elements that contain the POI. The other steps remain similar to the 3C protocol.

As the analysis is performed on a population of cells, 3C and 3C-based technologies provide information about the frequency, but not the functionality, of DNA interactions. Thus, additional, often genetic, experiments are required to address whether an interaction identified by 3C-based technologies is functionally meaningful for the cells. Moreover, it is important to note that because of the flexibility of the chromatin fibre, DNA elements on the same fibre are engaged in random collisions, with a frequency inversely proportional to the genomic distance between them. Therefore, the mere detection of a ligation product does not necessarily reveal a specific interaction.

Figure 3.17 represents the possibilities offered by the different 3C-based approaches in order to investigate the DNA interactions within the genome.

3.4   Large-scale proteomics

Proteomics investigates at a large-scale the various properties of the proteins including their sequence, quantities, Post-Translational Modifications (PTM), interactions between each other, cellular localisations and structures. As the proteins are effectors inside the cell, proteomics studies are essential to understand the function of the genes. Moreover, while messenger RNAs (mRNA) can be detected inside the cell, they are not necessarily translated into proteins. Even translated, the proteins might be present but inactive. Thus, proteomics studies give a complementary knowledge in addition to genomics and transcriptomics.

A protein has specific functions that depend on four properties: (1) the peptide sequence (the primary structure), (2) the local structure such as alpha helix or beta sheet (the secondary structure), (3) the tri-dimensional shape (the tertiary structure) and (4) its ability to form complexes with other proteins (the quaternary structure). In addition, PTMs (phosphorylation, ubiquitination, glycosylation, acylation, *etc.*) modify the secondary, tertiary or quaternary structures of proteins changing their activity and properties.

Inside the cell, the concentration of the different proteins can range from several pg/ml up to several mg/ml. This implies the use of highly sensitive techniques in order to identify and quantify the less abundant ones. In contrast to nucleotide-based experiments in which the Polymerase Chain Reaction (PCR) can produce enough material (for hybridisation in a microarray or for NGS), no amplification method exists for proteins. As a result, the sensitivity of the technique is all the more required in order to obtain sufficient signal for a reliable measurement. Moreover, the complementarity property in nucleotide sequences does not exist in proteins. Finally, proteomics requires the extraction and purification of proteins which is a difficult step. The aforementioned reasons make the complexity of the proteome huge and its analysis very challenging.

In this section, we will focus on techniques to identify, quantify the proteins, detect their PTMs and characterise their interactions while the study of their structure will not be addressed. The main approaches used to decipher the proteome are microarrays, mass spectrometry and two-hybrid systems (Johnson and Hunter, 2005).

3.4.1   Microarray-based proteomics

To tackle the complexity of the proteome, investigation methods based on immunoassay experiments have been largely developed. In that case, proteomics studies rely on antibodies which are naturally used by the immune system to recognise specific antigens (*e.g.* proteins) from foreign organisms (*e.g.* viruses, bacteria). The antibodies possess a specific, highly variable do-

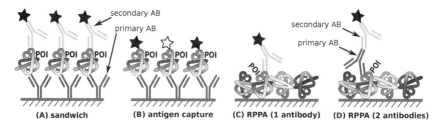

FIGURE 3.18 Protein arrays. (A) In a sandwich microarray, a specific primary antibody is tethered on the slide to capture the Protein of Interest (POI) which is then detected using a specific secondary antibody coupled with a fluorochrome. (B) In an antigen capture assay, the POI is similarly captured by tethered antibodies, but the captured proteins are detected directly (after chemically labelling the complex mixture of proteins). (C) In the Reverse-Phase Protein Array (RPPA), the mixture of proteins itself is tethered on the slide. The POI can be recognised by a primary antibody coupled with a fluorochrome or (D) with a primary antibody which recognises the protein and a secondary antibody coupled with a fluorochrome which targets the primary antibody. AB = antibody. Image adapted from (MacBeath, 2002).

main which can be adapted to recognise any potential antigen encountered during the life of the host organism. Biotechnological approaches exist to prepare antibodies capable of targeting a Protein of Interest (POI) but it is a delicate task to ensure both their specificity and sensitivity. As a consequence, only a few thousands of antibodies can target some proteins among the several millions of protein forms (including their PTMs) which may exist in humans. Microarray-based approaches have been combined with immunoassay methods in order to perform high-throughput quantification of proteins in biological samples. However, the throughput still remains lower than those obtained with nucleotide-based microarrays due to difficulty to prepare and use antibodies. Three microarray approaches allowing a relative quantification of protein levels exist and are described below (MacBeath, 2002; Tomizaki et al., 2010; LaBaer and Ramachandran, 2005; Spurrier et al., 2008).

Sandwich immunoassay. This technique is based on the Enzyme-Linked ImmunoSorbent Assay (ELISA) widely used as a diagnostic tool in medicine. First, specific antibodies are tethered on the slide and capture the POI which is then detected using a second specific antibody coupled with a fluorochrome (see **Figure 3.18A**). This technique requires that two specific antibodies are available to recognise the POI. For a single biological sample, several tens of proteins can be quantified simultaneously provided that no cross-reactions exist between the antibodies and the POIs.

Antigen capture immunoassay. The proteins from the sample are first subjected to a labelling procedure which adds fluorochromes to every protein (see **Figure 3.18B**). The use of two different fluorochromes allows the measurement of two different samples in one single experiment in the same way as the array Comparative Genomic Hybridisation (aCGH) protocol (see **Figure 3.3**). The POI is captured by tethered antibodies as in the previous technique and both fluorescences are measured. For a single sample, several hundred proteins can be quantified simultaneously.

Reverse-phase protein array (RPPA). The proteins from the sample are directly tethered on the slide surface. The POI can be recognised using either a specific primary antibody coupled with a fluorochrome (see **Figure 3.18C**) or both a specific primary antibody and a secondary antibody coupled with a fluorochrome (see **Figure 3.18D**). The primary antibody generally comes from another given species (*e.g.* rabbit when the proteome under study is human) and the secondary antibody is able to recognise any rabbit antibody. This strategy reduces the cost of primary antibody-coupled-fluorochrome preparation. While the two previous techniques allow the measurement of many different proteins for one or two samples in a single experiment, the RPPA allows the quantification of a single POI in several hundreds of samples in a single experiment. In this case, a spot on the microarray corresponds to the protein lysate from one sample.

❏ **BOX 3.4: Mass spectrometry**

A mass spectrometer consists of the following devices:

an ion source converts a gas, liquid or solid phase sample molecules under investigation into ions.

a mass analyser sorts the ions by their mass-to-charge ratio (m/z) by applying electromagnetic fields. Different technologies exist like time-of-flight, ion trap, quadrupole, Fourier transform mass spectrometry and orbitrap. They have their own specificities and are used for different applications.

a detector registers the number of ions at each m/z value.

The output result from a Mass Spectrometry (MS) or a Mass Spectrometry/Mass Spectrometry (MS/MS) experiment is a spectrum which consists of a series of peaks at given m/z values. The height of each peak indicates the ion abundance. The height of each peak is generally rescaled such that the highest peak is 100.

3.4.2 Mass spectrometry proteomics

Mass Spectrometry (MS) is an analytical technique to determine the composition of molecules or the list of compounds in mixture of many different molecules (see **Box 3.4**). This technique has been widely used in different scientific fields and particularly in biology for proteome study. While microarray-based proteomics target proteins known in advance, MS can identify any pro-

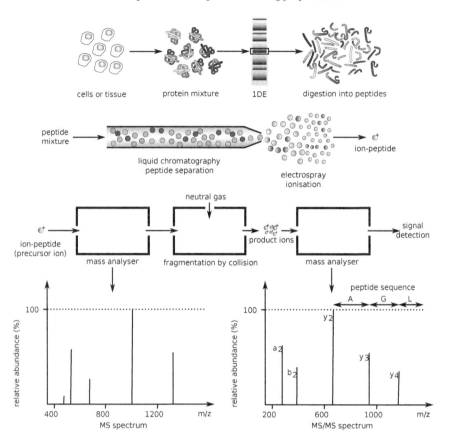

FIGURE 3.19 Mass spectrometry protocol. Proteins are extracted from cells or tissues and a sub-proteome is selected after one-dimensional electrophoresis (1DE). The proteins are digested into peptides, separated by Liquid Chromatography (LC) and ionised. The Mass Spectrometry (MS) spectrum and Mass Spectrometry/Mass Spectrometry (MS/MS) spectrum are obtained. (See colour insert.)

tein without any *a priori* knowledge. At the very beginning, MS was more convenient for small molecules and the analysis of macromolecules such as proteins was very challenging. In the 1980s soft ionisation techniques appeared such as Electrospray Ionisation (ESI) or Matrix-Assisted Laser Desorption/Ionisation (MALDI) allowing the production of ions from macromolecules without breaking their chemical bonds (John B. Fenn and Koichi Tanaka, Nobel prize chemistry 2002). This was a crucial step for the MS being used in biology especially to study proteins. Interestingly, a two step mass spectrometry has been developed to better characterise the molecules under investigation. This technique, called tandem mass spectrometry or Mass Spectrometry/Mass Spectrometry (MS/MS), involves two mass analysers. The combination of both MS and

MS/MS spectra ensures the identification of a protein. The investigation of PTMs is possible using MS and MS/MS as the PTMs change the peptide mass fingerprint. **Figure 3.19** describes the typical workflow for a MS experiment (see Aebersold and Mann, 2003; Patterson and Aebersold, 2003; Choudhary and Mann, 2010; Domon and Aebersold, 2010):

- The proteins are extracted from cells or tissue. Since the mass spectrometer cannot handle many proteins at the same time, a sub-proteome is extracted. This is generally done using a 1D or 2D **gel electrophoresis***. Moreover, the MS of whole proteins is less sensitive than peptide MS, a proteolysis reaction digests the proteins enzymatically by trypsin (or other enzymes) to produce small peptides. As a result, each protein has its own unique signature called peptide mass fingerprinting which is a series of peaks at given m/z values.

- In order to improve both the sensitivity and specificity of the MS, a Liquid Chromatography (LC) is used to separate the mixture of peptides from the sample. Briefly, the analyte's motion through the column is slowed by specific chemical or physical interactions with the stationary phase as it traverses the length of the column. How much the analyte is slowed depends on the nature of the analyte such as its hydrophobicity and on the compositions of the stationary and mobile phases. The time at which a specific analyte elutes (comes out of the column) is called the retention time. The retention time under specific conditions is considered a reasonably unique identifying characteristic of a given analyte. When the analyte comes out of the column, it is ionised by ESI and analysed by MS and MS/MS.

- After ionisation, a mixture of ions enter the first mass analyser, are sorted according to their m/z value and detected. A MS spectrum is produced for the ions which entered the mass analyser during a fixed time window called a survey scan.

- In MS/MS, the process continues as follows. From the MS spectrum, the ions in a survey scan are either specifically selected (*e.g.* from an ion list defined in advance by the user) or automatically selected (*e.g.* the precursor ions from the highest peaks). The precursor ions generally correspond to a unique peptide but can be contaminated by other peptides in some cases. Then, the precursor ions are fragmented into product ions after collision with a neutral gas. The product ions are termed a_i, b_i, c_i if they contain the N-terminus and x_i, y_i, z_i if they contain the C-terminus where i represents how much amino acids are present in the ion (see **Figure 3.20**). For stability, the cleavage preferentially occurs at the peptide bond such that b and y product ions are mostly observed. Finally, a second mass analyser sorts the product ions according to their m/z values. A MS/MS spectrum is produced which permits identification of the amino acid sequence.

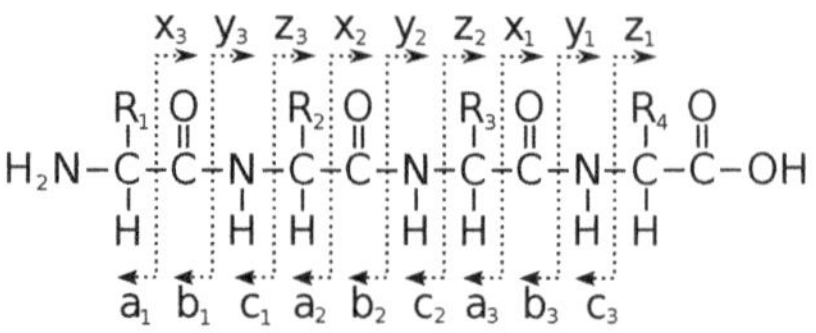

$$\begin{array}{c} x_3 \ y_3 \ z_3 \ x_2 \ y_2 \ z_2 \ x_1 \ y_1 \ z_1 \\ \end{array}$$

H$_2$N–C–C–N–C–C–N–C–C–N–C–C–OH

$$a_1 \ b_1 \ c_1 \ a_2 \ b_2 \ c_2 \ a_3 \ b_3 \ c_3$$

FIGURE 3.20 MS/MS peptide nomenclature. The nomenclature has been proposed by Roepstorff and Fohlman (1984).

- The MS/MS spectrum allows the peptide identification, querying reference databases with bioinformatic tools such as MASCOT or SEQUEST (Shadforth et al., 2005). The sequence of many genes have been identified both by traditional biological approaches and bioinformatic prediction models. As a result, about 25,000 genes are annotated in the human reference genome. Applying the genetic code to the DNA followed by *in silico* trypsin digestion permits to infer the exhaustive list of peptides which could be obtained. Databases which contain the list of m/z values for each peptide are finally queried in order to retrieve the candidate proteins which are very likely to be present in the sample.

Many biological or clinical questions require the comparison of two or even more different conditions. Therefore, there is a need for the MS to compare protein quantities or at least relative protein quantities between different conditions. The substantial discrepancy between the number of peptides present in a digest of a proteome and the analytical capacity of the LC-MS/MS system (*i.e.* the number of components that can be separated, detected and identified) prevents a perfectly reproducible set of peptides from being identified in repeated analyses of the same sample. As a consequence, quantitative proteomics by MS is a particularly challenging task. Different approaches have been developed to overcome these limitations (see Bantscheff et al., 2007; Elliott et al., 2009). The first method is Stable Isotope Labelling with Amino acids in Cell culture (SILAC) (see **Figure 3.21A** and Ong et al., 2002) and relies on a metabolic labelling using either light or heavy[4] arginine and lysine for the two different conditions. As the trypsin cleaves the protein after either arginine or lysine amino acids, it ensures that all the peptides obtained after proteolysis carry at least one labelled amino acid. As a result, a mass increment differentiates the same ions from both conditions. This technique requires culture cells as the labelling occurs during cell growth. Since this is not always possible, a second method called stable isotope incorporation via enzyme reaction (Enzymatic labelling) allows the labelling of the peptide C-terminus during proteolysis by trypsin as either light H_2O or heavy H_2O is used for the reaction (see **Figure 3.21D**). A third method is the Isotope-Coded Affinity Tag (ICAT) in which a reagent is added to cysteine residue (see

---

[4]light or heavy refer to the use of different isotopes.

Figure 3.21B and Gygi et al., 1999). From the MS spectrum, a shift will be observed and the ratio between peak heights represents the relative quantity between both conditions. For more reliable results, the MS/MS spectrum can be computed for validation. A fourth method was developed allowing more than two conditions to be compared simultaneously. This is the isobaric Tag for Relative and Absolute Quantitation (iTRAQ), which attaches a specific tag for each condition at the N-terminus of all peptides (see **Figure 3.21C** and Ross et al., 2004). Each tag consists of a reporter and a balance. As all the tags have the same weight, a mass combination for the reporter and the balance is chosen to be specific for each condition. The tags are cleaved into reporter and balance product ions during the fragmentation step. Therefore, from the MS/MS spectrum each condition can be identified by the reporter peak at its corresponding m/z value and the relative protein quantities can be inferred from the peak height ratio. As the sample preparation remains a tedious task in the aforementioned methods, bioinformatics algorithms have been proposed in order to correct inherent biases and sources of variations such that quantitative proteomics can be used in classical MS and MS/MS experiment (Griffin et al., 2010). Such approaches which rely on bioinformatics algorithms are called label-free quantitative MS.

3.4.3 Protein–protein interactions

In the cell, proteins are essential molecules involved in all biological processes, including the formation of macromolecular assemblies. Multiple proteins interact which each other thus conferring the cell and organism with specific functions and behaviours. The whole set of Protein–Protein Interactions (PPI) of a given organism is referred to as the *interactome*. Disruption of PPIs can result in the emergence of various diseases, including cancer. Besides, drug-based disruption of PPIs can be used for combating diseases. Therefore, the investigation of interacting partners and analysis of protein networks formed by PPIs is expected to have major implications in the understanding of diseases and the drug discovery. A number of high-throughput experimental methods have been developed to investigate PPIs. Genetic methods based on two-hybrid system were originally used by Fields and Song (1989) in yeast, *Saccharomyces cerevisiae*, to monitor PPIs. Yeast Two-Hybrid (Y2H) allows the determination of PPIs *in vivo*. It is based on the use of transcription factors characterised by a modular structure which consists of physically and functionally separable domains: a DNA-binding domain (DB) and a transcription activation domain (AD). Physical separation of DB and AD domains results in transcription factor inactivation (for a review, see Causier, 2004; Terentiev et al., 2009). In Y2H, a bait protein X is fused to the DNA-binding domain and a prey protein Y is fused to the activation domain resulting in two hybrid proteins (X-DB and Y-AD). The functional transcription factor is reconstituted upon the physical interaction between the bait and the prey (Suter et al., 2008). Physical association between the proteins X and Y in yeast cells genet-

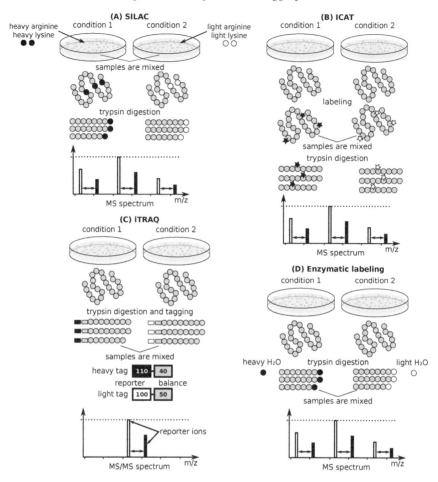

FIGURE 3.21 Quantitative mass spectrometry. Four different strategies allow specific labelling of peptides in order to distinguish the different conditions. For each peptide, a shift is observed for the m/z value either in the MS spectrum for (A) SILAC, (B) ICAT, (D) Enzymatic labelling or in the MS/MS spectrum for (C) iTRAQ. The ratio between peak heights indicates the relative difference between conditions.

ically engineered to express both hybrid proteins brings the DB and AD into proximity, thereby reconstituting the yeast transcription factor (*e.g.* GAL4). The DB of this functionally complemented transcription factor can bind a recognition site in the reporter gene promoter region called Upstream specific Activation Sequence (UAS). The AD interacts with the transcription machinery including RNA polymerase II, driving transcription of one or more reporter genes. The produced reporter proteins enable selection of those cells that harbour a pair of interacting proteins. Importantly, neither of these sub-domains alone can induce transcription. Yeast cells are transfected by two plasmids, the first one with X-DB protein and the second one with Y-AD protein such that both POIs can be produced within the cells (see **Figure 3.22A**). A system has been developed in order to improve the throughput (Jin et al., 2007). It is important to mention that Y2H can lead to false negative interactions (a steric hindrance can prevent the activation of the reporter gene) or false positive interactions (an overexpression of both the bait and the prey lead them to interact due to their high concentration; the bait can self-activate the reporter). Moreover, pseudo-interactions can be detected but not happen in native condition (interacting proteins can colocalise to cellular regions where the endogenous proteins normally never exist, thus enabling non-native interactions; the interacting proteins are coexpressed, whereas the corresponding endogenous proteins might never be present simultaneously). Such high risk of artifact is a real challenge to meaningful data interpretation.

Y2H generates mostly binary interactions. However, it has been extended to Yeast Three-Hybrid (Y3H) such that the reporter gene activity can only be detected if X and Y proteins interact with a third Z known protein (see **Figure 3.22C**). The Y3H was adapted in order to screen for prey proteins which interact with a small molecule displayed by the bait protein. Conversely, it might be interesting in drug discovery to assess if a small molecule can prevent two proteins from interacting. In this case, a counterselection marker is used with a gene reporter allowing the production of a toxic metabolite which leads to cell death. A small molecule which inhibits a bait-prey interaction is expected to rescue the cell viability (see **Figure 3.22B**).

The Y2H have been used to characterise the interactome in bacteria and metazoan model organisms (*Drosophila melanogaster* and *Caenorhabditis elegans*) and malaria parasite (*Plasmodium falciparum*). Stelzl et al. (2005) and Rual et al. (2005) used Y2H to investigate the human interactome. While the Y2H is obviously suitable for studying interactions in yeast it is not fully appropriate to study PPIs in mammals as the condition within the yeast might not be representative of what happens in mammal's cells. Indeed, yeast and mammalian cells differ in patterns of PTMs, as well as in the intracellular localisation of proteins. These types of protein modifications, as well as other unique factors or modulators present in mammalian cells, may influence the ability of proteins to interact. Therefore, it would be more reliable to investigate interactome in mammalian cells under the appropriate native cellular conditions such that the proteins have undergone the proper modifications to interact. Thus, not only should mammalian methods enable detection of a subset of interactions which might remain hidden using yeast-based approaches, they should also allow protein interactions to be tracked as a function of time, space (subcellular distribution) and physiological context (activation or inactivation of a cellular process induced by natural or synthetic stimuli). Whereas Y2H methods will probably remain unsurpassed in throughput and coverage, mammalian technologies could become essential tools for focused studies on the dynamics of (subsets of) the interactome. The Mammalian Two-Hybrids (M2H) system relies upon three plasmids which are co-transfected into mammalian cells (Lievens et al., 2009). Each plasmid has unique features. As in Y2H, the first plasmid contains the fused protein X-DD and the second one the fused protein Y-AD. The third plasmid contains a DNA binding site upstream of a specific reporter gene (Luo et al., 1997) (see **Figure 3.22D**). Such a technique was initially useful for a gene by gene validation thus limiting the genome-wide analysis. To overcome this limitation, Fiebitz et al. (2008) developed a cell array protein-protein interaction assay (CAPPIA). In the assay, mixtures of bait and prey expression plasmids together with an auto-fluorescent reporter are immobilised on glass slides in defined array formats. Adherent cells which grow on top of the microarray will become fluorescent only if the expressed proteins interact and subsequently trans-activate the reporter. This allows high-throughput investigation of PPIs in native condition within mammalian cells. Determination of physically interacting protein pairs makes it possible to design interactome map as graphs. Each node of the graph corresponds to a protein and an edge between two nodes indicates an interaction.

Two-hybrid approaches allow the investigation of a limited number of protein partners. However, in a cell, several different proteins can interact and form complexes which need to be discovered. The main limiting step in protein complex characterisation is the protein purification. To tackle this problem, the Tandem Affinity Purification (TAP) procedure was developed (Rigaut et al., 1999; Puig et al., 2001). It is an affinity purification technique based on Co-Immunoprecipitation (Co-IP). TAP was originally developed in yeast and enables the purification of protein complexes under close-to-physiological conditions. Protein complex composition is then determined by MS. TAP is a rapid and reliable technique which has been successfully applied in the analysis of PPIs in prokaryotic and eukaryotic cells such as yeast (Gavin et al., 2002). The method was improved in order to increase its sensitivity in mammalian cells (Bürckstümmer et al., 2006). The technique is based on the use of an affinity tag attached to a target protein. Genes which encode tag components and a target protein are incorporated using retrovirus into a host cell capable of maintaining the target protein expression at a level close to physiological. The standard tag, used in yeast, consists of two immunoglobulin-G-binding (IgG) fragments of *Staphylococcus aureus* protein A, a cleavage site for the tobacco etch virus (TEV) protease and a calmodulin-binding peptide. The target protein complex with the tag is isolated from the cell extract by a two-step procedure of affinity purification. The first step is based on binding of protein A to IgG-Sepharose beads, after which the complex undergoes action of the above-mentioned protease. The second step is based on partial binding of calmodulin-binding peptide, to calmodulin-Sepharose beads in the presence of calcium (see **Figure 3.23**). The use of affinity tags allows rather rapid purification of protein complexes from a small number of cells without preliminary elucidation of the protein composition of the complexes and func-

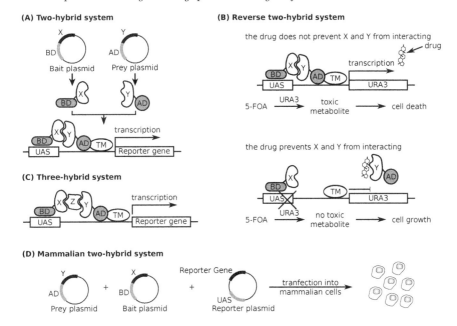

FIGURE 3.22 Principles of yeast and mammalian two-hybrid systems. (A) In the Yeast Two-Hybrid system, both prey and bait plasmids are transfected into the yeast. If proteins X and Y physically interact, the reporter gene is transcribed. (B) In the reverse Yeast Two-Hybrid system, the transcription of the reporter gene is lethal for the yeast. In this example, the protein coded the *URA3* gene transform the molecule 5-FOA into a toxic metabolite. If proteins X and Y interact, the yeast cells die in the presence of 5-FOA (top part) while if a drug prevents X and Y from interacting the cell growth is observed in the presence of 5-FOA. (C) In the Yeast Three-Hybrid system, the reporter gene is transcribed if both X and Y interact with a third known protein Z. (D) In Mammalian Two-Hybrid system, both prey and bait plasmids are transfected into the mammalian cells as in the Yeast Two-Hybrid system. Moreover, a reporter plasmid which contains the UAS region and the reporter gene which are naturally present in the yeast is also transfected. TM: Transcription Machinery; UAS: Upstream specific Activation Sequence. Image and Legend adapted from Suter et al. (2008); Causier (2004); Lievens et al. (2009); Luo et al. (1997).

tions of individual proteins. In combination with MS, this method allows the identification of proteins under study and their interactions. Many variations in the original tag and modifications of this method were proposed (Xu et al., 2010; Figeys, 2008) such that affinity purification combined with MS is widely used to study PPIs. For example, Ewing et al. (2007) identified more than 24 thousand PPIs in humans.

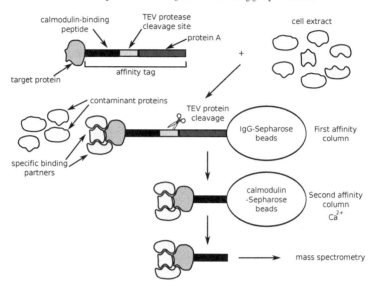

calmodulin-binding peptide
TEV protease cleavage site
protein A
cell extract
+
target protein
affinity tag
contaminant proteins
TEV protein cleavage
IgG-Sepharose beads
First affinity column
specific binding partners
calmodulin -Sepharose beads
Second affinity column
Ca^{2+}
mass spectrometry

FIGURE 3.23 Principle of tandem affinity purification.

3.5   Cellular phenotyping

The phenotype corresponds to any observable characteristics of an organism and is the result of the interaction between the genotype and the environment. Investigating the phenotype of living cells provides functional information regarding the biological processes involved under particular growth conditions. For instance, let us consider that a cancer cell line is grown with a potential anticancer chemical compound added within the culture medium. The characterisation of simple phenotypic traits such as cell viability versus cell death or growth rate indicate *in vitro* what could be the therapeutic efficiency of the anticancer agent *in vivo* (see **Figure 3.24**).

Pharmaceutical and biotechnological companies have developed large compound libraries which can exceed one million distinct chemical entities. The compounds of a library are referred to as perturbators. In drug discovery approach, potentially active compounds (*hits*) are first selected among the compound library and subsequently used in order to allow further development of compounds for pre-clinical testing (*leads*). Because the compound library is very large, a rapid and massive screening is required for the drug discovery process to be efficient. Major technical advances such as lab automation for sample preparation, assay miniaturisation, robotics, development of fast and automated microscopes combined with automated extraction of quantitative measurements from the acquired images have enabled microscopy to enter

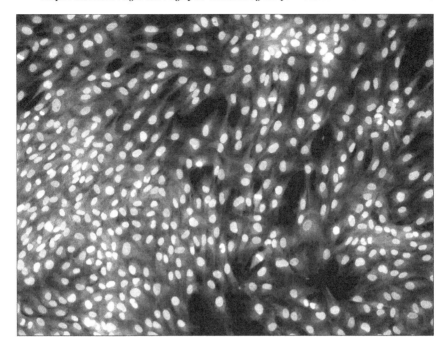

FIGURE 3.24 Characterisation of cell growth rate using cellular phenotyping. The cell nuclei appear in white or light grey. They are surrounded by the cell membrane. Image courtesy of Dr. Jacques Camonis. © 2012 Institut Curie.

the high-throughput era (Mayr and Bojanic, 2009; Mishra et al., 2008). As a result, large-scale screening can be performed to test rapidly a compound library in order to assess for each compound the phenotypic characteristics of the cancer cell line or any cellular model. The assessment of each compound activity is done by performing parallel assays in microtiter plates containing 96, 384, 1,536 or even 3,456 wells. In each well, cells are grown under the presence of one compound among the library. Depending on the plate size, using tens to several hundreds of plates allows a large-scale screening including control experiments.

The phenotypic traits which are considered generally correspond to an average biological response of thousands of cells present in the well. However, under a given condition, some phenotypic characteristics can be observed only for a subset of cells (for example, within heterogeneous or co-cultured cell cultures, stem cell subpopulations, *etc.*). Therefore, the monitoring of phenotypic characteristics at the level of each cell within a cell culture is essential to have a more accurate understanding of the perturbator effect. For example, the cell morphology, the spatial organisation of the organelles, their size and their number, and the subcellular location of a POI are phenotypic features which can be assessed (Zanella et al., 2010).

Major high-throughput technologies for cellular phenotyping are High-Throughput Screening (HTS) and High-Content Screening (HCS). HTS allows the measurement of a single feature, while HCS can record many different features simultaneously. HTS and HCS rely on the availability of fluorochrome used as biosensors to indicate physiological changes in the cell or to label specific organelles, including the nucleus, cytosol, mitochondria, endoplasmatic reticulum, Golgi apparatus and lysosomes. Typically, antibodies coupled to a fluorochrome or genetically encoded fluorescent proteins are used. However, the use of organic dyes in live cell imaging is often limited by their cytotoxicity and photobleaching.

Besides drug discovery, HTS and HCS are very useful in systems biology in order to decipher which genes and **signalling pathways*** are involved in a biological process. In that case, small interfering RNAs (siRNA) are used to specifically inactivate a gene of interest and figure out the resulting phenotype. The limitation of such an approach is obviously the siRNA design to ensure the most efficient depletion of its target gene. Therefore, each siRNA must be accurately validated in order to rely on downstream analysis. Instead of adding a chemical compound in each well, a siRNA is added. Interestingly, the interaction between siRNAs can be assessed in an epistatic study by adding siRNAs targeting two different genes.

3.6   Conclusion

The major high-throughput techniques used to characterise the molecular profiles in cancer have been presented in this chapter. Among them, the microarray has already provided significant improvments on the understanding of tumour progression and the classification of tumours. Today, NGS has dramatically increased our possibility to go deeper and deeper in the molecular investigation of cancer and cancer cells. Moreover, MS and cellular phenotyping are very valuable techniques. All these techniques offer insights to unravel the complexity and heterogeneity of cancer and are very likely to enter daily clinical practice in a near future. However, data processing based on sophisticated mathematical and statistical approaches is definitely required to extract the relevant biological and clinical information from the huge amount of data generated by these biotechnologies. This will be raised from **Chapter 4** to **Chapter 6**.

✎ **Exercises**

- In Ewing's **sarcoma***, you want to investigate what the possible target genes for the chimeric oncogenic transcription factor gene *EWS/FLI1* are (see **Page 29**). What kind of high-throughput technologies would you suggest for this purpose and which experimental design would you propose?

- Let us assume that you have sequenced the breast cancer cell line T47D using mate-pair sequencing with the Sequencing by Oligonucleotide Ligation and Detection (SOLiD^{TM})platform. How could it be possible from these sequencing data to produce the B Allele Frequency (BAF) profile as shown in **Figure 3.8**?

⇨ **Key notes of Chapter 3**

- A large variety of high-throughput technologies exist to study many different molecular levels.

- High-throughput technologies evolve very quickly.

- High-throughput technologies allow the identification and characterisation of molecular components and their interactions within biological systems.

- Microarrays can investigate oligonucleotide sequences or proteins which have to be known *a priori*.

- NGS can decipher previously unknown characteristics of the genome and improve the sensitivity with respect to microarrays.

- The study of proteins still remains complex thus limiting the throughput.

- The improvement of technologies offers the possibility to zoom in from cell population to single-cell behaviour and organisation.

Chapter 4

Bioinformatics tools and standards for systems biology

Systems biology relies heavily on a number of preliminary steps for preparing high-throughput experiments and making the results readily available for biological analysis and modelling. Though these steps are not *per se* part of what we commonly define as systems biology, they are essential for enabling the systems biology approach (Ghosh et al., 2011). Therefore, this chapter presents an overview of bioinformatics tools and standards used in a typical analysis workflow **Figure 4.1** which includes the following steps. Once the biological and/or clinical question is posed (❶), an experimental design is defined in order to efficiently answer the problem raised (❷). Then, the high-throughput experiments are performed (❸). A scanner generally analyses the **microarray***, sequencing slides or phenotyping screening, and produces images which are processed using appropriate algorithms to quantify the raw signal (❹). This step is followed by normalisation which aims at correcting the systematic sources of variability in order to improve the signal-to-noise ratio (❺). The quality of data is checked at the level of both the image analysis and the normalisation steps (❻). At this stage, the information provided after normalisation is still rough. The meaningful biological information relevant for biologists must be extracted from the data (❼). Once the relevant information is extracted, the data can be used in a transversal analysis to perform clinical biostatistics, classification or systems biology approaches (❽). Finally, the results need to be validated, interpreted and can lead to new experiments (❾). The bioinformatics workflow and computational systems biology approach are cyclical processes involving data acquisition and preprocessing, modelling and analysis. The integration and sharing of knowledge help to sustain the capabilities of this cycle to predict and explain the behaviour of biological systems. Therefore, to be successful, the workflow strongly relies on enabling processes to annotate (①), manage (②) and compute (③) the data. In this chapter, the steps ❷, ❺, ❻ and the processes ①, ② and ③ will be described. Steps ❼ and ❽ will be raised from **Chapter 5** to **Chapter 12**. The image analysis will not be addressed in the present book but the reader can refer to Fraser et al. (2010) and Novikov and Barillot (2007). Finally, this chapter illustrates how knowledge from the literature and databases can be extracted, and visualised using appropriate standards and software used in computational systems biology.

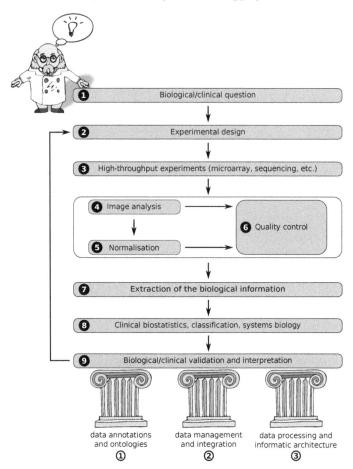

FIGURE 4.1 Bioinformatics workflow to analyse high-throughput experiments. A typical bioinformatics workflow generally includes steps ❶ to ❾ and strongly relies on enabling processes ① to ③.

4.1 Experimental design

Experimental design is an essential part of the scientific method. As such it is also part of the systems biology approach, but many of the techniques used are often considered out of the scope of systems biology, because they are part of another body of knowledge and attached to classical statistics. Though having a long tradition in industrial and agricultural trials since the last early century, the experimental design step is still too often neglected. Sir Ronald Aylmer Fisher, one of the pioneers in the field of experimental design, said in

1938 during his presidential address to the First Indian Statistical Congress: *"To consult the statistician after an experiment is finished is often merely to ask him to conduct a post-mortem examination. He can perhaps say what the experiment died of."* As an essential preliminary step, experimental design aims at two main objectives:

- Ensuring that the question of interest can be answered from a given set of experiments.

- Ensuring that the answer will be the most accurate for a confident statistical inference.

4.1.1 Choosing the optimal set of experiments

In practical situations, the limited amount of available biological material, the costs, *etc.* are strong constraints restricting the number of possible experiments which can be carried out. Given this number of experiments, the most efficient strategy has to be defined. In other words, the experimental design selects the most appropriate set of experiments among all the finite possibilities.

To illustrate this, let us assume that a gene expression study using two-colour microarrays (see **Section 3.1**) has quantified the effect of two drug treatments (factor with the two modalities D_0 and D_1) on two cancer cell lines (factor with the two modalities C_0 and C_1). As a multiplicative error model is generally assumed on the intensities, it is necessary to use a logarithmic transformation of the data in order to have an homoscedastic model (*i.e.* each observation is assumed to have the same variance). Let us consider Y_{ij} to be the quantity for a given gene in a condition S_{ij} which combines cell line i and drug j (with $i \in 0, 1$, $j \in 0, 1$). For the statistical inference, a two-way Analysis of Variance (ANOVA) model can be written as follows:

$$\log(Y_{ij}) = \mu + \gamma_i + \delta_j + (\gamma\delta)_{ij} + \varepsilon_{ij}, \text{with } \varepsilon_{ij} \sim \mathcal{N}(0, \sigma^2).$$

The terms γ_i and δ_j represent the effect of modality i (for the cell line factor) and the effect of the modality j (for the drug factor). The term $(\gamma\delta)_{ij}$ represents the interaction between the cell line and the drug factors. For the model to be identifiable (*i.e.* all the parameters of the model can be estimated), γ_0, δ_0 and $\gamma\delta_{00}$ are set to 0. The expected Y_{ij} value for the four possible conditions are given in **Table 4.1**.

In a typical two-colour microarray experiment, the relative difference between two conditions are compared. For example, let us consider the microarray with the condition S_{11} compared with S_{10}. For a given gene, the observed $\log_2$-ratio $L_{11 \text{ vs. } 10} = \log(Y_{11}) - \log(Y_{10})$ should be in average equal to $\mu + \gamma_1 + \delta_1 + (\gamma\delta)_{11} - \mu + \gamma_1 = \delta_1 + (\gamma\delta)_{11}$. Let us note θ the vector of parameters as follows:

$$\theta = \begin{pmatrix} \gamma_1 & \delta_1 & (\gamma\delta)_{11} \end{pmatrix}^\top.$$

cell line		drug	
		D_0	D_1
	C_0	μ	$\mu + \delta_1$
	C_1	$\mu + \gamma_1$	$\mu + \gamma_1 + \delta_1 + (\gamma\delta)_{11}$

TABLE 4.1 Expected values for the two-way ANOVA model.

Therefore, for one gene and one microarray, the expected $L_{11 \text{ vs. } 10}$ value can be formulated as the vector product between a design vector and the vector of parameters:

$$L_{11 \text{ vs. } 10} = \begin{pmatrix} 0 & 1 & 1 \end{pmatrix} \theta = X_{11 \text{ vs. } 10} \theta.$$

A given experimental design will exhaustively list the set of microarrays, each of them corresponding to a comparison between two different conditions. Since each microarray can be represented, as formulated previously, with a vector product, an experimental design can be summarised by both an X matrix and the vector of parameters θ. If we note L the vector which contains the observed $\log_2$-ratio $L_{ij \text{ vs. } i'j'}$ and X the matrix in which the rows are the corresponding $X_{ij \text{ vs. } i'j'}$ design vectors (with $i, i' \in 0, 1$ and $j, j' \in 0, 1$), we have:

$$L = X\theta + \epsilon, \text{with } \varepsilon \sim \mathcal{N}(0, 2\sigma^2 I).$$

The ANOVA model implies that $V(\theta) = 2(X^\top X)^{-1}\sigma^2$. As a result, different experimental designs can be compared based on the variance of the parameters of interest, the lower being the better. For example, let us consider design 1 in **Figure 4.2**. The top part of the figure indicates how the conditions are combined on the microarray: an arrow represents a microarray experiment and its direction defines which condition has been used as a test and which one has been used as a reference (see **Figure 3.3**). $S_{11} \rightarrow S_{00}$ means that S_{11} is the test condition while S_{00} is the reference one. The middle part of the figure shows the corresponding X design matrix. The bottom part indicates the variance for each parameter: for γ_1, its variance is $1 \times 2\sigma^2$, for $(\gamma\delta)_{11}$, its variance is $3 \times 2\sigma^2$, *etc.* If we compare design 1 and design 2, we clearly see that design 2 must be avoided. Increasing the number of experiments generally decreases the variance (for γ_1, the variance is the same in design 1 and design 4). The design 3 must be preferred over design 4.

4.1.2 Efficient statistical inference

In a design of experiments, the researcher investigates the effect of some factors (*e.g.* drug treatment) on *experimental units* such as cell lines, tumour patients, *etc.* in order to draw conclusions for the system under study. The procedure of statistical inference relies on the test of hypotheses, the estimation of

graphical representation of the microarray design

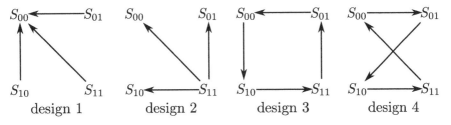

X matrix

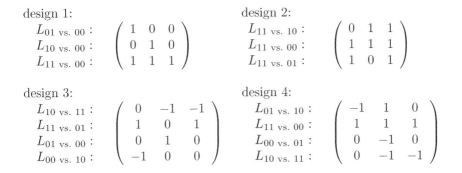

design 1:

$L_{01 \text{ vs. } 00}$:

$L_{10 \text{ vs. } 00}$:

$L_{11 \text{ vs. } 00}$:

$$\begin{pmatrix} 1 & 0 & 0 \\ 0 & 1 & 0 \\ 1 & 1 & 1 \end{pmatrix}$$

design 2:

$L_{11 \text{ vs. } 10}$:

$L_{11 \text{ vs. } 00}$:

$L_{11 \text{ vs. } 01}$:

$$\begin{pmatrix} 0 & 1 & 1 \\ 1 & 1 & 1 \\ 1 & 0 & 1 \end{pmatrix}$$

design 3:

$L_{10 \text{ vs. } 11}$:

$L_{11 \text{ vs. } 01}$:

$L_{01 \text{ vs. } 00}$:

$L_{00 \text{ vs. } 10}$:

$$\begin{pmatrix} 0 & -1 & -1 \\ 1 & 0 & 1 \\ 0 & 1 & 0 \\ -1 & 0 & 0 \end{pmatrix}$$

design 4:

$L_{01 \text{ vs. } 10}$:

$L_{11 \text{ vs. } 00}$:

$L_{00 \text{ vs. } 01}$:

$L_{10 \text{ vs. } 11}$:

$$\begin{pmatrix} -1 & 1 & 0 \\ 1 & 1 & 1 \\ 0 & -1 & 0 \\ 0 & -1 & -1 \end{pmatrix}$$

parameter variances

	design 1	design 2	design 3	design 4
γ_1	1	2	3/4	1
δ_1	1	2	3/4	3/4
$(\gamma\delta)_{11}$	3	3	1	2

FIGURE 4.2 Experimental design with a two-way ANOVA model. The top part provides four possible experimental designs using three two-colour microarray experiments. The middle part shows the design matrices for each design. The table in the bottom part indicates the variance of each parameter used in the two-way ANOVA model.

parameters of interest and the comparison of different mathematical models. As the experimental units are selected among a global population, *replication*, *blocking* and *randomisation* are three fundamental statistical principles to consider for an efficient and reliable statistical inference.

Replication consists of adding a number of replicates for each different condition, to take into account the fact that a given measurement is subject to variation. In the different experimental designs in **Figure 4.2**, we only had one microarray for each condition. Following this first principle, replicates for each condition is definitively required. We generally distinguish *technical replicates* from *biological replicates*. Technical replicates deal with the variability inherent to the technology while biological replicates account for the variability which exists within the population under study. Choosing whether technical or biological replicates (or both) should be preferred clearly depends on which variability is the most important. For example, we will favour biological replicates if the biological variability is greater than the technical variability. In general, both variabilities cannot be known in advance. Therefore, a pilot study is necessary in order to estimate these variabilities.

Since replicates are needed, the number of required microarrays can exceed the maximal number of experiments which can be performed in a single day by a single operator on the same laboratory. As a result, one possibility is to process the samples on different days but the day of experiment generally impacts on the signal measurement. Indeed, humidity, temperature, ozone concentration vary from one day to another and modify the measurements (Lander, 1999; Fare et al., 2003; Byerly et al., 2009). This effect is often termed *batch effect*. Imagine you would like to compare samples treated with drug D_0 and samples treated with drug D_1, then it would be a very bad idea to process all D_0 samples in a first batch (*i.e.* day 1) and all D_1 samples in a second batch (*i.e.* day 2). As the batch and drug are confounding effects, you will not be able to conclude whether the difference you may observe is due to the day of experiment or to the drug treatment. The same thing holds if you decide that two different operators can prepare the samples, or that the sample will be processed in different laboratories. Batch effect, operator effect or laboratory effect must not be confounding with the effect of interest in order to rely on the results. These factors are irrelevant with respect to the question of interest but have an impact on the signal measurement. Taking them into account is an essential consideration in experimental design and is called *blocking*.

The last important concept is *randomisation* which consists of affecting at random the experimental units between the different drug treatments in order to avoid any bias in the results. For example, in the case of a clinical trial which compares a new drug with respect to a standard drug, the patients are affected to either the new drug or to the standard drug control at random.

An abundant literature about experimental design for microarray experiments exists (Kerr and Churchill, 2001; Yang and Speed, 2002; Nguyen and Williams, 2006; Churchill, 2002). Although the illustration described here is devoted to two-colour microarray data, the same principles hold whatever the high-throughput technology used. For example, Auer and Doerge (2010) and Fang and Cui (2011) raised the problem of experimental design in **NGS**[*] experiments (RNA-seq). Obviously, each technology has its own specificities but the selection of the experiments, replication, blocking and randomisation are four essential preliminary steps which will ensure the reliability of downstream analysis.

4.1.3 Specific aspects in systems biology

In systems biology, one goal is to build and evaluate mathematical models taking into account the mechanistic and dynamic components of the biological system under study (see **Chapter 7**). In addition to the essential statistical principles we have previously introduced, other important aspects have to be specifically considered in systems biology (Ideker et al., 2000; Kreutz and Timmer, 2009). To be representative of the behaviour of the biological system in various experimental conditions, the model must be in agreement with observed measurements in real data. Therefore, the selection of a given set of perturbations to be applied on real experiments is necessary to challenge the model in real conditions. Defining the most relevant set of perturbations can be addressed using various techniques. Moreover, choosing sampling times, *i.e.* the times of consecutive measurements, is crucial as the dynamics of a system is an important component. The last aspect deals with the parameter values such as kinetics or affinity constants necessary in some mathematical models. Among all the parameters, some are critical as they can strongly impact on the model predictions. The identification of these critical parameters and their correct measurement is required for reliable and robust predictions.

4.2 Normalisation

Normalisation (also called low-level analysis) aims at correcting the systematic sources of variability in order to improve the signal-to-noise ratio for better biological and/or clinical interpretation. Historically, the normalisation started in the field of messenger RNA (mRNA) expression microarrays (see Quackenbush, 2002; Do and Choi, 2006; Irizarry et al., 2006; Stafford, 2007; Wu et al., 2003; Irizarry et al., 2003, for a review). The first method was the *Lowess normalisation* proposed for two-colour microarrays (Yang et al., 2002), followed by RMA (Irizarry et al., 2003) and GC-RMA (Wu et al., 2003) devoted to Affymetrix GeneChip®. Normalisation remains a very active field of research in any current high-throughput technologies.

As already mentioned in the previous section, there are inherent sources of variability which have a direct impact on the signal measurement. In some way, the *blocking* in an experimental design already integrates effects which have to be corrected. Typically, the correction of the batch effect is generally considered as part of the normalisation step. However, while necessary, *blocking* does not generally account for all possible sources of variability. Indeed, each experiment is singular and shows a specific variability which needs to be corrected. For example, spatial artefacts are frequently observed in microarray experiments and many spatial normalisation methods have been developed to correct them for gene expression (Workman et al., 2002), Comparative Genomic Hybridisation (CGH) (Neuvial et al., 2006) and DNA methylation (Sabbah et al., 2011) microarrays. **Figure 4.3** illustrates how the method called MicroArray NORmalisation (MANOR) improves the signal-to-noise ratio on array Comparative Genomic Hybridisation (aCGH) profile (Neuvial et al., 2006). Koren et al. (2007) suggested that normalisation methods which correct for spatial biases, such as MANOR, should be routinely applied when analysing microarray data.

Among other parameters, let us mention the GC-content (see **Box 3.3**) which affects the signal measurement in technology based on nucleotide sequence requiring **Polymerase Chain Reaction (PCR)**[*] amplification (microarrays and NGS) (Metzker, 2010). Rigaill et al. (2008) proposed the ITerative and Alternative normaLIsation and Copy number calling for affymetrix Snp arrays (ITALICS) which is based on multiple regression to correct, among other things, the effect of the GC-content which affects Affymetrix GeneChip® SNP arrays. Similarly, Boeva et al. (2011a) and Risso et al. (2011) proposed methods to correct the effect of GC-content on read counts in NGS data (see **Figure 4.4**).

While the effect of both spatial bias and GC-content can be clearly observed, the shape of the bias can differ from one experiment to another. As a result, the normalisation has to be adaptive, that means tailored for each experiment. Importantly, the identification of all parameters which can bias the signal has to be investigated and discussed with the platform provider and the operator in charge of the platform, as they have a strong experience in all the protocol steps affecting the signal measurement. The data normalisation is a critical step which needs to be considered carefully, as it will affect reliability, accuracy and validity of downstream analyses (Stafford, 2007).

4.3   Quality control

Any high-throughput technology, even if well standardised and carefully carried out, suffers from experimental bias or uncontrolled variations as we have discussed previously. Good laboratory practices require a quality control to be performed on a regular basis, and if possible on each sample analysed. For usage in clinical practice, the requirement for reliability of high-throughput technologies is of course even higher. Therefore, adequate quality control procedures need to be defined. Assessing the efficiency of any measurement protocol is based on different metrics (see **Box 4.1**).

The Food and Drug Administration (FDA) has initiated a huge quality control project called MicroArray Quality Control (MAQC). Initially devoted to gene expression microarrays, the project was further extended to NGS. The MAQC project is separated into three main phases (see **Box 4.2**). MAQC-I (MAQC Consortium et al., 2006) demonstrated both an intra-platform con-

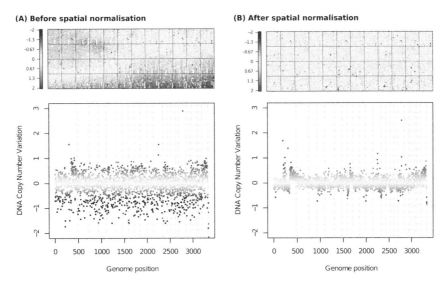

FIGURE 4.3 Spatial bias in aCGH experiment. (A) The top part represents the log-ratio values quantified on an aCGH. The experiment suffers from a spatial gradient from top left corner (abnormal high log-ratio values) to bottom right corner (abnormal low log-ratio values). As a result the DNA copy number profile plotted in the bottom part is very noisy. (B) The top part represents the log-ratio values after spatial normalisation with the MANOR algorithm which has removed the spatial gradient. The normalisation improves the signal-to-noise ratio of the DNA copy number profiles. Images from Neuvial et al. (2006).

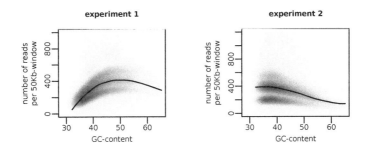

FIGURE 4.4 Effect of GC-content on the number of reads in NGS experiments. For two different NGS experiments, the effect of GC-content on the number of reads is plotted. The reference genome is split into contiguous windows of 50Kb-size. The number of reads which align within a given window (y-axis) is plotted as a function of the percentage of GC on the reference genome within this window (x-axis). The black curves represent the three-order polynomial fits of the data. As the patterns are very different in each experiment, an adaptive normalisation method such as control-FREE Copy number caller (FREEC) is required in order to correct the GC-content effect. Image adapted from Boeva et al. (2011a).

❏ **BOX 4.1: Quality control metrics**

precision: it evaluates the variance of repeated measurements under the same conditions. Precision can be decomposed into two variance components:

    repeatability: it corresponds to the variance of the measurements when the conditions remain identical (the same instrument, operator, *etc.*) and the measurements are repeated during a short time period.

    reproducibility: it corresponds to the variance of the measurements using the same measurement protocol but using different conditions (different instruments, operators, *etc.*) and the measurements are repeated during a long time period.

accuracy: it evaluates the bias between the measurement of a quantity to its true value.

---

sistency and inter-platform concordance in terms of genes identified as differentially expressed (platform means the provider such as Affymetrix, Illumina, *etc.*). MAQC-II (Shi et al., 2010) showed that the performance of predictive models (see **Chapter 6**) depended largely on the question being addressed. For example, in breast cancer data, predicting Oestrogen receptor status is a much easier task than predicting a pre-operative treatment response. MAQC-II showed that the lower the prediction performance, the lower the stability of gene lists. The last phase on the project is still ongoing.

While some experiments can pass through quality control thresholds they may present abnormal behaviour for unexplained reasons. Such experiments are called outliers and represent an observation which is significantly different from the rest of the data. Such experiments have to be discarded for reliable downstream analyses. In order to identify such outlier experiments, statistical methods are used, such as hierarchical clustering or Principal Component Analysis (PCA) (see **Section 5.3.2** and **Section 5.4.2**).

4.4 Quality management and reproducibility in computational systems biology workflow

The MAQC consortium pinpointed that not only the experiment but also the data analysis must be reproducible as reported by Ioannidis et al. (2009).

❑ **BOX 4.2: The MAQC project**

MAQC-I assesses the precision and comparability (intra/inter-laboratory, intra/inter-platform) of microarrays and develops guidelines for microarray data analysis.

MAQC-II assesses the capabilities and limitations of different data analysis methods in building microarray-based predictive models and provides best practices for development and validation of predictive models.

MAQC-III, also called Sequencing Quality Control (SEQC), aims at assessing the technical performance of NGS platforms and evaluating advantages and limitations of various bioinformatics strategies in RNA and DNA analyses.

To do so, Noble (2009) suggested that any bioinformatician must follow the simple guiding principle: *"Someone unfamiliar with your project should be able to look at your computer files and understand in detail what you did and why."* This requires practical guidelines such as the use of a standard folder tree (*e.g.* script, data, results, reports), the code documentation, the use of a lab notebook (possibly electronic) and the versioning of files. For example, tools such as Apache Subversion, Mercurial or GIT may be used as a versioning file system while the Sweave format (Leisch, 2002) can be used to produce reports which embed the code used to generate the analysis (*n.b.* a Sweave file is supplied in the accompanying materials). We strongly encourage bioinformaticians to adopt these tools for better reproducible *in silico* research in bioinformatics and computational systems biology.

4.5 Data annotations and ontologies

High-throughput technologies produce a huge amount of data which require reliable annotations (also termed metadata) in order to provide significant biological and/or clinical interpretations and be used in systems biology approaches. Therefore, for any experiment, the quality and availability of annotations is essential. The annotations can be separated into two categories:

sample annotations correspond to all the important information related to the samples under consideration. In this category fall the clinical annotations such as the treatment, culture condition, type of pathology, age of the patient *etc.* Moreover, all the characteristics related to the protocol which have been applied to prepare the samples need to be recorded, such as the type of technology used, version of protocol, *etc.* These annotations are generally stored in clinical databases and Laboratory Information Management System (LIMS).

feature annotations correspond to the description of the entities which are measured by the technology. Typically, on the microarray, different probes are quantified but it is required to know which gene/protein is targeted by the probe. Three levels of information are essential and presented in the next section.

4.5.1 *A priori* biological knowledge

Three levels of prior biological knowledge exist for the feature annotations and are available in different databases:

gene and protein sequences are generally based on the human genome sequence which is regularly updated as long as improvement in the human genome assembly is obtained by a continuous effort. The sequences are available from NCBI and UCSC Genome Browser databases. The major repositories include NCBI Nucleotide (formerly GenBank), DNA Data Bank of Japan (DDBJ), EMBL which are part of the International Nucleotide Sequence Database Collaboration (http://www.insdc.org/).

gene and protein biological information contains cellular localisations, biological processes, molecular functions, link between genes and diseases, *etc.* They are available in databases such as Gene Ontology (GO), Online Mendelian Inheritance in Man (OMIM), NCBI Gene, and Universal Protein Resource Knowledgebase (UniProtKB).

molecular interactions correspond to information about Protein–Protein Interactions (PPI), metabolic pathways, signalling pathways, gene regulatory networks, *etc.* (see **Box 4.3**). Among the main public databases are Database of Interacting Proteins (DIP), IntAct, Molecular INTeraction database (MINT), Human Protein Reference Database (HPRD), MIPS Mammalian Protein-Protein Interaction database (MPPI), Kyoto Encyclopaedia of Genes and Genomes (KEGG), BioCarta, Reactome, the Cancer Cell Map and Wikipathways. A large variety of pathway databases exists and the Pathguide resource provides a good overview (Bader et al., 2006). Besides the public databases, commercial tools propose comprehensive collections of information regarding molecular interactions such as Ingenuity®, TRANSFAC® and TRANSPATH® from BIOBASE (Krull et al., 2003, 2006) and ResNet® from ARIADNE (Nikitin et al., 2003).

While the first two levels are important, the third one is obviously the most valuable information to consider and serves as an important *a priori* biological knowledge in systems biology approaches. The identification and characterisation of components involved in regulation mechanisms on one hand, and the discovery of their interactions on the other hand have been permitted not only by classical biomolecular techniques but also by the advent of high-throughput technologies such as Chromatin Immunoprecipitation (ChIP) approaches, Yeast Two-Hybrid (Y2H), affinity purification combined with Mass Spectrometry (MS) and transcriptomics (see **Chapter 3**). As a result, many articles have yielded thousands of molecular interactions for human and for model organisms. Huge efforts have been devoted to build pathway databases which serve as repositories of current knowledge (Bauer-Mehren et al., 2009; Tsui et al., 2007). Importantly, one must distinguish between predicted interactions (which are deduced from computational studies) and those which have been experimentally established. Within the latter group, it must be indicated whether a single direct experiment or a high-throughput experiment has been used to identify the interactions.

❏ **BOX 4.3: Molecular interactions**

gene regulatory network: It is a set of DNA regions (coding and noncoding genes) which interact with each other either through their RNA or the coded protein. Transcription factors are the main players in this network. These interactions drive the gene transcription within the cell.

metabolic pathway: It is a set of biochemical reactions catalysed by enzymes that are connected by their intermediates. The reactants of one reaction are the products of the previous one, and so on. For example, the Krebs cycle is a major metabolic pathway in cellular respiration.

signalling pathway: It is set of molecules which control a cellular function (*e.g.* apoptosis). Once the first molecule in the pathway is activated in response to a *stimulus*, it activates another molecule. This process is repeated in an activation cascade until the last molecule is activated and the cell function involved is carried out. For a given cellular function, many different cascades can be connected with crosstalks.

Protein–Protein Interaction (PPI) network: It provides the indication whether two or more proteins are able to bind together within the cell.

protein-compound network: For a set of protein and chemical compounds (*e.g.* drugs), it indicates which compound can interact with a protein.

Molecular interactions available in databases allow the mathematical analyses of the **emergent properties**[*] of the network and the formulation of hypotheses that can be tested in the laboratory. Iterative cycles of prediction and experimental validation will result in the refinement of our knowledge in regulation mechanisms (such as as feedback loops or architectural features, see **Chapter 7**) and understanding of the robustness of the system (see **Chapter 9**). Molecular interactions will be a very valuable *a priori* biological knowledge to explore the diversity of cancer (see **Chapter 5**), to improve **prognosis**[*] and prediction (see **Chapter 6**) and to find new drug targets (see **Chapter 11**).

4.5.2   Standards for data and knowledge sharing

The reader would easily understand that all the annotations encompass a large amount of heterogeneous data types, both regarding the content and the source. This heterogeneity hampered the exchange and comparison of data, and the use of data analysis software, slowing down research. Offering the most efficient use of software and data resources will facilitate an in-depth understanding of biological systems. Beyond productivity improvements in each research group, common standards could potentially connect research groups globally. It is therefore crucial that the scientific community agrees on knowledge representation, data format standards and unique identifiers used to share information. The definition and the use of standards are still very challenging and this practice must be promoted.

In order to formalise knowledge representation, one has first to answer one question: what information needs to be recorded for an experiment? With this question in mind, the Functional GEnomics Data (FGED) Society (formerly known as the MGED Society) initiated different projects in order to define guidelines and minimal information standards to describe the high-throughput data to enable the unambiguous reproduction and interpretation of an experiment. Among others let us mention Minimum Information About a Microarray Experiment (MIAME) (Brazma et al., 2001), and Minimum Information About a Proteomics Experiment (MIAPE) (Taylor et al., 2007, 2008). This principle of minimal information does not hold only for high-throughput data but also for mathematical models published by researchers. For this reason, Minimum Information Required in the Annotation of Models (MIRIAM) has been proposed in systems biology to define the rules for model annotation (Le Novère et al., 2005). However, information about a model alone is not sufficient to enable its efficient reuse in other computational studies. Indeed, the minimal set of information regarding the description of a simulation experiment based on a mathematical model must be provided. Minimum Information About a Simulation Experiment (MIASE) (Waltemath et al., 2011a) defines the rules in order to reproduce numerical simulations.

Controlled vocabularies also called ontologies are required to describe this minimal information. An ontology is a formal representation of knowledge with definitions of the relevant semantic attributes, their hierarchy and their relationship using a well-defined logic. It includes a single identifier for each attribute and the terminology is augmented with synonyms, abbreviations and acronyms. Initiatives such as Open Biological and biomedical Ontologies (OBO) and BioPortal have produced controlled ontologies for shared use across different biological and medical domains. Gene Ontology (GO) and Biological Pathway Exchange (BioPAX) are widely-used ontologies that represent biological knowledge. GO provides terms to describe genes and their products, each entity being annotated according to three properties: the cellular localisation, the molecular function and the biological process. BioPAX is the ontology used to represent molecular interactions. While these ontologies are required to describe the biology, they are not sufficient in systems biology approaches. Indeed, information related to mathematical models need a formal representation too. Therefore, efforts have been initiated to develop ontologies in order to encode the semantics for models and simulations in systems biology (Courtot et al., 2011). Among them, Systems Biology Ontology (SBO) provides information about the components of the model, Kinetic Simulation Algorithm Ontology (KiSAO) describes existing simulation algorithms and their inter-relationships through their characteristics and parameters, and TErminology for the Description of Dynamics (TEDDY) supplies information about dynamical behaviours, observable dynamical phenomena, and control elements of the models. While not an ontology as such, it is important to mention that unique gene symbols and names are necessary and a gene nomenclature has been proposed by the HUman Genome Organisation (HUGO).

Formats allowing exchange between different software platforms and further processing by network analysis, visualisation and modelling tools have been proposed. In bioinformatics, eXtensible Markup Language (XML) has been widely used for storing information (Achard et al., 2001). The Web Ontology Language (OWL) is a family of knowledge representation languages for authoring ontologies based on XML syntax and Resource Description Framework (RDF) data model (Gedela, 2011). Among the most used formats in systems biology there are Proteomics Standards Initiative Molecular Interaction (PSI-MI), Systems Biology Markup Language (SBML), Simulation Experiment Description Markup Language (SED-ML) and Systems Biology Graphical notation (SBGN) (Le Novère et al., 2009; Waltemath et al., 2011b).

4.6 Data management and integration

Systems biology and integrative analysis approaches require the combination and comparison of different levels of information from heterogeneous data sources. Therefore, an essential prerequisite is that all the data can be accessible and queried using unified tools. Typically, a unified data management system which allows the user to access, browse and retrieve heterogeneous data would be ideal because the different pieces of knowledge required in analysis and modelling are scattered across many databases and repositories. Having a seamless computational tool platform is very challenging and even unrealistic. However, attempts have been proposed to facilitate data access and query for end-users. For example, BioMart provides a data integration system which involves the following four steps: (1) with *querying*, the user can filter the data according to criteria of interest; (2) a *configuration* step ensures the compliance between heterogeneous data in order to support the same structured and unified query system; (3) a *transformation* step prepares the data from the source in the expected XML format; and (4) *source data* contain available datasets in Structured Query Language (SQL) databases.

Data management and integration system are generally implemented as a three tier client-server architecture. The first tier corresponds to the databases storing the information in a structured form (*e.g.* SQL databases). The second tier embeds an Application Programming Interface (API) which is a program library using the set of configurations and the databases. The third tier provides a user-friendly query interface which uses the API.

4.7   Public repositories for high-throughput data

Bioinformatics standards and tools for data integration provide means to enhance cross-software interoperability and data exchange between laboratories. In addition to these prerequisites, gathering high-throughput data into public repositories is necessary to offer an access to this huge prospect of information to the scientific community. Researchers can, this way, reuse the data and reproduce the analysis, compare their results, and test and validate new hypotheses. Moreover, it provides a unique opportunity for integrative approaches and **meta-analysis*** across different datasets (see **Section 5.6** and **Section 6.6**).

Major public repositories are the Gene Expression Omnibus (GEO) (Edgar et al., 2002; Barrett et al., 2011) from NCBI and ArrayExpress (Brazma et al., 2003) from the European Bioinformatics Institute (EBI). In the context of cancer study, specific initiatives such as the cancer Biomedical Informatics Grid (caBIG), The Cancer Genome Atlas (TCGA) (Collins and Barker, 2007), the International Cancer Genome Consortium (ICGC) (International Cancer Genome Consortium et al., 2010) or Oncomine (Rhodes et al., 2004b) gather an important variety of molecular profiles (gene expression, DNA methylation, DNA copy number, *etc.*) using various technologies (microarrays, NGS) across many different cancers. These public data are available through web portals which are listed in the **Appendices**.

4.8   Informatics architecture and data processing

Today, high-throughput technologies enable data to be generated at unprecedented scales and transform life science research into *big data* science. This is particularly true for NGS which can produce terabytes of data within a few days or even less in a near future (Richter and Sexton, 2009; Baker, 2010). In all life science research institutes worldwide, bioinformatics core facilities are overwhelmed by this data tsunami. According to Moore's law, Kryder's law and Butter's law, costs are halved every 18, 12 and 9 months for processor, storage and data transfer respectively while 5 months is the rule for sequencing costs (Stein, 2010). These figures give an idea of how big the challenge is. Many years ago, high energy particle physics had to tackle the big data challenge and it is now the turn of life sciences. High-Computing Performance (HPC) infrastructure is definitely required for data storage, data transfer, data computation and access control. Typical needs are hundreds of CPUs, a few perabytes of storage, and 10 Gigabit networks. Different HPC solutions exist such as cluster, cloud or grid computing. Such massively parallel infrastructure allows many tasks to be run simultaneously in order to reduce the effective computation time. Efficient bioinformatics definitely relies on strong Information and Technology support.

Moreover, skills in low-level programming languages such as C or C++, parallel programming such as Message Passing Interface (MPI), Open Multi-Processing (OpenMP) or MapReduce (Dean and Ghemawat, 2008) and algorithm analysis are required to improve the efficiency of software used in downstream analyses. This is really important as integration of many layers of molecular information is necessary to understand the complexity of living systems.

Finally, the informatics architecture has to fulfil the requirement for end-users such as biostatisticians, systems biologists and biologists. As they have not necessarily skills in low-level programming, graphical tools are needed in order to manipulate and analyse the data (Gehlenborg et al., 2010; Nielsen et al., 2010). Among others, let us mention workflow management systems such as Konstanz Information Miner (KNIME), Taverna (Hull et al., 2006) or Galaxy (Goecks et al., 2010) which make it possible to design and execute scientific workflows and aid *in silico* experimentation. Among others, let us mention the R software (R Development Core Team, 2011) and Bioconductor (Gentleman et al., 2004) which have been widely used by the scientific community to process high-throughput data.

4.9   Knowledge extraction and network visualisation

As we previously mentioned, a lot of *a priori* biological knowledge has been gathered in various and heterogeneous databases. Although this information is based on experimental results obtained from published data, some valuable information still remains in numerous available scientific articles and has not been integrated into databases yet. The extraction of knowledge from this large number of valuable articles is very challenging. While text mining algorithms are very useful tools to query the literature, human reading, visual inspection and manual curation by experts are still needed to extract and formalise these pieces of information from articles. In order to tackle this challenge, knowledge representation is an essential requirement to describe and structure the information. The representation of biological knowledge as what we refer to as *maps of knowledge* offers a simple but efficient way to visualise the data. In this section, we detail the tools and standards useful for researchers involved in computational systems biology to represent biological knowledge.

4.9.1   Charting a map of knowledge

The databases listed in **Section 4.5.1**, such as Reactome, DIP, MINT, *etc.*, partly support the building of any type of map. Protein–Protein Interactions (PPI) networks, metabolic pathways, signalling pathways, gene regulatory networks, *etc.* are referenced with the standards in which they are provided (see **Box 4.3**). The ontologies and formats in which the pathways are provided are BioPAX, SBML, PSI-MI and SBGN (see **Section 4.5.2**).

More particularly, the goal of SBGN is the standardisation of pathway notation in a human-readable format. The notation defines the graphical representation of any network such that users can interpret it consistently. Three types of representations, called diagrams, (see **Box 4.4**) were proposed in SBGN: the process description diagram, the entity relationship diagram and the activity flow diagram (Le Novère et al., 2009). Process description diagrams are bipartite graphs and are often used to represent biochemical reaction networks specifying states and locations of proteins. Nodes are biochemical species (proteins, complexes, metabolites, *etc.*) and reactions. Arcs are molecular interactions. Entity relationship diagrams show the influences that an entity can have on a process (*e.g.* transport). Nodes can be species and arcs transformations. Activity flow diagrams are best suited for cascades of activity and concentrate on the influences between entities. Nodes can be entities and arcs influences. According to the diagram and what we wish to do with it, some software will be more appropriate than others. For instance, there exists an editor of SBGN, SBGN-ED (Czauderna et al., 2010) used with Vanted (Junker et al., 2006). CellDesigner is also widely used for drawing process diagrams

(Kitano et al., 2005). Public BioUML and commercial *geneXplain*^{TM} platforms support SBGN standard. Some plugins of Cytoscape (Shannon et al., 2003) are developed by the community to answer other specific graphical and data management issues, such as BiNoM (Zinovyev et al., 2008), best suited for manipulating, analysing and working on any type of diagrams.

❏ **BOX 4.4: Diagrams**

A diagram is a graphical and schematic representation aiming at demonstrating and explaining the relationships between parts of a whole. The parts are referred to as *nodes* and their relationships as *edges*. In our context, a network, a chart, a map, or a graph will be used as synonyms even though they have slightly different meanings. All of them refer to a more general term, a *map of knowledge.*

A considerable effort has been made in the construction of comprehensive maps of knowledge, either based on the description of biochemical reactions (Oda et al., 2005; Oda and Kitano, 2006; Calzone et al., 2008; Kohn, 1999) or on the description of influences of one species on the others (Schlatter et al., 2009). Both types of networks provide some valuable information, even without the dynamical mathematical model description. Using maps of knowledge developed by experts who spend months studying a pathway strengthens the relevance of mathematical models used in systems biology.

4.9.2 Example of a map of knowledge: RB pathway

In this section, an example of such maps of knowledge is presented. It describes the comprehensive map built around the *RB* gene (also referred to as *RB*), a tumour suppressor gene playing a major role in cell cycle entry (see **Section 2.2.2**, **Section 2.7.1** and **Box 2.1**). A reaction network (or process diagram) has been constructed to summarise published data around *RB* activity. The resulting map encompasses 78 proteins, 208 species, 165 biochemical reactions, and is based on more than 350 publications (Calzone et al., 2008). It describes the molecular players and the biochemical reactions that participate in RB phosphorylation and, more generally, in the G1 to S transition of the cell cycle. For more information on the biochemical information summarised in this reaction network, an interactive clickable and zoomable version is available at the following address: http://bioinfo.curie.fr/projects/rbpathway/. The map was built using CellDesigner software (Kitano et al., 2005).

It is possible to improve the readability of the map by proposing a modular view of the comprehensive *RB* pathway. Using BiNoM, the reaction network is translated into an influence network in which each node corresponds to a set of proteins or of genes closely related in the comprehensive map, and grouped into what we refer to as a *module* whereas each edge corresponds to the influence of one module onto another. The method of modularisation

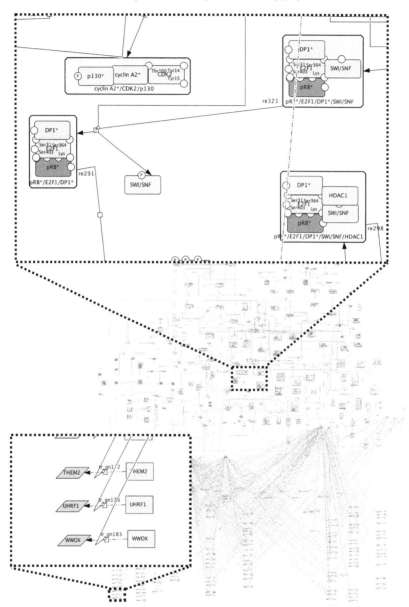

FIGURE 4.5 Comprehensive map of *RB* pathway. Lower part of the graph: the genes regulated by the *E2F* family genes (*E2F1* to 8) are represented as rectangular boxes; the corresponding mRNAs are represented as parallelograms. Upper part of the graph: a network of proteins linked by the biochemical reactions that are involved in the regulation of *RB* activity. A readable and interactive version of the map can be found at: http://bioinfo.curie.fr/projects/rbpathway/ (Calzone et al., 2008).

allows an abstraction of the initial complex and detailed map and is explained in details in Calzone et al. (2008). In **Chapter 5** we illustrate the use of this modular maps of knowledge (see **Figure 5.8**) in the context of bladder cancer and explain how it can help in the interpretation of *omics* data.

✎ **Exercises**

- Using the same ANOVA model as described in **Section 4.1**, compute the parameter variances for γ_1, δ_1 and $(\gamma\delta)_{11}$ as shown in **Figure 4.2** when 6 different microarray experiments are performed.

- Import and visualise in Cytoscape the RB/E2 pathway of **Section 4.9.2** (MODEL 4132046015 in Biomodels database). For that, on the Biomodels webpage, download the model in SBML L2 V1 (with Identifiers.org URLs), open Cytoscape and import the file using Cytoscape import function: Import → Network from Multiple file types.

⇨ **Key notes of Chapter 4**

- Experimental design is a prerequisite in any scientific approach and in particular in systems biology.

- Many parameters impact the measurement in high-throughput experiments. Normalisation procedures must correct these effects for a reliable downstream interpretation.

- Simple guidelines allow *in silico* research to be reproducible.

- Ontology and standard formats exist and must be promoted to share data and knowledge within the scientific community.

- Systems biology and bioinformatics rely on a substantial computation infrastructure.

- Molecular interactions can be visualised with maps of knowledge.

Chapter 5

Exploring the diversity of cancers

Cancer is not one disease, but a multitude of different diseases. Cancers can arise in different organs and cell types, and have different visual aspects under the microscope. They have different epidemiological risk factors, different patterns of progression, they respond to different treatments, and are associated to different risks of **relapse***. In a sense, each cancer is unique since it is intimately associated to the unique genetic background and somatic evolution of each individual (see **Section 2.1**).

The diversity of cancers has been recognised for a long time, and is a central issue in cancer clinical management since different cancers can be associated to different risks and be best treated by different treatments. Understanding and classifying the diversity of the disease is therefore a prerequisite not only to better understand the disease, but also to allow a more rational and personalised approach to cancer management and treatment.

In this chapter we quickly review how cancers are currently classified, and investigate how the omics revolution (see **Chapter 3**) has confirmed and expanded our understanding of cancer diversity and heterogeneity. We illustrate how new insight can emerge from the systematic analysis of large quantities of cancer molecular data. We discuss in particular the emergence of new molecular classifications of cancers, and how they can be related to particular biological processes. The computational and mathematical tools required to perform such analysis are numerous, and we provide a short introduction to some of them. The medical implications of this new molecular view of cancer diversity in terms of clinical management are postponed to **Chapter 6**.

5.1 Traditional classification of cancer

More than 200 types of cancers are commonly defined, based on the organ and cell type in which they start (National Cancer Institute, 2012). Within each cancer type, such as breast **carcinoma*** (breast cancer), patients are usually further stratified into sub-categories based on clinical information gathered from the patient, such as his/her age and possible previous cases of cancers in his/her family, and about the tumour, such as its location, size,

Type	ER/PR	HER2	Prolif	Recommended treatment
Luminal A	+	-	-	E
Luminal B	+	+/-	+	E+C (+H)
HER2 positive	-	+		C+H
Basal-like	-	-		C

TABLE 5.1 Systemic treatment recommendation for breast cancer subtypes. The four main subtypes of breast cancers are defined from the expression of marker proteins (ER, PR, HER2) and the tumour proliferation (usually quantified from Ki-67 assessment or grade). To each subtype corresponds a default recommended choice of systemic treatments, among endocrine therapy (E), cytotoxics (C) and anti-HER2 agents (H). Table adapted from Goldhirsch et al. (2011).

or histological type under the microscope. These informations are usually collectively referred to as **clinicopathological*** parameters. They are particularly useful for clinical management, and for most cancers guidelines exist to suggest the best therapeutic choices based on these parameters.

To be more precise, let us take as an example the case of breast cancer (Lønning, 2007; Cianfrocca and Goldstein, 2004), which is not restrictive as the principles we describe below can be extended to other types of cancer (Sawyers, 2008). The precise characterisation of a tumour is based on the observation by a **pathologist*** of a thin slice of tumour (*i.e.* a **histological section***), taken either from a biopsy or after surgery, through the microscope. Very different sorts of cancers can be identified visually (see **Figure 5.1**). The detailed characterisation of the histological section by the pathologist includes the assessment of parameters such as the appearance of the cells, the size and the shape of the cancer cell nuclei, the number of mitoses in the slice, or the invasiveness of adjacent tissues. These observations allow the determination of the so-called *histological type* of the tumour. In addition, the size of the tumour, the presence of cancerous cells in axillary lymph nodes, and the presence of a **metastasis*** elsewhere in the body are combined to define the **stage*** of the cancer, which quantifies its extension and size. The **grade*** of the tumour, which quantifies how abnormal the cancer cells look and how quickly the tumour is likely to grow and spread, is also assessed from the size of the nuclei, the proliferative activity within the tumour evaluated on ten high power field images, and the differentiation of the tumour (Ellis et al., 1992). In addition to these histological parameters, the presence of specific markers, such as oestrogen (ER), progesterone (PR) and human epidermal growth factor (HER2) receptors, is evaluated by immunohistochemical methods. Other clinical parameters such as the age of the patient can also be used. Taken together, these clinicopathological parameters currently determine the choice of the therapy proposed to the patient. For example, early breast cancers with no lymph node invasion can be spared surgery with axillary dissection (Goldhirsch et al., 2011), and the choice of systemic treatment options depends on the cancer subtype (see **Table 5.1**).

Although of tremendous help for patient management, the traditional classification of cancers based on clinicopathological criteria is not without drawbacks. First, the objective and consistent assessment of some clinicopathological factors is difficult to ensure. It may not only vary with the particular histological section studied, but also depend on the expert analysing the sample. For example, Billerey and Boccon-Gibod (1996) have shown on bladder cancer that the concordance for the grade and stage assessment between different pathologists is no more than 70%, a level which varies with cancers but can be even worse for example for gliomas. Second, this coarse classification lets unveiled many differences between patients that are important for therapeutic treatment and surveillance. Tumours with similar clinicopathological parameters frequently follow different clinical courses or respond differently to therapies, suggesting that a further level of variability exists within clinicopathological subtypes and calling for finer classifications.

5.2 Towards a molecular classification of cancers

The genomic revolution and subsequent development of several high-throughput omics technologies (see **Chapter 3**) have started to revolutionise

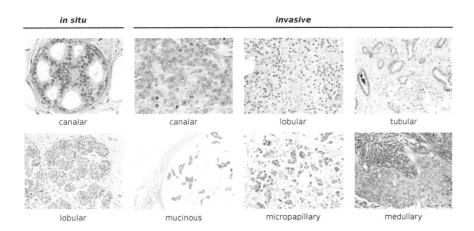

FIGURE 5.1 Histological sections of breast cancer. The histological sections are used by the pathologist to classify the tumours into histological types (Ellis et al., 1992) and to determine the stage and grade of the tumour. *In situ* tumours do not spread to the surrounding tissues while *invasive* tumours have started to break through normal breast tissue barriers and invade surrounding areas. Image courtesy of Dr. Anne Vincent-Salomon, Institut Curie. © 2012 Institut Curie. (See colour insert.)

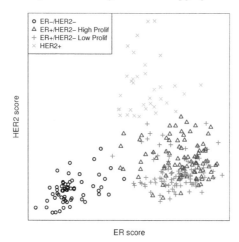

FIGURE 5.2 Breast cancer diversity in 2 dimensions. Global view of the 286 tumours in the Wang dataset, organised in terms of ER and HER2 status. The scores and subtype assignments were computed from gene expression data with the *genefu* R package (Haibe-Kains et al., 2011), mimicking the traditional histopathological analysis.

the way we apprehend cancer diversity. They now allow us to contemplate cancer cells at the molecular level, and to detect phenomena unobservable through the microscope. Furthermore, many omics technologies such as DNA **microarrays*** or **NGS*** ensure an unbiased and systematic collection of data, potentially paving the way to new discoveries in hitherto unexplored domains. Not surprisingly, the systematic profiling of various cancer types has been among the first applications of microarray-based transcriptomic studies in the late 1990s (*e.g.* Golub et al., 1999; Alizadeh et al., 2000; Perou et al., 2000), and has since remained at the forefront of applications targeted by new omics technologies.

Many questions related to cancer diversity can potentially be addressed when molecular omics data are collected on different tissues and patients. Can we observe at the molecular level the diversity we are familiar with at the macroscopic level or under the microscope? Can we define new, robust classification schemes based on molecular **biomarkers***? What biological insight can we get from comparing the molecular portraits of diverse samples?

As a preliminary answer to some of these questions, let us notice that several clinicopathological parameters such as the dosage of protein markers are directly related to measures that we can perform at the molecular level, such as the expression level of the corresponding or related genes. Let us illustrate this on a standard breast cancer dataset made public by Wang et al. (2005), which we use as a running example throughout the chapter and simply refer to as the Wang dataset. Note that the reader can reproduce most of the illustrations of this chapter using the Sweave file supplied online on the book web site. The Wang dataset consists of a cohort of $n = 286$ lymph node-negative **primary tumours*** of the breast, not treated by systemic **chemotherapy***, for which genome-wide gene expression data have been measured with the Affymetrix GeneChip® technology (see **Section 3.1.1**). Desmedt et al. (2008) showed that the ER and HER2 status usually measured by pathologists in the clinics can be recovered, with good accuracy, from the expression level of a few genes, allowing in principle the automatic classification of each tumour in one of the four classical subtypes (see **Table 5.1**). **Figure 5.2** shows the set of all tumours, as a function of their ER and HER2 status, together with their subtypes. We instantly see that the well-known diversity of breast cancers is confirmed in this simple plot, and that in spite of some overlap between the categories there are indeed well-defined subtypes easily detectable. We also see that the boundaries between subtypes do not seem to be clearly marked, and that the simplified binary scheme of **Table 5.1** hides a certainly more complex reality.

To go further in the analysis of cancer diversity, we need to go beyond these two-dimensional representations and investigate more finely how cancers vary in light of the thousands of molecular measures we have access to. This requires the use of various mathematical and computational models, which would be too long to exhaustively present here. Instead, we present a small selection of such tools in the next sections, illustrating on real data how they can contribute to the investigation of high-dimensional omics data. For the sake of clarity, we try to use a common language and constant notations to refer to the data we manipulate, as summarised in **Box 5.1**.

5.3 Clustering for class discovery

Capturing and understanding the diversity of cancers from omics data can be investigated from a variety of viewpoints, and with different computational tools. In this section, we investigate the possibility to automatically discover homogeneous subtypes within a collection of tumours characterised by high-dimensional measures, such as whole-genome expression profiling with DNA microarrays. Finding subgroups from molecular characterisation of tumours could indeed pave the way to new, robust and unbiased taxonomies of cancers, which could parallel and refine the classical classification schemes based on clinicopathological factors (see **Section 5.1**). It could also reveal new molecular factors underlying the classification, such as the activation of particular **signalling pathways*** or the deletion of specific genomic regions, increasing our understanding of the molecular biology of cancers.

Anticipating the description of how algorithms for subtype discovery work, let us first look at what they do. **Figure 5.3** shows a snapshot of the expression

---

❏ **BOX 5.1: Terminology and notations.**

We consider n **samples**, which correspond to patients, cell lines, or any biological sample analysed.

For each sample, we measure p **features**, which are quantitative properties of the sample. This can include, for example, clinical parameters, gene expression levels, DNA copy numbers, genomic or epigenetic markers.

We represent the set of p features measured on a sample by a p-dimensional vector $x \in \mathbb{R}^p$, and store the n vectors of features for all samples in a $n \times p$ **data matrix** X. With omics data the number of features used to characterise each sample is typically much larger than the number of samples available, *i.e.* $n \ll p$, as suggested in the following picture.

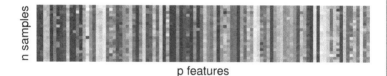

A **marker** is a feature, represented by its index $i \in [1, p]$, selected because it is interesting for some application. Typically a marker is selected because it is found to be associated to some properties of the sample, such as some phenotype of the sample or the risk of relapse of a patient.

A **signature** $M \subset [1, p]$ is a collection of markers, of size $|M| = q$, which have been selected because they allow together to predict a property of the sample. With omics data, we often look for small signatures, *i.e.* $q \ll p$.

---

data of the Wang breast cancer dataset. In this picture, each row corresponds to a gene, each column to a sample, and the intensity varies with the level of expression. We display the $n = 286$ samples, but only 400 genes among the 12,065 for which we have expression measures, taking those which vary the most across samples. The columns and rows have been reordered in such a way that rows (and columns) near each other tend to have similar patterns, making the visualisation more intuitive. In fact, they have been organised into a tree-like structure known as a *dendrogram*, which allows to visually capture the presence of homogeneous subtypes of cancers, and groups of genes. The dendrograms were obtained by Ward's hierarchical agglomerative clustering with correlation distance, explained in **Section 5.3.2**. This way to visualise a large data matrix is known as a *heatmap* and was popularised in biology by Eisen et al. (1998).

Interestingly, when we consider the structure organisation of samples in the dendrogram of **Figure 5.3**, and compare it to the subtypes based only on ER, HER2 and proliferation status (see **Figure 5.2**), we see that the classical categorisation in subtype is largely, although not perfectly, recovered: if we cut the branches of the dendrogram at a depth corresponding to 4 clusters, we see a strong enrichment, from left to right, in samples of subtypes basal-like, HER2+, luminal B and luminal A, respectively. This suggests that a new, completely automatic and biologically relevant classification of cancers from automatic analysis of omics data is possible.

Before discussing the biological and medical implications of cancer subtype discovery from omics data in **Section 5.3.6**, let us first focus on the algorithmic aspects of finding clusters in a dataset. Automatically finding subgroups within a collection of samples is the domain of *clustering* or *unsupervised classification* methods. There exist many such methods, which vary in their underlying hypothesis and algorithmic strategies, and which we briefly discuss below. We refer the interested reader to the many publications and textbooks on clustering for an in-depth coverage of this class of algorithms, (*e.g.* Jain et al., 1999; Gordon, 1999; Chipman et al., 2003), and only provide a brief account of how they work below.

5.3.1 Choosing a distance between samples

The starting point of any clustering method is the $n \times p$ data matrix X, which contains p features such as p gene expression measured in n samples (see **Box 5.1**). We assume that data have been pre-processed to remove technical artefacts and batch effects (see **Section 4.2**), and, for simplicity, that missing values have been imputed (although some clustering methods can also directly work with missing values in X).

Clustering methods attempt to organise samples into a small number of groups, also called clusters, in such a way that samples within a group tend to be similar to each other, while samples from different groups tend to be different from each other. A notion of similarity, or equivalently of dissimilarity or distance, between samples is therefore required to start any clustering analysis. Popular distances between samples $x, x' \in \mathbb{R}^p$ are the ℓ_q distances:

$$\| x - x' \|_q = \left(\sum_{i=1}^{p} | x_i - x'_i |^q \right)^{\frac{1}{q}},$$

in particular the ℓ_2 Euclidean distance and the ℓ_1 Manhattan distance. Alternatively, when we wish to compare samples not in terms of absolute values of their measures, but more in terms of relative values within the samples, it can be advantageous to prefer a similarity based on the *Pearson's correlation*

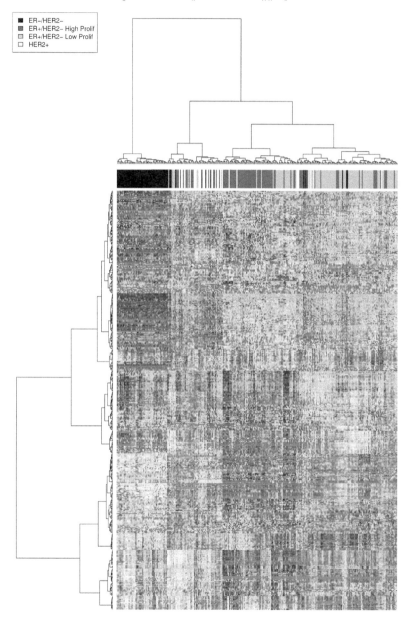

FIGURE 5.3 Molecular classification of breast cancer from mRNA expression profiles. This heatmap allows us to visualise a small part of the transcriptome of the 286 samples in the Wang breast cancer dataset. Dendrograms automatically cluster samples (and features) in a hierarchy of groups. We note a good, although not perfect correspondence between this clustering and the standard classification of breast cancers in four subtypes based on ER, HER2 and proliferation. (See colour insert.)

coefficient:

$$r(x, x') = \frac{\sum_{i=1}^{p} (x_i - \bar{x}) (x'_i - \bar{x}')}{\sqrt{\sum_{i=1}^{p} (x_i - \bar{x})^2 \sum_{i=1}^{p} (x'_i - \bar{x}')^2}} ,$$

where $\bar{x} = \left(\sum_{i=1}^{p} x_i\right)/p$ is the mean value of the measures for sample x. The Pearson's correlation coefficient ranges between $+1$ for identical samples (up to a global translation and scaling of the measures) and -1 for completely anti-correlated samples. It can be transformed into a dissimilarity measure for the purpose of clustering by considering $1 - r(x, x')$ or $1 - |r(x, x')|$, depending on how negatively correlated samples should be treated. When a similarity up to a nonlinear transform of the data is expected, other measures can be used: the *Spearman's rank correlation coefficient*, which is similar to the Pearson's correlation coefficient when the exact values of the measures are replaced by their rank in the list of p measures sorted by decreasing value for a sample; or more general notions of *mutual information* which capture more general relationships between the measures of two samples (Priness et al., 2007).

This short presentation of some of the most popular measures of similarity for clustering in bioinformatics is by no mean exhaustive, and defining a notion of similarity between samples is a research topic in itself. Indeed, the choice of similarity measure has a very strong influence on the result of clustering (Steuer et al., 2002; D'haeseleer et al., 2000), and few generic guidelines exist to guide its choice in practice. The choice of a particular similarity measure is often driven by prior knowledge we may have about the data considered, or can be optimised to reach some criterion, *e.g.* ensuring that biological replicates are similar to each other. For example, studying a set of gene expression profiles of breast cancer samples, Perou et al. (2000) defined a distance between samples based on only 496 genes (termed the *intrinsic* gene subset) out of 8,102 genes analysed. These genes were selected because they had significantly greater variation in expression between different tumours than between paired samples from the same tumours, and the resulting clustering consequently grouped together paired samples. This strategy to optimise a distance in order to fulfil constraints of similarity or dissimilarity between particular pairs of samples is even amenable to automatisation with the help of *metric learning* algorithms (Xing et al., 2003; Weinberger et al., 2006).

5.3.2 Hierarchical clustering methods

Once a distance or similarity measure between samples is chosen, a clustering method can be used to automatically organise the data into coherent groups according to the distance. By far, the most popular method for data clustering in bioinformatics is *hierarchical clustering*, which was in particular popularised among biologists and bioinformaticians by the seminal work of Eisen et al. (1998). Hierarchical clustering was, for example, applied to produce the dendrogram in **Figure 5.3**. It provides not only a clustering into K groups for a fixed K, if we cut the dendrogram at a particular depth, but also a full hierarchical organisation of the data into nested clusters with individual samples at the bottom leaves of the dendrogram and clusters of increasing size when we go up the tree toward the root. The length of a branch in the tree is related to how strong the separation at the upper part of the branch is, *e.g.* the long branches just below the root in **Figure 5.3** suggest that there is a clear separation of the full dataset into two large subgroups, corresponding to the separation between basal-like tumours and the rest. Cutting the tree at a given depth defines a clustering of the data into a finite number of groups.

There exist several algorithms to perform a hierarchical clustering, leading to different dendrograms. Hierarchical methods can be *agglomerative*, when groups are formed by a *bottom-up* strategy, iteratively joining the most similar groups into larger groups, or *divisive*, when groups are split in a *top-down* strategy, starting from a single group with all instances and iteratively splitting groups into two subgroups as separated as possible. In addition to a (dis-)similarity between individual samples, hierarchical clustering algorithms therefore depend on a linkage function, which defines how the distance between two groups is computed from the distances between the samples they contain (see **Box 5.2**).

❏ BOX 5.2: Linkage criteria for hierarchical clustering

In addition to a distance $d(x, x')$ between any two samples (see **Section 5.3.1**), hierarchical clustering algorithms depend on a linkage criterion in order to decide which clusters should be combined (for agglomerative clustering), or where a cluster should be split (for divisive clustering). The linkage criterion determines the distance between sets of observations and as a function of the pairwise distances between observations. Common linkage criterion $L(A, B)$ between two sets of observations A and B include:

- Maximum or complete linkage: $L(A, B) = \max_{a \in A, b \in B} d(a, b)$.

- Minimum or single linkage: $L(A, B) = \min_{a \in A, b \in B} d(a, b)$.

- Average linkage: $L(A, B) = \frac{1}{|A||B|} \sum_{a \in A} \sum_{b \in B} d(a, b)$.

- Centroid linkage: $L(A, B) = d\left(\frac{1}{|A|} \sum_{a \in A} a, \frac{1}{|B|} \sum_{b \in B} b \right)$.

- Ward's linkage: $L(A, B) = ESS(A \cup B) - [ESS(A) + ESS(B)]$ where the Error Sum of Squares (ESS) is $ESS(A) = \frac{1}{|A|} \sum_{a \in A} \left(a - \frac{1}{|A|} \sum_{b \in A} b \right)^2$.

As illustrated in **Figure 5.3**, hierarchical clustering methods have the advantage that they provide a visually appealing organisation of data, providing a multi-resolution view of groups within data and possibly suggesting biological interpretations. A drawback is that they will always output a nice dendrogram, even when samples have no reason to be organised into a tree hierarchy. A good way to assess the statistical significance of a clustering is to assess its stability, a problem related to the problem of choosing the number clusters which we briefly discuss in **Section 5.3.4**. They are also sensitive to possible errors made in the construction of the tree, since a wrong split or merge decision in the construction of the tree cannot be undone.

5.3.3 Partitioning methods

Although slightly less popular in bioinformatics than hierarchical clustering methods, partitioning methods try to solve more directly the problem of splitting a given set of samples into K groups, for a given number K. Roughly speaking, they directly try to optimise a partitioning of all samples into K groups, such that the similarity between samples tends to be large within each group, and small between different groups. Partitioning methods include in particular K-means, K-medoids, mixture of Gaussians, and Self-Organising Map (SOM).

Let us consider for example K-means. If we denote by $\mathcal{C}_K$ the set of all partitions of the n samples into K groups, it is therefore tempting to try to find the partition C which minimises the within-group sum of squares $W(C)$, or equivalently maximises the between-group sum of squares $B(C)$ (see **Box 5.3**), *i.e.* to solve:

$$W_K = \min_{C \in \mathcal{C}_K} W(C).$$ (5.1)

Unfortunately, **Equation 5.1** is a NP-hard problem which in practice cannot be solved exactly with more than a few tens of samples (Aloise et al., 2009). K-means is a computationally efficient algorithm which attempts to solve **Equation 5.1** approximately, *i.e.* to find a partition C with a small $W(C)$, although not the best one (MacQueen, 1967). It iteratively alternates between computing the centroid $\bar{x}_k$ of each cluster C_k for a fixed partition C, on one hand, and finding a new partition by minimising $W(C)$ over C with the centroids $\bar{x}_k$ fixed, on the other hand. The later minimisation is easily obtained assigning to cluster C_k all samples which are closer to $\bar{x}_k$ than to any other centroid $\bar{x}_i$, for $i \neq k$. The iterations are repeated until no change occurs from one iteration to the other. Although K-means does not in general converge to the global minimum, it always converges at least to a local minimum. In practice, it is useful to restart K-means several times with different initial conditions, and keep only the local minimum with the lowest score.

Many other partitioning methods have been proposed. The mixture of Gaussian model assumes that the data are random points generated by a mixture of Gaussian distributions, each Gaussian corresponding to a different cluster. When the covariance matrices of all Gaussians are all equal and spherical, the algorithm to fit the model to the data, called Expectation Maximisation (EM), is very similar to a version of K-means with soft assignment of each sample to all clusters. The K-medoids method is similar to K-means but replaces the centroid of a cluster, *i.e.* the vector which minimises the sum of squared Euclidean distances to the samples in the cluster, by its medoid, *i.e.* the sample of the cluster which has the minimal total sum of distances to the other samples in the cluster (Kaufman and Rousseeuw, 1990). K-medoids is usually more robust to outliers than K-means. SOMs is another variant where different clusters are organised into a user-defined grid, in such a way that samples in neighbouring clusters on the grid are forced to be more similar than samples in clusters far from each other on the grid (Kohonen, 1990). They have also been found useful to interpret gene expression data (Tamayo et al., 1999).

An advantage of partitioning methods over hierarchical methods is that they do not make strong assumption about the nature of the organisation of samples in the high-dimensional sample space, beyond being clustered into

---

❑ **BOX 5.3: Within- and between-group sum of squares**

For a given partitioning of the n points into K non-overlapping groups $C = (C_1, \ldots, C_K)$, of sizes $n_1 + \ldots + n_K = n$, the within-group sum of squares is by definition:

$$W(C) = \sum_{k=1}^{K} \sum_{i \in C_k} \| x_i - \bar{x}_k \|^2 \, ,$$

where $\bar{x}_k = \left(\sum_{i \in C_k} x_i \right) / n_k$ is the centroid of samples in the k-th cluster. It measures how concentrated the samples are within each group, and should therefore be small for a good clustering. Another interesting measure is the between-group sum of squares:

$$B(C) = \sum_{k=1}^{K} n_k \| \bar{x}_k - \bar{x} \|^2 \, ,$$

where $\bar{x} = \left(\sum_{i=1}^{n} x_i \right) / n$ is the global centroid of all samples. It measures how similar the cluster centres are to each other, and should therefore be large for a good clustering. In fact, both quantities are related by the simple equality

$$W(C) + B(C) = \sum_{i=1}^{n} \| x_i - \bar{x} \|^2 = T \, ,$$

where T denotes the total sum of squares, which is independent of C. This shows in particular that decreasing $W(C)$ is equivalent to increasing $B(C)$.

subgroups. On the other hand, they explicitly or implicitly make assumptions that should be kept in mind, such as the idea that data form clusters well modelled by spherical Gaussian distributions for K-means. A general drawback of partitioning methods is that the optimisation problem they try to solve to find the *best* partition is usually intractable, and the solution found by these methods generally depends on the initialisation of the algorithm. In practice, it is common to run these methods several times with different initialisation, and keep the best solution found.

5.3.4 Choosing the number of groups

A generic problem with clustering methods is to choose the number of groups K. In an ideal situation in which samples would be well partitioned into a finite number of clusters and each cluster would correspond to the various hypothesis made by a clustering method (*e.g.* being a spherical Gaussian distribution), statistical criteria such as the Bayesian Information Criterion (BIC) can be used to select the optimal K consistently (Kass and Wasserman, 1995; Pelleg and Moore, 2000). On real data, the assumptions underlying such criteria are rarely met, and a variety of more or less heuristic criteria have been proposed to select a good number of clusters K (Milligan and Cooper, 1985; Gordon, 1999). For example, given a sequence of partitions $C^1, C^2, \ldots$ with $k = 1, 2, \ldots$ groups, a useful and simple method is to monitor the decrease in $W(C^k)$ (see **Box 5.3**) with k, and try to detect an *elbow* in the curve, *i.e.* a transition between sharp and slow decrease (see **Figure 5.4**, upper left). Alternatively, several statistics have been proposed to precisely detect a change of regime in this curve (see **Box 5.4**). For hierarchical clustering methods, the selection of clusters is often performed by searching the branches of dendrograms which are stable with respect to within- and between-group distance (Jain et al., 1999; Bertoni and Valentini, 2008).

In **Figure 5.4**, we illustrate two of these methods, the elbow method and the CH index, on a toy clustering example with three well-separated groups of points in two dimensions. Both methods easily identify the correct number of clusters. However, running the same techniques to estimate the number of groups in the Wang dataset does not lead to a clear conclusion (see **Figure 5.5**). This suggests that common techniques to estimate the number of groups are not well adapted to real omics data, and that probably the intuition that the *space* of cancers can be well represented by a limited number of homogeneous subgroups may not be very accurate.

5.3.5 Clustering features and biclustering

While clustering samples allows investigation of the presence of subtypes of cancers within a heterogeneous collection of samples, it is mathematically possible to perform exactly the same analysis after reversing the rows and columns of the data matrix X in order to cluster features. This is what was

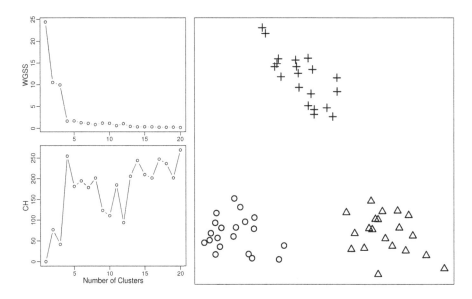

FIGURE 5.4 Choosing the number of groups. The K-means method was run on this toy example (right) for different values of K. The elbow method (upper left), based on the detection of a change of slope in the within-group sum of squares (WGSS), and the CH index of Calinski and Harabasz (1974) (bottom left), both easily identify that $K = 3$ groups is the best choice.

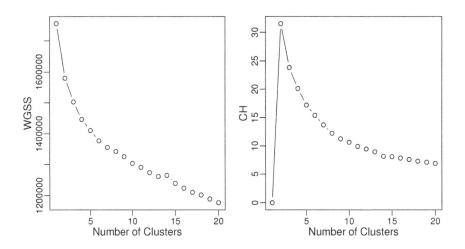

FIGURE 5.5 Number of groups in the Wang dataset. Compared to the toy clustering example (see **Figure 5.4**), no clear elbow is visible (left), and the CH index (right) does not clearly identify a clustering structure in the Wang breast cancer dataset.

❑ **BOX 5.4: Statistics for choosing the number of groups**
The CH index of Calinski and Harabasz (1974) is defined by:

$$CH(k) = \frac{B(C^k)/(k-1)}{W(C^k)/(n-k)},$$

It was shown by Milligan and Cooper (1985) to be the best among 30 other methods.
Another heuristics is the Hartigan's index (Hartigan, 1975):

$$Har(k) = \left(\frac{W(C^k)}{W(C^{k+1})} - 1 \right) / (n - k - 1),$$

The *gap statistics* which compares the observed decrease in the curve $W(C^k)$ to the expected from an appropriate null model is another possibility (Tibshirani et al., 2001).

performed in **Figure 5.3**, where both samples and genes are clustered. Clustering features allows the discovery of groups of genes which behave similarly across samples, suggesting that they may work together or at least be involved in the same biological processes. In fact, early work on gene expression data analysis mostly focused on gene coexpression clustering in order to help infer the function of poorly characterised or novel genes (Eisen et al., 1998). Inspecting gene clusters highly expressed or inhibited in particular samples clusters can also help the interpretation of the sample subtypes discovered: for example, Perou et al. (2000) identified groups of coexpressed genes which showed substantial variations in expression among the tumours, and which could be associated to specific signalling and regulatory systems, such as the interferon pathway or the activity of *HER2*.

An interesting variant of clustering samples and/or features is the notion of *biclustering*, where one wishes to capture subsets of samples which are similar on subsets of features only. In other words, a bicluster is a pair (I, J) where $I \subset [1, n]$ is a subset of samples and $J \subset [1, p]$ is a subset of features, such as the samples in I exhibit highly correlated values for the feature in J. Biclustering has been introduced for gene expression data analysis by Cheng and Church (2000), and has been investigated by many researchers (see Madeira and Oliveira, 2004, for a survey). The computational complexity of biclustering depends on the exact problem formulation, but most interesting variants are NP-complete, requiring either large computational effort or the use of heuristics to short-circuit the calculation (Madeira and Oliveira, 2004).

5.3.6   Applications: New molecular classifications of cancers

The possibility to define cancer subtypes by unbiased and systematic analysis of large quantities of omics data has been investigated by many researchers, and already has had a profound impact on our understanding of cancer heterogeneity. For some cancers such as acute **leukaemia*** , known subtypes were automatically recovered from automatic analysis of gene expression data (Golub et al., 1999), paving the way to automatic and reproducible classification of patients. For many other cancers such as diffuse large B-cell **lymphomas*** (Alizadeh et al., 2000), **melanomas*** (Bittner et al., 2000), lung cancers (Bhattacharjee et al., 2001; Garber et al., 2001) or breast cancers (Perou et al., 2000; Sørlie et al., 2001), new molecular classifications were proposed.

The case of breast cancers is perhaps the most illustrative in terms of how molecular subtypes have emerged as gold standards for cancer classification during the last decade. Perou et al. (2000) proposed in their seminal work a classification of invasive breast cancers into four subtypes, from the automatic clustering of gene expression for 39 invasive breast cancers and 2 normal tissues: there was one ER-positive subtype (luminal-like), and three ER-negative subtypes (basal-like, HER2+ and normal-like). The basal-like samples were characterised by high expression of keratins 5/6 and 17, while the oncoprotein HER2 was relatively overexpressed in the HER2+ group. The normal-breast-like resembled normal breast tissue samples, while the luminal-like samples expressed ER and had breast luminal cell markers relatively overexpressed. The luminal subtype was later subdivided into two subclasses (luminal A and B), as more samples were analysed (Sørlie et al., 2001); a third luminal subgroup was also proposed (luminal C), but was not supported by the subsequent analysis of an expanded dataset (Sørlie et al., 2003), leading to the classification mentioned in **Table 5.1**. This molecular classification of breast cancer on the basis of gene expression was validated on large meta-analysis involving thousands of patients (Wirapati et al., 2008), and validated on more recent technological platforms (Hu et al., 2006; Parker et al., 2009). Despite criticisms related to the instability of the subtypes defined, in particular their dependence on the original set of samples and genes used for the analysis (Kapp et al., 2006; Weigelt et al., 2010), this molecular classification has entered the common language in breast cancer research. It allowed reconsidering cancer classification based on immunohistochemical markers as surrogates of molecular classification (Blows et al., 2010; Voduc et al., 2010). As more data are available, more detailed and robust subtypes are starting to emerge (Guedj et al., 2011).

The success of these attempts to define new molecular classification of cancers should not hide the fact that clustering high-dimensional data remains a challenging task from a methodological viewpoint. In particular, we have seen that many parameters influence the classification obtained by clustering methods, including the features and metric used to compare samples, the clustering algorithm itself, and the procedure to select the number of clusters. Notwithstanding these limitations, it is fair to say that the new molecular classifications of cancers obtained by automatic clustering of omics data, in particular gene expression profiles, have started to revolutionise the way we apprehend cancer heterogeneity. As larger collections of samples are being analysed, it is likely that finer classifications into well-specified and robust subtypes will emerge from clustering methods, and allow a more precise stratification of patients into subcategories which would not be captured by clinicopathological parameters only. As different subgroups can have different prognosis or respond differently to different treatments, a more precise and robust classification of patients can improve clinical management, a question further addressed in **Chapter 6**.

5.4 Discovering latent processes with matrix factorisation

Classifying cancers into several distinct subgroups is useful for human understanding, but may however, be too limited to capture the intrinsic nature of cancer heterogeneity. On the one hand, the observation of omics cancer data does not obviously call for an organisation into well-separated subgroups (as suggested by **Figure 5.2** and **Figure 5.5**), and is not well adapted to describe continuously varying parameters such as the progressive activation of a pathway. On the other hand, it is likely that samples in different clusters may share common features, such as the high activity of the ER proteins in the luminal A and B subtypes commonly accepted in breast cancer classification (see **Figure 5.2**). It may then be more pertinent to describe a sample not by a cluster assignment, but as a superposition of a few well-defined molecular properties, just like classical clinical parameters, to unravel the inherent complexity of variations between tumours. In this section we discuss several computational strategies to automatically detect and quantify such decompositions, using in particular the concept of matrix factorisation.

5.4.1 From clustering to matrix factorisation

Clustering can be thought of as discretising and quantising the complex high-dimensional space where we represent tumours with thousands or millions of molecular parameters, into a finite number of K groups. The underlying motivation is that there may exist K basic subtypes of cancers, sufficient to describe the diversity of cancers. Mathematically, if we denote by C the $n \times K$ binary matrix which indicates the cluster assignment of each sample ($C_{ij} = 1$ if sample i is in cluster j, 0 otherwise) and by M the $K \times p$ matrix of cluster centroids (*i.e.* the k-th row of M is the p-dimensional centroid $\bar{x}_k^\top$ of cluster k), then the $n \times p$ matrix CM has in each row the cluster centroid of each sample. Consequently, denoting by $\| A \|_2^2 = \sum_{i,j} A_{ij}^2$ the classical Frobenius norm for matrices, we see that the within-group sum of squares $W(C)$ (see **Box 5.3**) can be expressed in matrix form as follows:

$$W(C) = \| X - CM \|_2^2,$$

and that the objective function of partitioning methods (see **Equation 5.1**) becomes:

$$\min_{C \in \mathcal{C}(n,k), M \in \mathbb{R}^{k \times p}} \| X - CM \|_2^2, \tag{5.2}$$

where $\mathcal{C}(n,k)$ is the set of $n \times k$ binary matrices with exactly one 1 in each row. This matrix formulation shows that clustering by partitioning methods, in particular K-means, can be thought of as attempting to *factorise* the data matrix X as a product $X \approx CM$, with particular constraints on C and M.

In this section we explore other popular *matrix factorisation* methods, which also attempt to approximately factorise the data matrix X as a product $X \approx CM$, although with different constraints on C and M. A constraint common to all methods we will discuss is the size of the matrices C and M, respectively $n \times K$ and $K \times p$, implying that the rank of the approximate matrix CM is at most K. The general motivation behind such low-rank matrix factorisation methods is to decompose any p-dimensional sample x_i as a superposition of K basic processes (sometimes called *atoms*, *dictionary elements*, or *metagenes* in the context of biological data). Indeed, by denoting μ_j the j-th row of M (for $j = 1, \ldots, K$), the factorisation $X \approx CM$ means in particular that the i-th sample x_i located in the i-th row of X is approximated by

$$x_i \approx \sum_{j=1}^{K} C_{ij} \mu_j. \tag{5.3}$$

In other words, the μ_i's serve as a set of K basis vectors, optimised to represent the collection of biological samples. The matrix C (often called the *score* matrix) contains the mapping of each sample as a linear combination of these basis vectors. A low-rank matrix factorisation therefore provides a K-dimensional representation of each sample, attempting to capture as much as possible the variations among samples in this low-dimensional representation. This is particularly attractive to reduce complex biological data into a small combination of basic factors, which may then pinpoint basic biological processes underlying the diversity of cancers.

As we have seen, clustering by partitioning methods such as K-means is a form of matrix factorisation, albeit a particular one where each sample is only represented by a single basis element (its cluster's centroid). In the rest of this section, we discuss more matrix factorisation techniques, obtained by varying the objective function and the constraints in **Equation 5.2**. From a computational viewpoint, it should be pointed out that the objective function of **Equation 5.2** is convex in C for M fixed, and also in M for C fixed. However, it is not jointly convex in C and M, meaning that except in very particular cases, the exact optimisation of the objective function in matrix factorisation is intractable, and many techniques find only a local minimum by alternating the minimisation over C and M, just like K-means.

5.4.2   Principal and independent component analysis

Principal Component Analysis (PCA) (Pearson, 1901; Jolliffe, 1996) is a very popular statistical technique used to visualise and capture variations between multi-dimensional samples, by projecting them to a low-dimensional space where most variations are retained. PCA is particularly useful for visualisation of high-dimensional data, by plotting the projection of the samples to the 2- or 3-dimensional subspace which captures the largest amount of variations, and for data understanding, by highlighting the key factors responsible for the diversity of samples. Unsurprisingly, PCA is a popular technique in bioinformatics to visualise high-dimensional biological data such as microarray data (Raychaudhuri et al., 2000; Alter et al., 2000; Misra et al., 2002). For example, **Figure 5.6A** shows the projection of the Wang breast cancers dataset onto the 2-dimensional subspace which captures the largest amount of variations. We observe that the first axis (the first principal component) separates the samples based on the activity of *ER*. The second direction is less obvious to interpret at first sight. Comparing to **Figure 5.2**, we recover the clear separation between the basal-like samples and the rest of the samples along the first axis, which captures the largest amount of variance. This confirms that the ER+/ER− split is predominant and responsible for global differences of transcriptome, explaining also why the basal-like samples are easily separated from the other ones by clustering (see **Figure 5.3**). On the other hand, the clear separation of HER2+ samples from the rest when we look specifically at the expression of HER2 (see **Figure 5.2**) disappears when we only look at the two main directions of variability among samples, as captured by PCA. This suggests other major sources of variations between samples on top of the four major subtypes discussed earlier.

Mathematically, PCA starts by centering each column of the data matrix X to obtain a centred matrix $\bar{X}$ where on average each feature is 0, and factorises the centred matrix by solving the following optimisation problem:

$$\min_{C \in \mathbb{R}^{n \times K}, M \in \mathbb{R}^{K \times p}} \| \bar{X} - CM \|_2^2. \tag{5.4}$$

Although nonconvex, this problem is easily solved with a singular value decomposition of $\bar{X}$. The optimal matrix M contains a metagene in each row, which defines the directions of projections, while the optimal C matrix contains the coordinates of the samples in the subspace defined by the K metagenes. For example, **Figure 5.6A** was obtained by solving **Equation 5.4** for $K = 2$ and plotting the 2-dimensional coordinates of each sample stored in the C matrix.

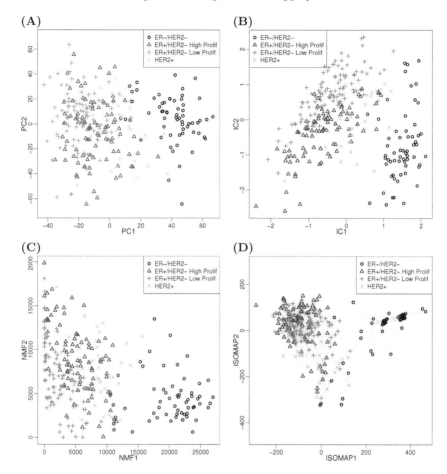

FIGURE 5.6 Breast cancer diversity in 2 dimensions. The 286 tumour samples of the Wang dataset are visualised in 2D using various matrix factorisation and projection methods: (A) PCA, (B) ICA, (C) NMF and (D) ISOMAP. (See colour insert.)

The biological interpretation of PCA metagenes is not always easy, since it requires analysing p-dimensional vectors which can often not be easily visualised (*e.g.* a 20,000-dimensional vector of gene expression). Furthermore, a mathematical property of the singular value decomposition which solves **Equation 5.4** implies that the successive columns of C are orthogonal to each other, and similarly the successive rows of M (metagenes) found by PCA are orthogonal to each other. If biological samples are linear combinations of basic processes which are not orthogonal to each other (we can think of different pathways, activated independently from each other, but which may share some genes), then PCA will fail to identify them and instead force them to merge into the same *super-metagenes* which will be difficult to deconvolve into independent components. Some variants of PCA, such as Independent Component Analysis (ICA) (Hyvärinen et al., 2001), try to overcome this limitation by decomposing the data onto a series of statistically independent processes. Although less popular than PCA among bioinformaticians, ICA has successfully been applied to the analysis of genomic data (Liebermeister, 2002; Lee and Batzoglou, 2003). For illustration, we plot on **Figure 5.6B** the projection of the Wang dataset onto the first two independent components computed by ICA. In spite of its overall similarity to PCA (see **Figure 5.6A**), the second component of ICA seems to better capture the difference between low and high proliferation luminal samples.

5.4.3   Nonnegative, sparse and structured matrix factorisations

PCA and ICA decompose each sample as a linear combination of metagenes, with positive and negative weights. This strategy to decompose a data matrix as a linear combination of basic signals is generic and does not incorporate any information specific to the genomic data analysed. In some cases, in can be useful to integrate specific prior knowledge we have about the data, in order to capture metagenes and decomposition with more biological relevance. The general framework of matrix factorisation is well adapted for that purpose, by keeping the objective to find a factorisation of the form $X \approx CM$ but modifying the constraints we put on C and M, *i.e.* by solving the problem:

$$\min_{C \in \mathcal{C}, M \in \mathcal{M}} \| X - CM \|_2^2, \qquad (5.5)$$

where particular choices for $\mathcal{C}$ and $\mathcal{M}$ induce various constraints on the solution. Unfortunately, this problem is usually not jointly convex in C and M, meaning that we usually can not find its global minimum in general. In practice, many approaches ensure that $\mathcal{C}$ and $\mathcal{M}$ are convex sets, so that at least an optimisation strategy based on alternatively minimising **Equation 5.5** in C and M is tractable (because it is then a convex optimisation problem in each variable), leading to a local optimum. A limitation of this strategy is that it can be very dependent on the initial choice of C and M.

A first example of such a strategy is Nonnegative Matrix Factorization (NMF), which enforces a decomposition as *a superposition of metagenes assumed to have nonnegative weights* (Paatero and Tapper, 1994; Lee and Seung, 1999). For example, if a pathway needs to be activated then the corresponding genes may get positively expressed. However, when the pathway is not activated, the corresponding genes may get silenced (no expression), but will certainly not be *negatively* expressed. This asymmetry may suggest that a more realistic model for a sample expression profile could be to express it as a nonnegative linear combination of metagenes with nonlinear loadings. Mathematically, NMF solves **Equation 5.5** with $\mathcal{C}$ (resp. $\mathcal{M}$) representing the sets of nonnegative $n \times K$ (resp. $K \times p$) matrices. Both $\mathcal{C}$ and $\mathcal{M}$ are then convex, and an optimisation strategy is to alternatively optimise in C and M (Lee and Seung, 1999). NMF was popularised in bioinformatics by Brunet et al. (2004) to construct metagenes from a gene expression matrix. In **Figure 5.6C** we show the result of NMF applied to the Wang dataset, where we have fixed $K = 4$ metagenes to decompose the samples and visualise the coefficients of the first two metagenes for each sample. We see that the first metagene (horizontal axis) is mostly characteristic of basal-like samples, while the second metagene (vertical axis) is more difficult to associate to a unique subtype.

Interestingly, the NMF solution usually produces a *sparse* representation of the data, in the sense that both C and M are not only nonnegative, but also contain many zeros. Intuitively, this is due to the fact that the nonnegativity constraint blocks to 0 the coefficients that would improve the objective function if they were allowed to be negative. This sparsity can be helpful to interpret the data and the metagenes, since only nonzero coefficients need to be analysed. However, the number of nonzero components is not easily controlled by NMF, and we may end up with too many nonzero components to easily interpret the metagenes. For example, the first NMF component of **Figure 5.6**(C), characteristic of the basal-like subtype, has no more than 10% of the genes with zero weight.

In order to improve the control of the sparsity, and increase the interpretability of metagenes at the expense of their ability to trustfully approximate the data, several authors have proposed to further constrain the sets $\mathcal{C}$ and/or $\mathcal{M}$ in **Equation 5.5**. A popular way to increase the sparsity of a vector or matrix is to constrain its ℓ_1 norm, *i.e.* to consider the sets:

$$\mathcal{C}_\lambda = \left\{ C \in \mathbb{R}^{n \times K} : \| C \|_1 \leq \lambda \right\} \quad , \quad \mathcal{M}_\mu = \left\{ M \in \mathbb{R}^{K \times p} : \| M \|_1 \leq \mu \right\} ,$$

where the ℓ_1 norm of a matrix is the sum of absolute values of its elements:

$$\| A \|_1 = \sum_{i,j} | A_{ij} | . \tag{5.6}$$

Intuitively, reducing λ (resp. μ) shrinks the entries of C and M towards 0, and leads to sparser solutions. When such constraints are used in **Equation 5.5**, we can enforce more 0 in C (resp. in M) by decreasing λ (resp. μ) and obtain a form of sparse PCA (Zou et al., 2006; Witten et al., 2009; Mairal et al., 2010). When in addition we constrain the values of C and M to be nonnegative, we obtain a sparse version of NMF, which was shown by several studies to be an efficient way to capture biologically relevant and robust metagenes (Hoyer, 2004; Gao and Church, 2005; Kim and Park, 2007).

Instead of promoting the presence of zeros anywhere in the C and M matrices, in order to improve their interpretability, one may sometimes know in advance which coefficients should be zero, and let the algorithm only optimise the nonzero coefficients. This is, for example, the case in Network Component Analysis (NCA) (Liao et al., 2003), where each row of M is constrained to have only nonzero coefficients on a subset of the genes, corresponding typically to the targets of particular transcription factors or the elements of particular pathways. Enforcing zero coefficients at particular positions in C or M leads to particular constraint sets $\mathcal{C}$ and $\mathcal{M}$, which can again be plugged into the general **Equation 5.5**. After optimisation, the nonzero weights of each row of M can be thought of as the contribution of each feature in the biological process considered (typically, activity level of a given transcription factor), and the coefficients in the C matrix quantify the activity of the process in each sample.

Let us conclude this section by mentioning that the general idea to include prior knowledge in matrix factorisation by choosing particular sets $\mathcal{C}$ and $\mathcal{M}$ can also be exploited to derive specific methods when the data analysed have particular structures. For example, when the data analysed for each sample are DNA copy number profiles (see **Section 3.1.2**), there is an obvious linear structure along each chromosome that can be exploited in matrix factorisation (see **Figure 3.5** and **Figure 3.8** for examples of such profiles). In order to capture metagenes which are not only sparse but also *piecewise constant* along each chromosome, several authors have proposed to consider the following constraint set for metagenes (Witten et al., 2009; Mairal et al., 2010):

$$\mathcal{M}_{\mu,\nu} = \left\{ M \in \mathbb{R}^{K \times p} : \| M \|_1 \leq \mu, \| M \|_{TV} \leq \nu \right\},$$

where the Total Variation (TV) semi-norm is given by:

$$\| M \|_{TV} = \sum_{i=1}^{K} \left(\sum_{j=1}^{p-1} | M_{i,j+1} - M_{i,j} | \right). \tag{5.7}$$

Similarly to the ℓ_1 norm which promotes sparsity, the TV semi-norm promotes piecewise constant profiles and is therefore well adapted to regularise metagenes which have a one-dimensional sequential structure (Rudin et al., 1992; Tibshirani et al., 2005).

5.4.4 Nonlinear methods and manifold learning

Matrix factorisation methods are by essence linear, in the sense that the factorisation $X \approx CM$ directly implies that each sample is represented by a linear combination of metagenes (see **Equation 5.3**). Although such linear methods are widely used in exploratory analysis of cancer data, in some situations the distribution of samples may be better approximated by nonlinear rather than linear objects. The objective of nonlinear data approximation is to construct such an approximation, typically a manifold embedded *in the middle* of the cloud of samples and which possesses some regularity properties. There exist a number of approaches that can construct such nonlinear approximations. Methods of data approximation constructing manifolds embedded in the multidimensional data space are collectively called *manifold learning*. Together with manifolds, other types of data approximations are suggested such as *principal graphs* or *principal trees* for branching data distributions. Review of the methods themselves and application of these methods to analysis of high-throughput data, in particular in cancer bioinformatics, can be found in Gorban et al. (2008); Lee and Verleysen (2007); Gorban and Zinovyev (2009).

Nonlinear methods of data approximation can be classified accordingly to three criteria:

- Those methods that assume the existence of a global probability distribution from which the data sample is generated, such as self-consistent principal curves and manifolds, and those who do not require it, such as elastic principal manifolds suggested by Gorban et al. (2008).

- Those methods that construct the nonlinear approximators as explicit objects, such as SOM, and those who do not construct them explicitly, such as methods of multidimensional scaling or kernel PCA (Schölkopf et al., 1999).

- Those methods that are most suited to the situation when the data points are located in close vicinity of some hidden intrinsic low-dimensional manifold, such as ISOMAP (Tenenbaum et al., 2000) and Locally Linear Embedding (LLE) (Roweis and Saul, 2000), and those methods that do not assume this.

Figure 5.6D illustrates the nonlinear mapping in 2D of the Wang dataset by the ISOMAP method. The mapping was performed in such a way that distances between each sample and its three most similar other samples in the original sample space (represented by light grey edges) tend to be conserved. Because long-range distances are not conserved, it is potentially possible to *unfold* any manifold structure that would describe the density of samples in the original *p*-dimensional space. Here we see that, again, the basal-like samples have been recognised as very different from the rest, while all ER+ luminal samples tend to overlap in the same region.

Each nonlinear method can be more or less suitable to analyse a given dataset. For example, methods requiring estimating probability distribution from data sample usually require a large number of samples. The methods which assume that the points are located on a low-dimensional manifold with little noise can be unsuitable to noisy microarray data.

When complex data are approximated by a nonlinear object, a usual problem of avoiding overfitting arises. In principle, a sufficiently flexible nonlinear object can approximate data with zero approximation error; it can simply pass through all data points. However, such an approximant is usually of little practical use; it can be as complex as data itself. Therefore, any nonlinear method, implicitly or explicitly, contains a parameter whose meaning is the trade-off between approximation accuracy and regularity properties. Tuning this parameter allows to switch between approximations with very regular properties (for example, close to the linear ones) and very irregular (for example, non-smooth) approximations.

One of the first nonlinear methods of data approximation which was applied to interpreting microarray data was SOM (Tamayo et al., 1999), a clustering method (see **Section 5.3.3**) which can also be considered as a nonlinear data approximation approach (because the clusters can be organised into a 2D or 3D grid). The benefits of application of nonlinear methods for cancer data analysis were explicitly shown and quantified (Gorban and Zinovyev, 2010). It was shown that constructing nonlinear principal manifolds for cancer-related microarray data can better reproduce the structure of distances between data points after projection onto the manifold. As a result, visualisation of multidimensional data based on such a projection sometimes gives more information about data clusters and classes.

5.5 Interpreting cancer diversity in terms of biological processes

Purely data-driven analytical methods such as clustering (see **Section 5.3**) and matrix factorisation (see **Section 5.4**) are very useful to quantify and visualise the relative similarity between samples, and investigate their diversity. What biological insight can we get from such analysis? Which biological processes underpin the large diversity of tumours at the molecular level? As often with omics data, extracting such biological information from raw high-dimensional data is not always obvious and requires automatic computational methods. In this section, we focus particularly on samples represented by gene expression data, and discuss several possible directions put forward recently to capture important biological processes underpinning the diversity of cancers.

5.5.1 From individual genes and proteins to higher-order biological functions

Most biological functions involve the coordinated actions of many genes and proteins, and can impact the activity of an even larger number of other genes. For example, the map of knowledge which summarises the molecular players involved in *RB* phosphorylation and G1 to S phase transition contains 78 proteins (see **Section 4.9.2**), and similarly all hallmarks of cancers involve the coordinated action of many genes (see **Chapter 8**). On a global scale, each protein can be seen as a node in several strongly interconnected networks (see **Box 4.3**): it forms complexes with other proteins, interacts with various ligands, transmits signals, *etc.*

Concretely, the scientific community has already collected large amounts of information about the functions and interactions of genes and proteins, and made them easily available on various databases (see **Section 4.5.1**). In this section we focus primarily on two types of knowledge we would like to exploit:

- On one hand, collections of *gene sets* (or modules) which group together genes sharing the same functional annotations or regulatory motifs, belonging to the same pathway, localised in the same subcellular compartment, or forming a modules of systematically coexpressed genes. For example, as of March, 2012, the v3.0 release of the MSigDB database (Subramanian et al., 2005) contains a heterogeneous collection of $G = 6,769$ sets of human gene groups by various functional annotations.

- On the other hand, *networks* of genes/proteins which may be directed or undirected graphs, and describe various physical, functional or genetic interactions among genes and proteins (see **Box 4.3**).

Exploiting this existing information in the analysis of high-dimensional omics data offers an exciting opportunity to unravel the biological phenomena involved in cancer biology and underlying its diversity. We discuss in the next section a few computational methods to perform such analysis. For simplicity, we first only consider the case where our prior knowledge is represented as a collections of G groups of genes $g_1, \ldots, g_G$, where each group is a subset of the genes, *i.e.* $g_i \subset [1, p]$.

5.5.2   Detecting biological functions in metagenes

Both clustering (see **Section 5.3**) and matrix factorisation (see **Section 5.4**) can be seen as low-dimensional projections of biological samples represented by high-dimensional omics data. We use the term *metagenes*, in a rather broad sense, to designate the basis vectors $\mu_1, \ldots, \mu_K$ defining this projection, be they the cluster centroids or the rows of the projection matrix M in matrix factorisation. Mathematically, each metagene μ is therefore a p-dimensional vector which assigns a weight to each gene. In order to interpret the variations among samples in terms of biological functions, it is therefore natural to first exploit the metagenes themselves and try to capture biological meaning from the weights they assign to each gene.

For example, one may look at gene weights one by one, and further inspect the genes with the largest absolute weights in a metagene. Except if a metagene has only a very small number of nonzero weight (*e.g.* using sparse matrix factorisation techniques as discussed in **Section 5.4.3**), this is however, likely to be a painful task missing the real biological phenomena underlying the metagene. Indeed, the real strength of working with high-dimensional genomic data is to capture signals hidden in tens or hundreds of individual genes, corresponding to biological functions involving the coordinated actions of many individual genes or more generally to phenomena impacting many genes.

An alternative to per-gene analysis of metagenes is to interrogate them in the light of known gene sets and gene networks, *e.g.* to see if particular gene sets or collections of interacting genes are particularly overrepresented among genes with large weights in a metagene, or collectively tend to have more important weights than what would be expected by chance. Technically, this functional analysis of metagenes is similar to the functional analysis of molecular signatures that will be discussed in more detail in **Section 6.4.2**.

5.5.3 Quantifying biological processes

Section 5.5.2 discusses the possibility to extract biological information from the metagene weights, after a first clustering or projection of the samples to a low-dimensional space. An alternative strategy is to first quantify the activity of each gene set in which we are interested in each sample, before carrying out further clustering or projection tasks. Indeed, as discussed in **Section 5.3.1**, the choice of features used to describe the samples can have a strong influence on the later processing of data, and one may conjecture that representing a sample as a G-dimensional vector of gene set activity is a better starting point than as a p-dimensional vector of gene expression. Conceptually, this means that if can be better to exploit our prior knowledge early rather that late in the analysis pipeline.

Given a sample $x = (x_1, \ldots, x_p) \in \mathbb{R}^p$ with the activity of p genes, this raises the question of how we can quantify the *activity* of a biological function represented by a set of gene $g \subset [1, p]$ of size $|g|$. Arguably, the simplest idea is to average the expression of individual genes, *i.e.* to define the activity $a_g \in \mathbb{R}$ of g as:

$$a_g = \frac{1}{|g|} \sum_{i \in g} x_i . \tag{5.8}$$

Such quantification of gene set activity was for example performed by Breslin et al. (2005), who called it the *group sample score*, in order to assess whether each of 29 cancer relevant pathways was significantly active or inactive in each sample of a collection of breast cancers and leukaemias. Guo et al. (2005) also averaged the expression levels of genes sharing the same Gene Ontology (GO) annotation to define the activity level of each GO term in a collection of cancer cell lines and lymphomas.

While the arithmetic mean (see **Equation 5.8**) is a natural choice when the different genes in a group all contribute similarly to the activity of the group, it may be too simplistic to capture more realistic phenomena such as the fact that some genes may be more important than others to define the activity of the group, or that some genes may even be negatively correlated to the activity of the group (*e.g.* genes negatively regulated in a given pathway). An alternative is to consider a more general linear combination of individual gene activity to define the activity of the group:

$$a_g = \sum_{i \in g} w_i x_i , \tag{5.9}$$

where the weight w_i may be nonuniform and even have different signs. For example, Tomfohr et al. (2005) propose to take for weights the first principal component of the expression matrix restricted to the genes of the group (after normalising each gene to zero mean and unit variance across samples). Intuitively, the first principal component defines a metagene that captures the largest amount of variations across samples (see **Section 5.4.2**); restricted to the genes of the group, this can correspond to the major variation in the biological activity of the group. In particular, a strength of this approach is that genes can contribute either positively or negatively to the activity of a group. Bild et al. (2006) follow a similar strategy to define the activity of 5 important pathways for cancer on several large collections of human cancers: *MYC*, *RASA1* (*i.e. RAS*), *SRC*, *Wnt/β*-catenin and loss of *RB* function. An interesting byproduct of this approach is to identify "hot spot" genes within a group of genes as the ones which contribute the most to the group activity.

Another promising approach is the use of Multiple Factor Analysis (MFA) (Escofier and Pages, 1994; de Tayrac et al., 2009) which can be viewed as a generalisation of the PCA. Instead of considering each feature individually as the PCA does, MFA works on groups of features (*e.g.* gene sets). For example, features can be grouped according to different pathways of interest. This way, it permits one to assess the contribution to each gene set with respect to the overall structure of the data. In the Wang dataset, the **Figure 5.7** shows the results of the MFA where RB pathway (*i.e.* the cell cycle), NFkB pathway (*i.e.* the inflammation) and TGF*β* pathway (*i.e.* the immune response) have been considered as gene sets in addition to all the other genes. The overall structure of the data in **Figure 5.7A** is as expected very similar to the PCA results in **Figure 5.6A**. Regarding the contribution of the RB pathway, **Figure 5.7B** suggests an elongated cloud, which indicates that the tumours can be ranked according to linear score representing very likely the level of activation of the cell cycle. The NFkB pathway (see **Figure 5.7C**) and the TGF*β* pathway (see **Figure 5.7D**) show qualitatively a similar behaviour with respect to the overall structure (see **Figure 5.7A**) while the RB pathway (see **Figure 5.7A**) definitively shows a peculiar behaviour with respect to other gene sets.

5.5.4 Detecting important gene sets and pathways

Instead of quantifying the activity of thousands of pre-defined groups of genes for each sample, another line of thought to get biological insight from a collection of samples is to try to directly detect which groups are *important* in the observed samples. While the notion of importance is vague and subjective in general, a fruitful definition is to assess the importance of a group of genes by their level of *coexpression* across samples (Gerstein and Jansen, 2000). The underlying assumption is that high expression correlation between genes implies some form of interaction between the proteins they code under the investigated conditions.

Following this idea, Jansen et al. (2002) measure for example the coexpression of a group of genes as the mean correlation coefficient between all

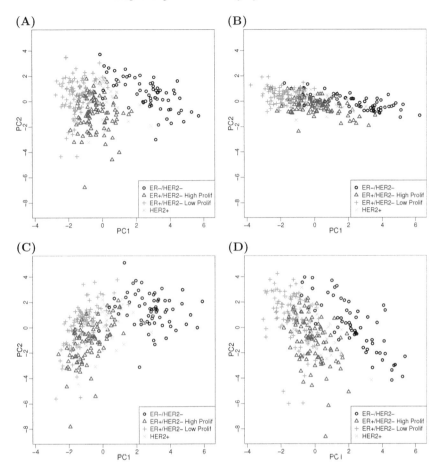

FIGURE 5.7 Multiple factor analysis on breast cancer data. (A) The 286 tumour samples of the Wang dataset are visualised in the first two components of the MFA when all the genes have been used. The tumour samples are projected on the same two components using only the genes in (B) the RB pathway (C) the NFkB pathway or (D) the TGFβ pathway. (See colour insert.)

pairs of genes in the group. Interestingly, when they analyse groups of genes whose products form a protein complex, they observe a difference between permanent complexes, such as the **ribosome*** and **proteasome*** which have a particularly strong coexpression, while transient complexes have not. Following a similar line of thoughts, Pavlidis et al. (2002a) also proposed to compute the mean average correlation between genes of a group, and derived a p-value for each group (called the *correlation score*) to assess how significant the mean correlation is by comparing it to the mean correlation among random groups. Similarly, Breslin et al. (2005) assessed the mean correlation between all pairs of genes downstream of a signalling pathway, as a hint for the fact that the pathway could be a functional unit.

An interesting variant of this problem of detecting important gene sets is the situation where, instead of a collection of groups of genes, one has a collection of gene networks representing different pathways (such as, *e.g.* the RB map of **Section 4.9.2** or the networks representing the hallmarks of cancer discussed in **Chapter 8**), or even a global network connecting all or most of genes (see **Box 4.3**). How can we then assess the importance of a particular network taking into account its particular structure, or even detect important "subnetworks" within a given large network?

Several ideas have been proposed in the literature to exploit the network structure. A first approach, proposed by Han et al. (2004), is to measure for each node in a protein network the average correlation coefficients it has with interacting partners. A large average correlation may suggest that the node somehow controls the expression of its local interactome. Focusing on **hubs***, they observe a clear bimodal distribution, suggesting that hubs can be divided into *party* hubs, which have a large average correlation with their neighbours, and *date* hubs, which have not. A second approach, proposed by Rahnenführer et al. (2004), tries to generalise the mean correlation score used to assess the importance of a group when the group is itself a network, typically a signalling or metabolic pathway. They propose to quantify the importance of the subnetwork by computing a weighted average of the correlation coefficient between all pairs of genes of the pathways, where the weight decreases with the shortest-path distance between the genes on the graph. This is a way to focus only on the important correlations within the pathway.

Finally, let us mention the work of Guo et al. (2007) who proposed a method to detect subnetworks within a large given network, responsive to some investigated gene expression data. They define the score of a connected subnetwork with k edges as the sum over the edges of the covariance of the connected genes; the score is then normalised as a Z-score by comparing it to the scores of random sets of k edges. Since finding the highest-scoring subnetwork in the entire network is a NP-hard problem (Ideker et al., 2002), Guo et al. (2007) implement a strategy based on simulated annealing to find a high-scoring subnetwork. This approach was tested on a prostate cancer dataset, and was able to select a subnetwork containing many genes known to be important in prostate cancer, in particular many genes of the NFkB and MAPK signalling cascades.

5.5.5 Example: Analysing RB pathway activity in bladder cancer

To illustrate the possibility to exploit pathway information in the analysis of gene expression data, let us take the example of a collection of 55 bladder cancers comprising noninvasive (Ta, T1 stages) and invasive (stages above T2) tumours, as well as 5 normal urothelium samples (Stransky et al., 2006). Since

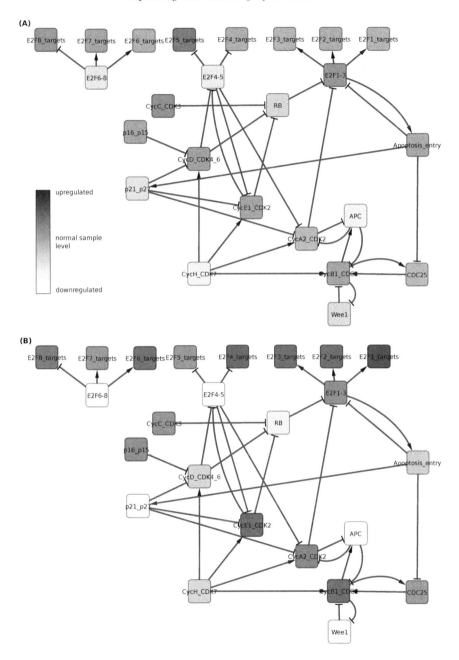

FIGURE 5.8 Modular map of RB pathway. *Coloured modular map of RB pathway*: (A) Module activity for noninvasive cancers; (B) module activity for invasive cancers. Red modules show global upregulation for the genes that compose each module. Green modules show global downregulation. (See colour insert.)

alterations in the *RB* pathway are frequent in bladder cancer (Mitra et al., 2007), we wish to analyse the difference between noninvasive and invasive tumours in the light of the *RB* pathway.

As explained in **Section 4.9.2**, the detailed description of the *RB* pathway is a complex network involving 78 proteins, which can be simplified by grouping together subsets of proteins into modules. The simplified version of the *RB* pathway we use, shown in **Figure 5.8**, contains 24 modules. We estimate the activity of a module in each sample using the first principal component of the expression matrix restricted to the genes of the module and identify *hot spot* genes within each module, as explained in **Section 5.5.3**. As expected, many of the hot spot genes correspond to genes known to have modified expression in cancers, including cyclins and cyclin-dependent kinase inhibitor genes.

Figure 5.8 shows the mean module activity of noninvasive (top) and invasive (bottom) tumours, and provides a general overview of differences in the pathway activity amenable to biological interpretation. The results are consistent with our understanding of the molecular mechanisms of bladder cancer progression. There are three modules of transcription factors: E2F1-3, the activating ones that are sequestered by RB and that play a role in apoptosis entry; E2F4-5, the inhibiting ones that are too sequestered by RB; and E2F6-8 that do not interact with RB at all. RB is inhibited through phosphorylation by CycC/CDK3, CycD1/CDK4,6 and CycE1/CDK2. There are two modules of cyclin-dependent kinase inhibitors: p16/p15 and p27/p21 that specifically bind to some of the Cyclin/CDK complexes. CycA2/CDK2 and especially CycB1/CDC2 are intervening in late phases of the cell cycle. CycB1/CDC2 activity is controlled by the kinase Wee1 and the phosphatase Cdc25 but both CycA2/CDC2 and CycB1/CDC2 are negatively regulated by the complex Anaphase Promoting Complex (APC). CycH/CDK7 is necessary for the activation of all Cyclin/CDK complexes. Finally, eight modules regroup all the gene targets of the E2F transcription factors.

Red modules correspond to overexpressed modules and green modules to underexpressed modules. The modules in black are close to the expression of normal samples. Cyclin expressions seem to play an active role in invasive cancers (activation of four cyclin modules: CycC/CDK3, CycE1/CDK2, CycA2/CDK2, CycB1/CDC2) as well as the activation of the E2F1-3 targets, even though the activity of E2F1-3 module at the transcriptional level is not strongly manifested. **Figure 5.8** also shows a particular behaviour of CycD1/CDK4/6 module, which is, in average, less activated in invasive than in noninvasive tumours, in agreement with the literature (Tut et al., 2001). Amplification (and overexpression) of *CCND1* (referred to as CyclinD1) can occur in invasive tumours but is a rare event (3 out of 30 tumours in this series).

Remarkably, all but one module that inhibit cell cycle progression (RB, E2F4-5, E2F6-8, p27/p21, Wee1, APC and Apoptosis entry) were downregulated in invasive tumours compared to noninvasive tumours. These results were already known for some genes (*RB*, *p27KIP1*) and point out new genes

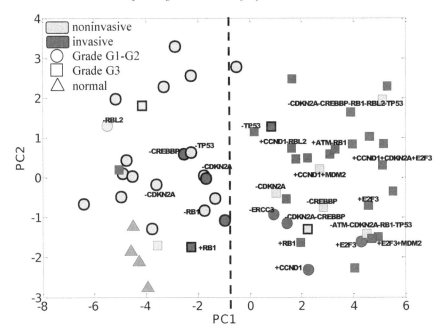

FIGURE 5.9 PCA of the modules activity. Two clusters can be identified, highlighting the existence of two paths to invasiveness, one corresponding to the Ta pathway (left cluster) and the other one to the carcinoma *in situ* pathway (right cluster). Alterations of genes of the network and for each sample are annotated on the graph: "+" means that the gene is amplified, "−" means that the gene is lost. (See colour insert.)

or mechanisms likely to be involved in bladder tumour progression. Downregulation of E2F4-5 and E2F6-8 modules is consistent with observed upregulation of their target gene modules. It should be underlined that a downregulation in the activity of a module, for example E2F4-5, does not necessarily correspond to a significant change of *E2F4* or *E2F5* expression itself, but rather to that of some components involved in the module.

Another way to look at these data is to consider the PCA plot of the modules in terms of their expression level. In the 24-dimensional space of the module activities (see **Figure 5.9**), tumours form two distinctive clusters separated along the first principal component. This separation is completely consistent with the two progression pathways which exist in bladder cancer: the Ta low-grade pathway (round shapes) and the carcinoma *in situ*/invasive tumour pathway (squared shapes) (Stransky et al., 2006). The left cluster of **Figure 5.9** contains mainly low grade noninvasive tumours, with 5 invasive tumours. Most of the invasive tumours belonging to this cluster (4/5) carry *FGFR3* **mutations**[*], a hallmark of the Ta pathway being *FGFR3* mutations (Billerey et al., 2001). Therefore, we can conclude that these tumours belong to the same progression pathway despite the difference in stage. The right cluster contains only invasive tumours or high grade (G3) noninvasive tumours. Most tumours of this cluster do not carry *FGFR3* mutations (29/31, compared to 3/24 in the other group).

We also labelled some of the samples according to the most significant genomic alterations (DNA copy number) in the regions of genes known to be involved in cancer and participating in the *RB* pathway. Thus, one can notice a higher frequency of genomic alterations in the aggressive tumours, and an amplification of E2F3 transcription factor that drives module activities towards the extremity of the right more aggressive cluster. Taken together, all these observations give a consistent picture of the alterations of the RB pathway in bladder cancers that seem to be different in the two pathways of bladder tumour progression.

It is important to note that the analyses of array Comparative Genomic Hybridisation (aCGH) and gene expression levels illustrate what could happen at the protein level. However, it does not guarantee that the corresponding proteins will be expressed, post-transcriptionally modified, active or inactive. High-throughput techniques such as Reverse-Phase Protein Array (RPPA) would make it possible to better assess the activation of pathways at the protein level (see **Chapter 3**).

5.6   Integrative analysis of heterogeneous data

The strength of data-driven methods normally increases when more data are analysed jointly. It becomes increasingly possible and easy to collect large quantities of cancer omics data from a public online database, but their analysis raises several challenges in integrative analysis:

- because we would like to jointly analyse data of different natures, *e.g.* gene expression data and DNA copy number profiles;

- because we could like to combine several datasets measuring the same type of data, *e.g.* gene expression, but on different cohorts, and potentially with different technologies.

In this section we briefly review a few computational strategies which have been proposed to carry out such integrative analysis, in the context of exploratory data analysis. We postpone to **Section 6.6** the presentation of methods for integrative analysis in the context of predictive modelling.

5.6.1   Joint analysis of data from different molecular levels

It is increasingly frequent to collect several heterogeneous genomic data about the same samples, such as gene expression data and genotype or DNA copy number profiles. It is then interesting to jointly analyse the different datasets, in particular to try to capture relationships between processes at different molecular levels such as the link between DNA amplification and overexpression (Hyman et al., 2002; Pollack et al., 2002; Morley et al., 2004).

In order to directly answer questions such as detecting genes which are differentially expressed in samples which exhibit a specific genomic alteration, the most direct method is to stratify the samples based on the genomic alteration and analyse the gene expression in relation to this stratification (Bungaro et al., 2009; Hyman et al., 2002; Pollack et al., 2002). Another approach which does not require explicit stratification is to compute the correlation between each DNA copy number and each gene expression, or formulate the problem as a regression problem, in order to detect genomic loci whose amplification is strongly associated to variations in expression of some genes (Peng et al., 2010).

Methods based on matrix factorisation can also be extended to capture low dimensional representations of two heterogeneous datasets associated to each other. For example, Canonical Correlation Analysis (CCA) searches for projections maximally correlated to each other (Hotelling, 1936). In the high-dimensional setting, CCA needs to be regularised to prevent overfitting, and many forms of regularised CCA have been proposed in the context of heterogeneous genomic data comparisons (de Tayrac et al., 2009; González et al., 2009). Like other matrix factorisation techniques discussed in **Section 5.4.3**, CCA can also be modified to capture sparse or structured projections (Waaijenborg et al., 2008; Witten et al., 2009; Lê Cao et al., 2009; Parkhomenko et al., 2009; Soneson et al., 2010).

5.6.2 Joint analysis of several datasets

The strength and robustness of statistical methods increase when the number of samples increases. Due to the practical limits of cost and access to large numbers of fresh frozen tumour samples, most individual studies collecting cancer omics data are limited in size from a few tens to a few hundreds of samples. Since the data of many of these studies are publicly and easily available, an efficient strategy to increase the number of samples analysed jointly is to pool several studies together and perform a **meta-analysis***. However, this meta-analysis task is often challenging because the biological diversity common to all studies is easily masked by batch effects, due in particular to differences in experimental protocols, cohorts and technologies between studies.

Batch effects are inevitable with high-throughput technologies, like for example expression microarrays which are very sensitive to many non-biological factors including the technicians who performed the experiment and even the atmospheric ozone level (see **Section 4.1.2**). Several methods have been proposed to remove batch effects in microarray experiments, as recently reviewed by Scherer (2009). For example, Distance-Weighted Discrimination (DWD) is a multivariate analysis technique to correct systematic differences between two datasets by performing a global shift of each dataset to ensure an optimal overlap between the two sets of samples (Benito et al., 2004). DWD was for example used by Hu et al. (2006) to show that the classification of breast cancers proposed earlier by Sørlie et al. (2001) was reproducible and conserved across different microarray platforms. In a recent benchmark of six adjustment methods for removing batch effects in expression data, Chen et al. (2011) concluded that an empirical Bayes method called Combating Batch Effects When Combining Batches of Gene Expression Microarray Data (ComBat) was particularly performant for this task (Johnson et al., 2007).

5.7   Heterogeneity within the tumour

Besides diversity of cancers across patients, another important source of heterogeneity and variability in tumour samples characterised by omics technology is the heterogeneity of cells within the tumour itself. Indeed, as discussed in **Sections 2.3, 2.6** and **3.2.6**, we know that a tumour consists of a variety of cells, including normal and cancer cells. Deciphering this heterogeneity within the tumour could rely on sophisticated biotechnological approaches, *e.g.* the ability to analyse samples at the single cell level. In parallel, we briefly discuss in this section how dedicated computational approaches can help capture this intra-tumoural diversity.

The characterisation of the heterogeneity within the tumour using high-throughput combined with statistical methods was first carried out on a DNA copy number profile using aCGH. Unravelling the DNA copy number profile of the subpopulations and also estimating what the percentage of each subpopulation is in a given tumour can be formulated straightforwardly as in Wang et al. (2009). Consider a mixture of $K + 1$ clonal subpopulations, with respective percentages in the full populations denoted by $p_0, p_1, \ldots, p_K$ with $\sum_{k=0}^{K} p_k = 1$. We assume that first population, with percentage p_0, is a normal sample with 2 DNA copies along the genome. At a genomic position i, the expected DNA copy number C_i can be expressed as $\sum_{k=0}^{K} p_k C_{ik}$ where C_{ik} is the DNA copy number of the kth subpopulation. A Bayesian framework based on a mixture model was used to solve this problem and estimate the most likely number of subpopulations. While very intuitive, this statistical model may fail in practice to predict reality as, on one hand, contamination by the normal tissue exists and on the other hand, the overall ploidy of the tumour can be greater than two (*e.g.* some tumours are triploid, tetraploid, *etc.*). With the advent of Single Nucleotide Polymorphism (SNP) arrays, the availability of both DNA copy number and B Allele Frequency (BAF) profiles (see **Section 3.1.3**) made it possible to refine the characterisation of tumour complexity and heterogeneity. For example, Popova et al. (2009) proposed a pattern recognition algorithm in which the observed DNA copy number and BAF profiles are fitted with respect to their expected values, shown in **Figure 3.7**. Their algorithm allows both the estimation of normal contamination and ploidy and the detection of clonal subpopulations. To fully benefit SNP arrays, Yau et al. (2010) extended the approach proposed by Wang et al. (2009) in order to jointly consider DNA copy number and BAF information as a set of different states in a unified statistical framework to better characterise heterogeneity within the tumour. Though initially developed on SNP arrays, these methods could also be applied on NGS data as DNA copy numbers and BAF profiles can be obtained from such technologies (Boeva et al., 2011b).

5.8 Conclusion

The large heterogeneity of cancers has been known for a long time, and has been a major focus of cancer research for decades. Is has direct bearing on clinical management, since apparently similar cancers may be better treated by different treatments. While stratification of patients based on clinicopathological parameters is well established in the clinics nowadays, it remains insufficient since patients in the same category can still have very different clinical evolution.

The possibility to observe cancer diversity at the molecular level with omics technologies has drastically changed and improved our understanding of cancer heterogeneity in the last decade. Hundreds of studies have now collected detailed molecular information about the genome, transcriptome and epigenome of diverse collections of cancer samples and cell lines, and systematically investigated the biological underpinnings of cancer diversity at the molecular level. New molecular classification of cancers have emerged and are now widely accepted as the gold standard classification. The biological processes underlying the diversity of cancers have also started to emerge from the systematic analyses of these datasets.

Besides expected discoveries, such as the importance of *ER* and proliferation pathways in breast cancer diversity, or the division of acute leukaemias into known subtypes, a multitude of other intriguing findings often emerge from these analyses. The field is still in its infancy, both in terms of computational methods to translate the wealth of omics data into useful biological findings, and in terms of capacity of the biological and medical community to investigate and validate new findings. After almost 15 years since the early work on genome-wide analysis of cancer samples, the analysis and modelling of cancer heterogeneity at the molecular level remains a fast-moving field which strongly benefits from technological developments. It will certainly benefit a lot from the upcoming availability of even larger quantities of molecular cancer data generated by large projects such as the International Cancer Genome Consortium (ICGC) and The Cancer Genome Atlas (TCGA). It also continues to raise important theoretical and computational questions regarding the best way to analyse these data, classify them and extract useful information from them.

✎ **Exercises**

- Compare the clusters obtained by different hierarchical or partitioning methods on the Wang breast cancer dataset. Which methods lead to the most similar and the most different classifications?

- Why do the ℓ_1 norm (see **Equation 5.6**) and TV semi-norm (see **Equation 5.7**) lead to respectively sparse and piecewise constant metagenes? Check it experimentally.

- Why do we assess the importance of a group of genes for a set of samples by the average correlation between the genes across the samples? What are we trying to quantify? Can you propose other ways to quantify the importance of gene sets using PCA?

⇨ **Key notes of Chapter 5**

- Cancer is a heterogeneous disease; cancer classification based on clinicopathological data is not precise enough for good clinical management.

- Automatic discovery of cancer subtypes by systematic analysis of omics data with clustering methods has imposed itself as a new gold standard taxonomy for several cancers, often referred to as molecular classification of cancers.

- There exist, however, many computational techniques to derive molecular classifications of cancers, which influence the results; current classification schemes are likely to evolve as larger sample sets are analysed and more robust analytical methods are developed.

- Automatic analysis of high-dimensional data with matrix factorisation or similar techniques allows to capture, to some extent, the basic biological processes underpinning cancer diversity.

- Integrating omics data analysis with prior knowledge we have about gene sets or gene networks can help understand the heterogeneity of cancers at the level of biological processes and pathways, and identify important pathways.

Chapter 6

Prognosis and prediction: Towards individualised treatments

The heterogeneity of cancers discussed in **Chapter 5** has immediate bearing on cancer clinical management. The diversity of cancers at the molecular level entails important differences at the clinical level, in particular in **prognosis**[*] (the risk of **relapse**[*] after an initial treatment) and in terms of responsiveness to different treatments. It is therefore important to characterise as finely as possible the disease in order to take the best therapeutic options for each particular case. In practice, when confronted with a newly detected cancer, physicians must first identify as precisely as possible the type and particular characteristics of the disease, a problem called *diagnosis*. Then, when several therapeutic options are available, they must choose the most suitable one based on whatever information they have about the patient and cancer to treat. Two important questions to be addressed when a treatment is available are (see **Figure 6.1**):

- *Does the patient need a treatment, and how aggressive should it be?* Many solid tumours can be removed by surgery and radiotherapy. However, while some patients are definitively cured by such operations, others will experience an often more aggressive relapse within a few years. It is therefore tempting to systematically treat patients by **adjuvant therapy**[*], aimed at preventing secondary tumour formation. Most existing adjuvant treatments for cancer, usually **chemotherapy**[*] with cytotoxic drugs, have however, strong deleterious side effects. Limiting aggressive treatments only to patients who would benefit from them is therefore important. In many cases, such as the decision to give or not give adjuvant chemotherapy for operable breast cancers, this decision boils down to the problem of *prognostication, i.e.* to estimate the risk of future relapse if no adjuvant treatment is given to the patient. This cannot only spare the morbidity of a treatment that offers no benefit to low-risk patients, but can also justify a more aggressive adjuvant treatment to patients with a bad prognosis (Clark, 1994).

- *Which treatment will work? Which treatment is the best among several possible choices?* For patients with poor prognosis, or with a tumour that cannot be removed by surgery or radiation only, the second question is to choose an adequate treatment. Since each treatment is typically effective only on a subset of all cancers, this question boils down to the problem of *predicting response*, *i.e.* to assessing the probability that the cancer will react to a specific treatment in order to select the best one. Factors which influence the probability of response to a treatment are usually simply called *predictive factors*.

In this chapter, we discuss current practices and new hopes generated by the omics revolution (see **Chapter 3**) in the assessment and clinical management of cancers through better prognosis and drug response prediction. This is the occasion to discuss some of the popular statistical and machine learning methodologies underlying **biomarker*** detection and predictive modelling, and to highlight the increasingly important role played by systems biology in this context. As in **Chapter 5**, we illustrate several of these techniques on the Wang dataset for breast cancer prognosis from gene expression data, and provide a vignette on the book's companion website to help the reader reproduce the analysis.

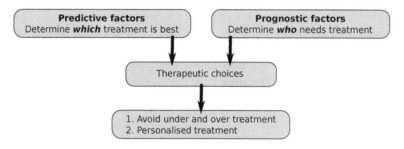

FIGURE 6.1 Prognostic and predictive factors. This illustration from Lønning (2007) highlights the clinical importance of prognostic and predictive factors in breast cancer.

6.1 Traditional prognostic and predictive factors

Many prognostic and predictive factors have been known for years. The type and location of the tumour, and its **clinicopathological*** characterisation including its **stage*** and **grade***, are currently the main prognostic factors used in decision-making for clinical management. For example, it is known that a high-grade breast cancer with cancer cells detected in lymph nodes has in general a higher risk of relapse than a small tumour of low grade, and requires a more aggressive adjuvant treatment to prevent relapse. Other factors, such as the detection of the steroid hormone receptors ER and PR in breast cancer, are both prognostic and predictive for the response to hormone and endocrine therapies. Some decision support tools have been developed to combine these factors, including the popular **Adjuvant! Online*** which is widely used by physicians and patients to predict mortality and **recurrence*** risks, as well as the benefit of adjuvant therapy for women with early-stage breast cancer (Ravdin et al., 2001). Adjuvant! Online makes these predictions automatically from six inputs that are well established as powerful predictors of mortality and recurrence: patient age, tumour size, grade, hormone receptor status, number of positive lymph nodes, and **comorbidity*** level. It is based on a statistical model that we present in **Section 6.2.3** (logistic regression), trained on thousands of cases.

However, the clinicopathological characterisation for prognostication remains subject to several limitations. First, tumours with similar clinicopathological criteria may still have very different prognosis and respond differently to different treatments. For example, recurrence is likely in 20–30% of young women with early-stage (lymph node-negative) breast cancer who only undergo surgery and localised radiation treatment (Van't Veer and Bernards, 2008). Yet, in the United States, 85–95% of women with this type of cancer receive adjuvant chemotherapy, mostly because conventional clinicopathological criteria fail to identify reliably those patients who are likely to relapse. Therefore, 55–75% of women with early-stage breast cancer in the United States undergo a toxic therapy from which they will not benefit but from which they will experience the side effects, because the clinicopathological characterisation of the tumour is not sufficient to differentiate them from those who need the treatment. This problem could be explained by both the existence of unknown cancer subtypes within clinicopathological categories, and by the lack of efficient clinicopathological prognostic factors. As a consequence, we need new prognostic factors in order to better stratify patients, and in particular identify more low-risk patients to spare them the deleterious effects of a treatment that brings them no benefit (Clark, 1994).

Second, while the identification of reliable predictive factors has the potential to spare patients ineffective treatment and unnecessary side effects, the reverse (that a factor may guarantee therapeutic success) may be more difficult to achieve. For example, while ER negativity is associated with lack of response to endocrine treatment, not all patients with ER positive tumours may benefit from such therapy. Similarly, while the absence of *HER2* (*ERBB2*) overexpression has been established as a predictive factor for nonresponsiveness to trastuzumab therapy, not all *HER2*-overexpressing tumours are trastuzumab sensitive, reflecting the complexity of breast cancer genetics.

Third, the objective and consistent assessment of some clinicopathological factors is sometimes difficult to ensure. As explained in **Section 5.1**, they may not only vary with the particular **histological section*** studied, but also depend on the expert analysing the sample. Therefore, depending on the pathologist, the patient might not be given the same therapy and thus the reproducibility for the clinicopathological parameter assessment, in particular grade and stage, needs to be improved.

There is therefore a need for new prognostic and predictive factors, with better reproducibility and better discriminatory power between different prognosis and drug responsiveness groups. Recent technologies to interrogate tumours at the molecular level (see **Chapter 3**) offer extraordinary opportunities to search for new biomarkers, and to build prognostic and predictive models from these biomarkers. As millions of candidate markers can nowadays easily and robustly be collected from patients and tumours, including genomic, epigenomic, transcriptomic and proteomic markers, major questions to be addressed include (1) how to select good biomarkers among all candidates, and (2) combine them into accurate predictive models. We start by addressing the second question in the next section, before coming back to the question of biomarker selection in **Section 6.3**

6.2   Predictive modelling by supervised statistical inference

As no single factor is usually perfectly prognostic or predictive, accurate models need to combine several factors for better risk assessment. In this section, we present a few computational methods to combine several factors into a single model, and postpone the question of how to select factors among many candidates to **Section 6.3**. For the sake of coherence, we follow the terminology of **Box 5.1**, calling *samples* the patients available to estimate a model and *features* the different factors available for each patient which we can use to build a predictive model.

6.2.1   Supervised statistical inference paradigm

Most prognostic and predictive factors have been found by empirical evidence of association with the response of interest (respectively cancer relapse and response to a treatment) over large populations of patients. Similarly, the main statistical paradigm to combine several features into a single model for risk prediction is to use empirical evidence: by observing both the features and the response of interest in a population of patients, we try to infer a rule to combine the features which is empirically associated to the response. Mathematically, we formalise this paradigm by modelling the observed population as a set of n pairs $\mathcal{S} = \{(x_1, y_1), \dots, (x_n, y_n)\}$, where the $x_i \in \mathcal{X}$ denotes the i-th sample and $y_i \in \mathcal{Y}$ his/her response. We restrict ourselves to the case where we have access to p features on each sample, meaning that x_i can be represented by a p-dimensional vector $(\mathcal{X} = \mathbb{R}^p)$. The response can take different forms. In many cases the response of interest is binary, such as good/bad prognosis or response/no response to a drug, in which case we represent the two alternatives by the arbitrary numbers -1 and 1, *i.e.* $\mathcal{Y} = \{-1, 1\}$. Other common settings include the cases where the response can be one among $K > 2$ classes, such as different subtypes of cancers, a continuous value, such as a quantitative measure of how a treatment works, or survival data, where we observe the time until an event such as relapse or death occurs. These situations can be modelled by different sets $\mathcal{Y}$.

Based on the observation of the population $\mathcal{S}$, the objective of the modeller is to infer a function $f : \mathcal{X} \to \mathcal{Y}$ which can then be used to predict the response $y \in \mathcal{Y}$ for any new sample $x \in \mathcal{X}$ by the value $f(x)$. This inference problem is often referred to as *supervised learning* in statistics, as opposed to *unsupervised learning* problems where we only observe the samples $x_1, \ldots, x_n$ without any response of interest; this was for example the case in **Chapter 5** where we investigated the diversity among the x_i's without trying to predict any particular property or response variable. The field of supervised learning is a mature field in statistics and computer science, with many well-understood and efficient methods and algorithms. It is however, also a field of active research, triggered in particular in the last decade by the particular challenges raised by genomic data. We present next a short selection of popular methods for supervised learning, and refer the readers to a number of excellent textbooks on the topic for a more in-depth presentation of the statistical machine learning (Devroye et al., 1996; Vapnik, 1998; Duda et al., 2001; Hastie et al., 2001; Bishop, 2006).

6.2.2   Supervised inference by generative models

In the case where the phenotype of interest is discrete $\mathcal{Y} = \{1, \ldots, K\}$, *i.e.* when we want to categorise the samples into K pre-defined categories, a possible strategy to infer a classification function is to model the sample and its phenotype as a random vector (X, Y) with values in $\mathcal{X} \times \mathcal{Y}$ and to learn the random distribution from the data. This approach, which is often referred to as learning a *generative model* of the data, uses the observed population $\mathcal{S}$ in order to infer the distribution $P(X, Y)$. Using the chain rule $P(X, Y) = P(Y) \times P(X \mid Y)$, this is often done in practice by inferring the class distribution $P(Y)$ on the one hand, and the class-conditional probabilities $P(X \mid Y)$ on the other hand, assuming the observed samples are randomly and independently sampled according to P. Once P is estimated from the observed samples, we can predict the probability of each possible response $Y \in \mathcal{Y}$ for a new sample $X \in \mathcal{X}$ using Bayes' rule:

$$P(Y \mid X) = \frac{P(X \mid Y)P(Y)}{P(X)},$$

and predict the response with largest probability:

$$f(X) = \underset{Y \in \mathcal{Y}}{\operatorname{argmax}} \ P(Y \mid X) = \underset{Y \in \mathcal{Y}}{\operatorname{argmax}} \ P(X \mid Y)P(Y). \tag{6.1}$$

This generic modelling framework underlies several popular methods. The class probabilities $P(Y = y)$ for $y \in \mathcal{Y}$ are usually simply estimated as the fraction $\hat{\pi}_y = n_y/n$ of each response class in $\mathcal{S}$, where n_y denotes the number of samples with response y in $\mathcal{S}$. Different methods make different assumptions on the class-conditional distributions $P(X \mid Y)$. For example, when samples are multidimensional real vectors $\mathcal{X} = \mathbb{R}^p$, possible models for the class-conditional distributions are Gaussian distributions $\mathcal{N}(\mu_u, \Sigma_y)$ with densities:

$$p(x \mid Y = y) = \frac{1}{(2\pi)^{p/2} |\Sigma_y|^{1/2}} \exp\left(-\frac{1}{2}(x - \mu_y)^\top \Sigma_y^{-1} (x - \mu_y)\right),$$

where $\mu_y \in \mathbb{R}^p$ and $\Sigma_y \in \mathbb{R}^{p \times p}$ are respectively the mean and covariance matrix of the distribution of samples in class $y \in \mathcal{Y}$. In that case, inferring the class-conditional distributions boils down to estimating μ_y and Σ_y from the observed samples. The class-conditional mean μ_y is usually estimated by the empirical average of samples with response class y, *i.e.* $\hat{\mu}_y = \sum_{i:y_i=y} x_i/n_y$. The covariance matrices Σ_y involve more parameters and can be estimated in different ways:

- Quadratic Discriminant Analysis (QDA) estimates each covariance matrix Σ_y independently from the others, as the empirical covariance matrix of the samples of class y:

$$\hat{\Sigma}_y = \frac{1}{n_y - 1} \sum_{i:y_i=y} (x_i - \hat{\mu}_y)(x_i - \hat{\mu}_y)^\top.$$

  The corresponding decision function (see **Equation 6.1**) for a new sample $x \in \mathbb{R}^p$ predicts $f(x) = \operatorname{argmax}_{y \in \mathcal{Y}} g_y(x)$, where $g_y(x)$ is the quadratic function (hence the name QDA):

$$g_y(x) = -\frac{1}{2}\log|\hat{\Sigma}_y| - \frac{1}{2}(x - \hat{\mu}_y)^\top \hat{\Sigma}_y^{-1}(x - \hat{\mu}_y) + \log \hat{\pi}_y. \quad (6.2)$$

- Linear Discriminant Analysis (LDA) assumes that all class-conditional covariance matrices are equal and estimates them jointly from all samples:

$$\hat{\Sigma} = \frac{1}{n - |\mathcal{Y}|} \sum_{y \in \mathcal{Y}} \sum_{i:y_i=y} (x_i - \hat{\mu}_y)(x_i - \hat{\mu}_y)^\top.$$

  In that case, the quadratic terms in **Equation 6.2** cancel and the decision function for a new sample x maximises over y linear functions:

$$g_y(x) = x^\top \hat{\Sigma}^{-1} \hat{\mu}_y - \frac{1}{2}\hat{\mu}_y^\top \hat{\Sigma}^{-1}\hat{\mu}_y + \log \hat{\pi}_y.$$

- Naive Bayes (NB) assumes that the p different sample features are independent given the response, *i.e.* that each covariance matrix is diagonal. It boils down to a simplified version of LDA, sometimes called Diagonal Linear Discriminant Analysis (DLDA), where the non diagonal entries of the estimated covariance matrix are set to zero. This method was used for example in the seminal work of Golub et al. (1999), with a particular way to estimate the inverse covariance terms, in order to learn a predictor to discriminate different subtypes of leukaemia from their transcriptome; or by Hess et al. (2006) to develop a model to predict breast cancer response to neoadjuvant chemotherapy from gene expression data.

Generative methods are very easy to implement, and frequently work surprisingly well on real problems although the various assumptions they make about the distributions of samples are often unrealistic. Simpler methods like LDA or NB are often at least as good as QDA, in particular when the number of descriptors p increases, since the loss in modelling power when we use a simpler model can be compensated by the gain in statistical power when we estimate less parameters.

6.2.3   Linear discriminative methods

Generative models necessarily make restrictive assumptions on the class conditional sample distributions. However, our goal in predictive modelling is not to model correctly the sample distributions $P(X \mid Y)$, but only to infer a good rule to predict the response class Y given a sample X. In many cases, it is therefore advantageous to focus our efforts on estimating $P(Y \mid X)$ only. For example, in the binary case $\mathcal{Y} = \{-1, 1\}$, Logistic Regression (LR) assumes a model for $P(Y \mid X)$ of the form

$$P(Y = y \mid X = x) = \frac{1}{1 + \exp(-yw^\top x)} , \tag{6.3}$$

where the weight vector $w \in \mathbb{R}^p$ is estimated from the observed population $\mathcal{S}$ by maximising the conditional log-likelihood:

$$\hat{w} = \operatorname*{argmax}_{w \in \mathbb{R}^p} \sum_{i=1}^{n} \log P(y_i \mid x_i) = \operatorname*{argmin}_{w \in \mathbb{R}^p} \sum_{i=1}^{n} \log \left(1 + \exp(-y_i w^\top x_i)\right) . \tag{6.4}$$

Once $\hat{w}$ is estimated, the class of a new sample x is predicted based on the value of the function $f(x) = \hat{w}^\top x$, using **Equation 6.3** to estimate the probability of each response; the resulting decision function is therefore linear in x. LR was for example used to estimate the predictive models of mortality and relapse risks in Adjuvant! Online from six prognostic clinicopathological factors (Ravdin et al., 2001).

LR is a well-established method for supervised classification, particularly powerful when the number of samples n is large compared to the number of features p. When this is not the case, which is often the rule rather than the exception with omics data where p can easily range in the thousands or more, it is recommended to regularise the maximisation of the conditional log-likelihood of **Equation 6.4** to limit overfitting, *i.e.* to prevent the estimation of weights w which fit very well the observed population but have little generalisation power on unseen samples. The most common way to regularise LR is to add a ridge term equal to the squared Euclidean norm of w in the objective function of **Equation 6.4**, leading to the following minimisation problem:

$$\hat{w} = \underset{w\in\mathbb{R}^p}{\operatorname{argmin}} \sum_{i=1}^{n} \ell_{logistic}(y_i w^\top x_i) + \lambda\| w \|^2 , \qquad (6.5)$$

where we use the shorthand

$$\ell_{logistic}(u) = \log(1 + e^{-u}) \qquad (6.6)$$

for the logistic loss. The $\lambda \geq 0$ parameter in **Equation 6.5** controls the amount of regularisation, and must be fixed by the user (see **Section 6.2.5**). Intuitively, the larger the number of features p, the more we need to regularise by increasing λ.

While the final objective function optimised by LR (see **Equation 6.5**) was motivated by a probabilistic model of $P(Y\,|\,X)$ (see **Equation 6.3**), one can also interpret LR as a method that attempts to minimise the mean logistic loss $\ell_{logistic}(yw^\top x)$ over the training set of observations, up to the regularisation terms. For a given sample $x \in \mathcal{X}$ with response $y \in \mathcal{Y}$, the value $yw^\top x$ is called the *margin* of the sample with respect to the classifier w. The margin is positive when $w^\top x$ and y have the same sign, *i.e.* when the sign of the prediction function $f(x) = w^\top x$ is correct. The margin is large when the prediction $f(x)$ has not only the correct sign, but also a large absolute value which can be thought of as a large confidence in the prediction. Because the function $\ell_{logistic}(u)$ is decreasing in u, LR can be thought of as a *large-margin classifier*, which focuses on finding a classifier $f(x) = w^\top x$ with large margin on the observed training set.

Other popular methods follow the same principle, although with technical differences. The best-known large-margin classification method is probably the linear Support Vector Machine (SVM), which also estimates a predictor $f(x) = w^\top x$ by minimising **Equation 6.5**, with the logistic loss replaced by the so-called hinge loss:

$$\ell_{hinge}(u) = \begin{cases} 1 - u & \text{for } u \leq 1, \\ 0 & \text{otherwise.} \end{cases} \qquad (6.7)$$

Figure 6.2 illustrates the losses used by LR (see **Equation 6.6**) and SVM (see **Equation 6.7**). Since both loss functions are convex, the problem (see 6.5) has a unique minimum which can be found efficiently by various optimisation algorithms.

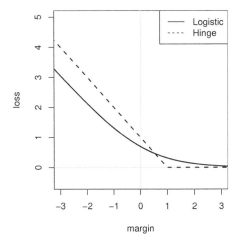

FIGURE 6.2 Losses of large-margin classifiers. LR and SVM both learn classifiers by enforcing large margins with different loss functions.

6.2.4   Nonlinear discriminative methods

Generative and linear methods are often sufficient to estimate good classifiers, in particular when we have many features. Nonlinear methods are also commonly used in supervised statistical inference, and can be more powerful than linear methods when the function that relates the response to the sample features is clearly nonlinear. However, one should keep in mind that more realistic, nonlinear models are not always better since they can be more difficult to infer from a limited number of observations.

One popular nonlinear discriminative method is the k-Nearest Neighbour (k-NN) algorithm. k-NN estimates the probability of a response y for a sample x as the frequency of samples with response y among the k training samples most similar to x. The predicted response $f(x)$ by k-NN is therefore the most frequent response among the k nearest neighbours of x in $\mathcal{S}$.

Alternatively, it is possible to extend the formalism of discriminative linear models presented in **Section 6.2.3** to nonlinear models by first performing a nonlinear transformation of the data through a mapping $\Phi : \mathcal{X} \to \mathbb{R}^q$, and then learning a linear function of the form $w^\top \Phi(x)$ which can be nonlinear in x. A particularly ingenious instantiation of this idea is the set of *kernel methods*, which consider mappings Φ such that the so-called kernel $K(x, x') = \Phi(x)^\top \Phi(x')$ is easily computed without the need for explicitly computing $\Phi(x)$ for each x (Burges, 1998; Vapnik, 1998; Schölkopf and Smola, 2002; Schölkopf et al., 2004; Shawe-Taylor and Cristianini, 2004). One can then show that LR (see **Equation 6.3**) and SVM (see **Equation 6.7**) can be solved efficiently and lead to nonlinear functions of the form:

$$f(x) = \sum_{i=1}^{n} \alpha_i K(x, x_i),$$

which can be arbitrarily nonlinear functions of x depending on the choice of the kernel K. A popular kernel is for example the Gaussian kernel $K(x, x') = \exp\left(-\gamma\| x - x' \|^2\right)$.

We conclude this short and nonexhaustive description of methods for supervised statistical learning by mentioning the Random Forest (RF) method proposed by Breiman (2001), which is increasingly popular in bioinformatics (Jensen and Bateman, 2011) for its good empirical performance on many different learning tasks. RFs are based on decision trees, which predict the label of a sample by asking a succession of questions with binary answers (such as *"is the expression of gene X larger than a threshold T?"*). A RF aggregates hundreds or thousands of simple classification trees, each built from random subsamples of $\mathcal{S}$ and restricted to asking questions among a random subset of questions.

6.2.5   Estimating performance and setting parameters

It is useful to be able to estimate the *generalisation performance* of a predictive model $f : \mathcal{X} \to \mathcal{Y}$, *i.e.* how well it will predict the response of future, unknown samples. Given a test set of samples, with observed responses, several metrics can be used to assess how good the predictions made by f are (see **Box 6.1**). However, whatever the metric used, when a predictor is trained using information on some samples, it is important to assess its performance on completely independent test data in order to consistently estimate its generalisation ability.

Several strategies exist to assess the performance of a predictive model from a set $\mathcal{S}$ of observed samples with their responses (MacLachlan, 1992; Molinaro et al., 2005). A popular one is to split the set of annotated samples $\mathcal{S}$ into two non-overlapping subsets $\mathcal{S}_{train}$ and $\mathcal{S}_{test}$, to train a model $\hat{f}$ using only the samples in $\mathcal{S}_{train}$, and to assess its performance on $\mathcal{S}_{test}$ using any performance metric. Since the result depends on the particular train/test split performed, it is recommended to repeat this procedure many times in order to estimate not only the expected level of performance, but also its sensitivity to the training set (usually measured by the average and standard deviation of the performance on the test set across many train/test splits). k-fold cross-validation is a particular strategy to create k train/test splits, by randomly splitting the full set $\mathcal{S}$ into k non-overlapping subsets $\mathcal{S}_1, \ldots, \mathcal{S}_k$, and consider in turn $\mathcal{S}_{test} = \mathcal{S}_i$ and $\mathcal{S}_{train} = \mathcal{S}\backslash\mathcal{S}_i$ for $i = 1, \ldots, k$.

Estimating the generalisation performance of a predictor is not only useful *per se*, but also to compare different predictors. In particular, for machine learning which depends on one or several parameters (*e.g.* the λ parameters in

Equation 6.5 for LR and SVM), it is often recommended to set the parameters as the ones with maximum estimated generalisation performance over a grid of candidate parameter values.

6.2.6   Application: Breast cancer prognosis from gene expression data

Let us illustrate the use of the supervised machine learning technique on the Wang breast cancer dataset, which we used several times already in the previous chapter (see **Chapter 5**). It contains genome-wide gene expression data for 286 early-stage breast cancers, together with follow-up and relapse information. Among the 286 patients included in the study, 93 showed evidence of relapse within 5 years, while 183 lived at least 5 years without relapse. The remaining 10 patients had no evidence of relapse, but either were followed less

---

❏ **BOX 6.1: How good is my predictor?**
In order to assess how good a binary predictor $f : \mathcal{X} \to \{-1, 1\}$ is, we can consider a test set of examples $x_1, \ldots, x_N$ (not used to infer the predictor f) and compare the predictions of f to the true binary labels of the test cases to compute the number of good positive and negative predictions, respectively called true positives (TP) and true negatives (TN), and the number of wrong positive and negative predictions, respectively called false positives (FP) and false negatives (FN). Note that $TP + TN + FP + FN = N$. We can compute different measures of performance:

- Accuracy: $ACC = (TP + TN)/N$,
- True positive rate (sensitivity, recall, hit rate): $TPR = TP/(TP + FN)$
- False positive rate (fall-out): $FPR = FP/(FP + TN)$
- Positive predictive value (precision) : $PPV = TP/(TP + FP)$,
- True negative rate (specificity): $SPE = 1 - FPR$

When the predictor outputs a continuous score, typically an estimate of the probability of the label being 1, we can furthermore construct the receiver operating characteristic (ROC) curve which plots TPR as a function of FPR when we vary the thresholds above which a label 1 is predicted, and summarise how good the ROC curve is by the area under it (AUC). AUC ranges usually between 0.5 for a random prediction to 1 for a perfect prediction (see for example **Figure 6.3**).

than 5 years, or died within 5 years; we remove them from the subsequent analysis.

The exploratory analysis of the Wang dataset (see **Chapter 5**) highlighted the heterogeneity of breast cancers, and suggested that at least 4 well-defined subtypes can be defined (see **Table 5.1**). Interestingly these different subtypes have different average risks of relapse; on the Wang dataset, we can estimate that the risk of relapse within 5 years increases from luminal A (19%) to HER2+ (34%), basal-like (35%) and luminal B (40%) subtypes. Although the precise number varies between studies, this observation has been recognised many times, in particular the important risk difference between luminal A and B subtypes (Sørlie et al., 2001). This suggests a first prognostication strategy: given a new breast cancer sample, map it to one of the 4 subtypes and predict its risk of relapse as the frequency of observed relapses in the subtype. Using the *genefu* R package (Haibe-Kains et al., 2011) to predict the subtype of any sample of the Wang dataset from its expression profile, we can estimate the performance of this prognostication in terms of ROC curve in 5-fold cross-validation. As shown on **Figure 6.3**, the ROC curve is clearly above diagonal, with an AUC of 0.63 ± 0.09, confirming that the subtype is a prognostic factor.

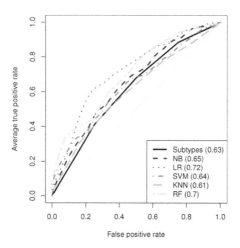

FIGURE 6.3 Breast cancer prognosis performance. This shows the ROC curves and AUC (in parenthesis) reached by different classification methods to predict 5-year metastasis in the Wang breast cancer dataset, in 5-folds cross-validation.

As an alternative to perform an unsupervised exploratory analysis of the data in order to define subtypes, and checking *a posteriori* that different subtypes have different prognoses, one can instead implement a supervised statistical learning strategy: given a training set of samples with known relapse information, train a machine learning model to predict the risk of relapse from gene expression data. Any of the methods discussed in **Section 6.2.2**–

Section 6.2.4 can be used for that purpose. For illustration, **Figure 6.3** shows the ROC curves reached in 5-fold cross-validation (repeated twice) by the generative model Naive Bayes (NB), the regularised linear discriminative models LR and SVM, and the nonlinear models k-NN and RF, trained on the $1,000$ most varying genes. We observe that most models outperform the subtype-based prognostic, with for example the regularised LR reaching an AUC of 0.72 ± 0.06. This is significant improvement over the simple subtype-based prediction, confirming the superiority of supervised statistical learning strategy when the goal is to build a predictive model from examples.

This simple example also shows that, even though the performance of the different methods is in the same range, there can be significant differences. We observe in our case that regularised LR and RF are the best methods, but the literature is not extremely consistent in this regard. For example, Dudoit et al. (2002a) compared nine methods on three cancer type and subtype prediction problems, and concluded that simple methods like Linear Discriminant Analysis (LDA) and k-NN performed remarkably well. Lee et al. (2005) found that SVM was often the best method, while Haury et al. (2011) pointed out the good performance of the Nearest Centroid (NC) method. Overall there is no single method which outperforms all others, and the relative performance of different methods depends on many factors including the data analysed, the number of samples and of features, and the experience of the programmer. For any particular application, it is therefore generally a good idea to test and compare a few representative methods before choosing one in particular.

6.3   Biomarker discovery and molecular signatures

The predictive modelling framework discussed in **Section 6.2** assumes that we have already chosen a representation of all sample vectors of p features, *i.e.* that we have decided which markers to use to build a predictive model. These markers may be anything from classical clinicopathological parameters (as used by Adjuvant! Online) to gene expression levels or DNA copy numbers, as long as we can measure them on a patient for whom a prediction is needed. In an era where we can easily measure thousands to millions of parameters on a biological sample with various high-throughput technology, the question of which features to use to design predictive models can be crucial. Indeed, as many of these parameters are likely to be irrelevant for the inference task considered, it may be a good idea to *select* only a few of them among the millions of candidates in order to design a predictive model, for at least three reasons:

1. From a statistical viewpoint, inferring a predictive model with many parameters from a limited number of observed samples is difficult and

may lead to poor models in terms of ability to predict the phenotype of interest on future samples. Reducing the number of parameters to describe each sample is a way to limit this difficulty, often referred to as the *curse of dimensionality*, and may lead to more accurate models (Donoho et al., 2000).

2. Reducing the number of markers used in a predictive model allows the design of dedicated devices for *cheaper and faster* prediction. It is for example still easier to measure the expression of a few genes than to completely sequence an individual nowadays.

3. Finally, predictive models based on only a few markers can suggest *biological interpretation*, and potentially lead to better understanding of the molecular underpinnings of prognosis or response to a treatment.

There is therefore an incentive to select a subset $M \subset [1, p]$ of size $q < p$, which we call a *signature* (see **Box 5.1**), to train a predictive model. Mathematically, the selection of a signature in the context of supervised statistical inference is called a problem of *feature selection*, which has deserved much attention in the statistics and machine learning communities (Guyon and Elisseeff, 2003). In this section, we briefly review some of the most popular feature selection methods and discuss the challenges that remain to be solved in the context of biomarker discovery from high-dimensional genomic data.

6.3.1   Feature ranking by univariate filter methods

The simplest way to select a signature M of q features among p candidates is to compute a score for each candidate, which assesses how relevant the candidate is for the response to predict, and to select the q candidate features with the largest scores to form the signature. Various scores have been proposed, depending particularly on the type of the response to be predicted (categorial or continuous).

In the case of binary classification where we want to predict a binary response, *i.e.* $\mathcal{Y} = \{-1, 1\}$, the score of a feature typically measures how differentially distributed the feature is between the two subpopulations of samples $\mathcal{S}_{-1}$ and $\mathcal{S}_1$, of respective sizes n_{-1} and n_1, with different responses. While a simple procedure is to simply compute for each descriptor the difference between its mean on both populations $\bar{\Delta} = |\bar{\mu}_1 - \bar{\mu}_{-1}|$, such as the Fold Change (FC) of each gene when we compare two populations by gene expression data, this is usually not a good idea since it does not correct for the variability of the score which may differ between different features (Allison et al., 2006). Instead, it is recommended to use normalised scores of the form

$$t = \frac{\bar{\Delta}}{S}, \tag{6.8}$$

where S is the variance of $\bar{\Delta}$. The classical estimate of S is

$$S_1 = \bar{\sigma} \sqrt{\frac{1}{n_1} + \frac{1}{n_{-1}}},$$

where $\bar{\sigma}$ is an estimate of the class-conditional variance of the feature. In that case **Equation 6.8** boils down to a classical Student's t-statistics, which is directly related to the F-test used in ANOVA modelling (Kerr and Churchill, 2001). Since the simultaneous estimation of S for many candidate markers from a limited number of samples is a difficult statistical task, many methods have been proposed to borrow information across features by shrinking the estimated variance S_t towards a predicted variance S_0 through the combination $S = \alpha S_t + (1 - \alpha) S_0$. The predicted variance S_0 is typically estimated from all markers jointly, and the coefficient α differs between methods such as Significance Analysis of Microarray (SAM) (Tusher et al., 2001), regularised t-tests (Dobbin et al., 2003), the randomised variance model (RVM, Wright and Simon, 2003), or limma (Smyth, 2004), to name just a few. Alternatively, nonparametric statistics such as the AUC (see **Box 6.1**) reached by a feature when used alone to predict the phenotype (which is equivalent to the U-statistics of the Mann-Whitney test) can be used, as well as other measures of distances between conditional distributions (Guyon and Elisseeff, 2003).

When the output is continuous, *i.e.* $\mathcal{Y} = \mathbb{R}$, a simple ranking score is the absolute value of Pearson's correlation coefficient, given for the i-th feature by

$$R(i) = \frac{\sum_{k=1}^{n} (x_{i,k} - \bar{x}_i)(y_k - \bar{y})}{\sqrt{\sum_{k=1}^{n} (x_{i,k} - \bar{x}_i)^2 \sum_{k=1}^{n} (y_k - \bar{y})^2}}, \tag{6.9}$$

where $\bar{x}_i = \left(\sum_{i=1}^{n} x_{i,k}\right)/n$ and $\bar{y} = \left(\sum_{i=1}^{n} y_i\right)/n$. Since Pearson's correlation is restricted to detecting linear relationships between the feature and the output, it is sometimes advised to use instead Spearman's correlation obtained by replacing the $x_{i,k}$ values by their rank (from 1 to n) in **Equation 6.9**. Since it is only based on the relative values between the feature value on different samples, and not on their absolute values, Spearman's correlation can capture more general monotonic relationships between the feature and the response than Pearson's. In cases where a more complex relationship is expected, one can also consider estimating the mutual information between the feature and the output, defined by

$$I(i) = \int_{x_i} \int_y p(x_i, y) \log \frac{p(x_i, y)}{p(x_i)p(y)} dx dy,$$

where $p(x_i)$, $p(y)$ and $p(x_i, y)$ are respectively the densities of x_i, y, and the joint density of (x_i, y). Estimating these densities needed empirically is however often not obvious, and again many methods have been proposed (*e.g.* Battiti, 1994; Torkkola, 2003)

6.3.2 Feature subset selection by wrapper methods

Univariate filter methods discussed in the previous section attempt to detect features related in some way to the response of interest. It is, however, not always a good idea to learn a predictive model using the top q features according to this criterion, particularly in the presence of correlations among features. Indeed, once a first feature is selected to enter a predictive model, adding a second feature strongly correlated to the first one can be less interesting than adding a third feature that adds complementary information to the first one, even though its univariate association with the response is lower than that of the second feature. In other words, instead of testing the features one by one and then taking the top q among them to form a signature, it seems more interesting to directly attempt to select a subset of k features which, together, allow the inference of a good model. This is exactly what *feature subset selection* methods try to achieve.

Feature subset selection is usually phrased in terms of a particular inference algorithm, able to estimate a model $f : \mathcal{X} \to \mathcal{Y}$ from a training set of samples $\mathcal{S}$ (see **Section 6.2**). By varying the set $M \subset [1,p]$ of features to be used in the model, the learning algorithm can estimate different functions which we denote by $f_M : \mathcal{X} \to \mathcal{Y}$, to insist on their dependencies on the subset of features used. In order to compare different sets of features in the context of predictive modelling, we need to be able to estimate how "good" the functions f_M are. Although the prediction accuracy on future samples cannot be estimated directly, different proxies for it are commonly used, including the performance on a validation set of observations left aside during inference, or the value of the objective function on the training set when the inference algorithm minimises an objective function. Denoting by $R(f_M)$ this estimated measure of goodness for a model estimated on the features in M, the best subset selection problem consists in finding the subset M which leads to the "best" model:

$$\hat{M} = \underset{M \subset [1,p]}{\mathrm{argmin}}\ R(f_M)\,. \tag{6.10}$$

Unfortunately, the number of candidate subsets with q features increases exponentially with q, and even with the efficient *leaps and bounds* algorithm (Furnival and Wilson, 1974), it is in general computationally impossible to exhaustively search over subsets of more than a few features when p is larger than 40. For example, there are more than 10^{12} ways to select only three genes among $20,000$ candidate genes.

Instead of solving **Equation 6.10** exactly, wrapper approaches usually follow greedy optimisation procedures which find a "good" subset, but not the best one. They start from an initial candidate set of features M_0, which could be the empty set or the full set of features, and iteratively add or remove features that maximally decrease $R(f_M)$, until it cannot improve any more by simple addition or removal of features. Starting from no feature and iteratively adding them one by one is referred to as *forward stepwise selection*, while starting from the full set of features and iteratively removing them is called *backward stepwise selection.* For example, in bioinformatics, a popular backward stepwise selection algorithm is the SVM recursive feature elimination (RFE) method which starts from all features and iteratively removes features with small weights estimated by a linear SVM (Guyon et al., 2002). Other more complex schemes which alternate between feature addition and deletion are obviously also possible, although the quest for more efficient methods to find solutions closer to the best subset is questionable since the risk to overfit M is difficult to control when we compare so many candidate subsets.

6.3.3   Embedded feature selection methods

Instead of using the statistical learning method only to assess the goodness of a candidate signature M, as in wrapper methods, some machine learning algorithms can directly both estimate a function $f : \mathcal{X} \to \mathcal{Y}$ and select features. This is in particular the case of methods for *sparse* learning, which estimate functions $f(x)$ which only depend on x through a limited number of features. For example, most linear methods estimated by minimising a regularised empirical risk (see **Equation 6.5**) can be turned into sparse inference models when the Euclidean norm $\| w \|$ is replaced by the so-called ℓ_1 norm as a penalty:

$$\| w \|_1 = \sum_{i=1}^{p} | w_i | .$$

This results in the following family of sparse linear discriminative methods:

$$\hat{w} = \underset{w \in \mathbb{R}^p}{\operatorname{argmin}} \; \frac{1}{n} \sum_{i=1}^{n} \ell(f_w(x_i), y_i) + \lambda \| w \|_1 , \qquad (6.11)$$

where ℓ is a loss function such as the logistic loss function (see **Equation 6.6**) or the squared error. The most famous example of sparse linear method is the lasso (Tibshirani, 1996; Chen et al., 1998), when the square loss $l(t, y) = (t - y)^2$ is used in **Equation 6.11**. Other variants using different loss functions in **Equation 6.11** include sparse logistic regression for categorial response data (Roth, 2004; Shevade and Keerthi, 2003) or sparse Cox regression for survival data (Tibshirani, 1997). Adding the ℓ_2 norm of w as a second additional penalty in **Equation 6.11** leads to the elastic net (Zou and Hastie, 2005), which can be more robust than the lasso in case of high correlations between features.

6.3.4   How many features should we select?

All feature selection methods discussed in **Section 6.3.1**–**Section 6.3.3** allow the selection of a signature of q markers among the initial p candidate features. The size of the signature is usually implicitly or explicitly controlled by the user. For example, the filter methods discussed in **Section 6.3.1** rank the features from the most interesting to the least interesting in terms of association with the response, and a signature of q markers is obtained by taking the top q features in this list. The wrapper methods discussed in **Section 6.3.2** can explicitly optimise greedily a set M of q features, where q is user-defined, and the embedded methods (see **Section 6.3.3**) control q by some other user-defined parameter, such as the regularisation parameter λ in sparse linear discriminative methods (see **Equation 6.11**).

How should the user choose q, the size of the signature? There are roughly two main types of procedures used to choose q, motivated by two different objectives: the *discovery* of markers significantly associated with the phenotype, on the one hand, and the choice of a subset of descriptors that will lead to optimal *prediction* of the response of interest, on the other hand.

In the case of marker discovery, we want to select all features which are truly associated with the response, independently of their ability to form a good predictor. q should then be the number of significantly associated features, and its estimation is related to hypothesis testing in statistics. Indeed, for each feature j, we need to test the hypothesis H_j that it is not statistically associated with the response, and q will be the total number of rejected hypotheses (corresponding to features associated with the response). When a single feature is tested, classical statistical procedures can be used to test its association with the design. In the case of binary response, for example, a two-sample t-test allows to control the probability of false positives (the so-called type I errors) at a desired level, typically 5%. When many features are tested simultaneously, one should, however, perform a correction for multiple testing in order to control the number of false positives, otherwise we will strongly overestimate the number of markers significantly associated with the response (see **Box 6.2**).

In the case of predictive modelling, the goal is to choose a signature M that leads to the most accurate model. If the size of the signature is too small, the predictive model may be too poor to correctly estimate the response of interest from the markers selected. But if it is too large, it may lead to statistical difficulties to estimate a correct model, potentially leading to sub-optimal performance. The size of the optimal signature is therefore often a trade-off between these two alternatives. In practice, it is common to estimate the performance of the predictive model for different values of q, using the techniques discussed in **Section 6.2.5**, and select the size q which has the best estimated performance. For example, **Figure 6.4** shows the performance of regularised LR to predict relapse on the Wang dataset, as a function of signature size. To select genes in the signature, we compare a random selection of genes to a selection by decreasing significance according to a two-sample t-test performed on the training set. The performance is the mean AUC reached in 5-fold cross-validation repeated 10 times. We see that, surprisingly, feature selection only degrades performance, and the best performance is reached when all genes are used (recovering the results already seen in **Section 6.2.6**). We also see that gene selection by t-test is better than random selection when we have to select few genes, although both strategies obviously converge to the same signature when we increase its size.

6.3.5   Molecular predictors in breast cancer

Many groups have applied the paradigm of supervised statistical inference coupled with feature selection in order to propose predictive models based on molecular signatures, particularly gene expression. In the particular case of breast cancer, some models have for example been proposed to predict the histological grade in breast cancer from gene expression data, defining a new

---

❑ BOX 6.2: Corrections for multiple testing

When a two-sample t-test is performed to assess whether a given gene is differentially expressed between two conditions, the evidence for differential expression is usually quantified by a p-value, the probability to reach the same level of significance only by chance. If the p-value is below a threshold, typically 5% the gene is considered significant, meaning that we control the risk of false positives (type I error) at 5%. When 10,000 genes are tested simultaneously with the same procedure, 500 genes are expected to have a p-value below 5% by chance alone. This represents too many false positives for many applications. It is therefore important to correct the individual p-values to account for multiple testing (Dudoit et al., 2002b; Slonim, 2002).

The Bonferroni correction multiplies the p-values of all genes by the total number of genes tested. If the adjusted p-value is still below the error rate, the gene is deemed significant. This correction allows to control the Family Wise Error Rate (FWER), *i.e.* the probability to make one or more false discoveries. For 10,000 genes, this means that a gene needs a p-value below 0.000005 to be considered differentially expressed. A less stringent correction to control the FWER is the Holm-Bonferroni correction, which ranks the genes by increasing p-values and multiplies each p-value by the number of genes with larger p-values (Holm, 1979). In genomics, it is often more interesting to allow false positives, but to control the False Discovery Rate (FDR), *i.e.* the proportion of false positives among selected genes (Benjamini and Hochberg, 1995). For p genes tested with individual p-values $P_1, \ldots, P_p$, the procedure of Benjamini and Hochberg (1995) to control the FDR below a level α is to order the p-values in increasing order $P_{(1)}, \ldots, P_{(p)}$ and select the largest k such that $P_{(k)} \leq k\alpha/p$. If the genes tested are strongly dependent, these procedures have, however, low power to detect true positives. The dependence between tests can be taken into account by using permutation-based strategies (Westfall and Young, 1993), such as SAM (Tusher et al., 2001).

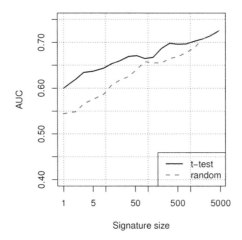

FIGURE 6.4 Influence of signature size on breast cancer prognosis performance. A regularised LR classifier using a signature of varying size is trained on the Wang expression dataset to predict relapse within 5 years. The genes in the signatures are selected either randomly, or by decreasing significance according to a *t*-test. The performance is estimated by 5-fold cross-validation, averaged over 10 repeats. In this example, it is better to keep all genes to train the classifier.

notion of *genomic grade* to quantify tumour differentiation (Sotiriou et al., 2003; Loi et al., 2007). In addition to tumour differentiation assessment, this genomic grade was shown to be prognostic. Several prognostic molecular predictors have also been proposed, including the 76-gene MammaPrint® signature developed at the Netherlands Cancer Institute in Amsterdam (van't Veer et al., 2002) and the 76-gene Rotterdam signature of Wang et al. (2005). Investigators from the University of Texas M. D. Anderson Cancer Center developed DLD30, a 30-gene signature to predict the response of a tumour to preoperative chemotherapies (Hess et al., 2006). The Oncotype DX® assay combines the expression of 21 genes to evaluate the risk of relapse and the benefits of chemotherapy for patients with early-stage, lymph node-negative, ER+/HER2- breast cancers (Paik et al., 2006; Paik, 2007). Several of these molecular predictors have reached the level of clinical trials, and are now being tested on large cohorts of patients. We can already foresee their routine use in the clinics within few years.

6.3.6   Pitfalls and challenges in biomarker discovery

Although an attractive strategy to improve the performance of predictive modelling in high-dimension and simultaneously identify biologically relevant markers, the automatic data-driven identification of new markers remains challenging for several reasons.

First, it is rarely the case that feature selection leads to drastic improvement in terms of prediction performance compared to methods using all p descriptors to train a model, at least when we consider a model designed to be able to learn in high-dimension such as SVM, RF or regularised LR. For example, we observed in **Section 6.3.4** (see **Figure 6.4**) that feature selection is only detrimental to the performance of regularised LR for relapse prediction in breast cancer from gene expression data, at least on the Wang dataset. The assumption that a model based on only a few markers should better capture the biological complexity of the prediction problem than a model based on all features should therefore be taken with caution. It is often not obvious that selecting markers from genomic data is the best way to infer the most accurate models.

Second, the trust we can have in the biological value of the selected markers, for example to suggest new therapeutic targets (see **Chapter 11**), should be taken with extreme caution. Indeed, the robust and consistent selection of good biomarkers is extremely challenging from a statistical viewpoint when the number of samples n is much smaller than the number of features p. This difficulty was quickly noticed in cancer genomics data analysis, when the first molecular signatures for operable early breast cancer prognosis were developed by feature selection on gene expression data, and different groups proposed independently lists of markers with barely any overlap. More precisely, only 3 genes are common in the now famous 70-gene signature of van't Veer et al. (2002) and the 76-gene signature of Wang et al. (2005). Although this intriguing lack of robustness could be attributed to differences in cohorts or technology, it was quickly understood that the major reason is purely statistical, and is related to the fact that many different sets of genes with little overlap can collectively have the same predictive power (Ein-Dor et al., 2005). Several studies analysed empirically (Michiels et al., 2005; Haury et al., 2011) or theoretically (Ein-Dor et al., 2006) the difficulty to select robustness of signatures, and overall concluded that unless we could gather a much larger number of samples than we currently have, it is an illusion to believe that we can robustly identify a unique set of a few markers to train good predictive models. An unfortunate consequence is that the automatic discovery of key markers which may, for example, lead to novel therapeutic targets as discussed in **Section 11.1**, remains out of reach nowadays.

Finally, let us mention a frequent methodological error in biomarker discovery, which can easily lead to overoptimistic opinions about the accuracy and stability of feature selection. When we want to estimate the predictive performance of a model involving feature selection by cross-validation of bootstrap procedures (see **Section 6.2.5**), it is important to carry out *both* feature selection and predictor learning on the training set only of each train/test split of the data, as we did to produce **Figure 6.4**. Selecting features using all data and then performing cross-validation to estimate the performance of a prediction algorithm by cross-validation overestimates the performance of the method because it does not account for the difficulty to select features, a problem known as *selection bias* (Ambroise and McLachlan, 2002).

6.4   Functional interpretation with group-level analysis

We saw in **Section 6.3** that much hope and effort has been devoted to the development of efficient methods for automatic biomarker identification and the inference of predictive models. However, the construction of robust and accurate models with direct biological interpretation remains challenging, in particular because of the relatively small number of samples analysed compared to the large number of candidate markers. Instead of expecting to overcome this challenge with a purely agnostic and data-driven approach, one can alternatively try to include prior knowledge about the candidate markers considered, such as the functions of the genes interrogated or their interactions, in order to drive the data-driven approach to more biologically relevant models — hopefully increasing their robustness and performance.

In this section we discuss different strategies to integrate in predictive modelling some prior knowledge we have about the existence of *groups* of features. We discussed a similar framework in **Section 5.5.1** in the case of unsupervised analysis. For example, we may know groups of genes sharing the same function when we analyse gene expression data, or groups of loci localised in the same genomic region when we analyse genotypes or DNA copy number profiles. Working at the level of groups can help provide a biological interpretation to the problem investigated, *e.g.* identify functional groups, **signalling pathways*** or gene regulatory network perturbed in a given prediction or predictive for a phenotype. Working at the group level may also be a way to reduce the dimension of the statistical problem and increase the robustness of signatures, in the sense that signatures with apparently different genes may in fact correspond to similar functions or pathways.

To fix notations, we therefore model the prior knowledge as a set of G groups $\mathcal{G} = \{g_1, \ldots, g_G\}$, where each $g_i \subset [1, p]$ is a subset of features. Note that in general, some groups may overlap, *e.g.* a gene may belong to several functional groups.

6.4.1   Detecting important groups in a signature

Once a small set of markers has been identified by any of the methods discussed in **Section 6.3**, it can be useful to compare them to the predefined groups $\mathcal{G}$ in order to capture overrepresented groups, resulting in potential biological interpretation of the signature. For example, when a short list of genes is selected from gene expression data analysis, it is helpful to detect which biological functions they collectively represent by assessing which functional groups are overrepresented in the signature. Another example of interesting grouping is the analysis of genes potentially regulated by the same transcription factors, whose binding sites can be predicted by computational approaches and experimentally characterised by technologies such as ChIP-on-chip and ChIP-seq.

Different statistical tests can be performed to assess the evidence that a group $g \in \mathcal{G}$ is overrepresented in the signature (see **Box 6.3**). When many groups are tested simultaneously, a correction for multiple testing is required, which is not without difficulty when groups overlap because the different tests are dependent and the correlation structure between groups may bias the significance of tests (Tian et al., 2005). Good discussions of such marker list analysis, its variants and limitations, are available in Khatri and Drăghici (2005); Rivals et al. (2007).

❏ **BOX 6.3: Assessing the overrepresentation of a gene group in a signature**

Suppose that the signature contains q markers out of a total of p possible features. Let now a predefined group g of k features, out of which $r \leq k$ are in the signature. Under the null hypothesis that the k features of the group are drawn randomly out of the p total features, the probability to find r of them in the signature follows the hypergeometric distribution:

$$P_{hyper}(r) = \frac{\binom{p}{q}\binom{k}{r}}{\binom{p+k}{q+r}},$$

where $\binom{p}{q} = \frac{p!}{q!(p-q)!}$ is the binomial coefficient. In order to assess the evidence that r is larger than what would be expected by chance, the *one-tailed exact Fisher test* computes the probability under the null hypothesis that the number of features from the group selected in the signature is at least r, *i.e.*

$$score_{Fisher}(g) = \sum_{k=r}^{p} P_{hyper}(k).$$

In practice, the hypergeometric distribution may be computationally difficult to compute when p is large, and it can be approximated by the simpler binomial distribution in this setting. Alternatively, a simple χ^2 test for equality of proportions (q/p versus r/k) can be performed. In most cases, the differences between the models will not be dramatic.

Using this strategy, Rhodes et al. (2005a) assess which transcription factors are significantly enriched in the promoter regions of genes in molecular signatures stored in the Oncomine database. They observe for example that genes regulated by the archetypal cancer transcription factor, E2F, are disproportionately overexpressed in a wide variety of cancers, whereas other transcription factors such as Myc-Max have this property only in specific cancers.

6.4.2 Biomarker group analysis

The signature analysis discussed in **Section 6.4.1** has several shortcomings. First, in order to detect an overrepresented group in the signature, enough individual markers of the group should be in the signature, *i.e.* should be individually strongly predictive for the phenotype of interest to pass the individual selection in the signature. This rules out the possibility to capture modest but consistent changes of markers within a group. Second, the signature is considered as an unordered set of markers, although many signatures construction methods score or rank the individual markers within the signature (see **Section 6.3**). Taking into account the order of markers in the signature may increase the power of the group detection method. Typically a group of markers among the strongest in the signature should be more rewarded than a group of markers of the same size among the weakest of the signature. This is particularly important when large signatures are considered.

An alternative to overcome these limitations is to bypass the construction of a signature and directly assess the importance of a group $g \in \mathcal{G}$ from a scoring or ranking of *all* candidate markers. Such a scoring or ranking can be obtained by various methods for signature inference (see **Section 6.3**), *e.g.* by statistical tests to assess the association of each marker with the phenotype of interest. Instead of selecting a signature from this list by choosing the top q markers in the ranking (see **Section 6.3.4**) and then analysing each group $g \in \mathcal{G}$ in light of these top q markers (see **Section 6.4.1**), we discuss in this section, methods which directly score a group g from the scores or rankings of its members. By inspecting them, one can typically assess whether the markers in g tend to be ranked towards the top of all markers — although not necessarily in the top q ones. This can allow, in particular, to detect small but consistent changes between two conditions of groups of related markers.

A simple implementation of this approach was proposed by Pavlidis et al. (2002a) when a p-value $P(i)$ is computed for each feature $i \in [1, p]$. A group of markers $g \in \mathcal{G}$ is then scored by the average of the negative log P values of the markers within the group:

$$score_P(g) = -\frac{1}{|g|} \sum_{i \in g} \log P(i).$$

The significance of the corresponding score can then be assessed empirically by comparing it to the scores obtained on random groups of genes of the same size.

Another very popular method is the Gene Set Enrichment Analysis (GSEA) tool (Subramanian et al., 2005). GSEA focuses mainly on the identification of known functionally related groups of genes tested at the level of expression for different conditions or phenotypes. GSEA starts by ranking all genes from the most to the least differentially expressed ones, using a statistical score like those discussed in **Section 6.3.1**[1]. For each predefined set of genes $g \in \mathcal{G}$, a one-sided Kolmogorov-Smirnov statistic is computed to assess whether the ranks of set genes are significantly concentrated towards the top (or the bottom of the list). The significance of this statistic, called the Enrichment Score (ES), is then assessed by repeating the same process after random permutation of the sample labels and comparing the ES obtained for the top-ranking gene sets with those of the top-ranking gene sets after random permutation. Alternatively, variants have been proposed to better correct for correlations in the expression matrix (Tian et al., 2005). The power of GSEA compared to classical signature analysis (see **Section 6.4.1**) was demonstrated in particular in Mootha et al. (2003), who found that genes involved in oxidative phosphorylation are coordinately downregulated in diabetic muscle while no individual gene shows significant deregulation.

6.4.3 Discriminative group analysis

Instead of scoring a group $g \in \mathcal{G}$ in terms of the scores or the ranking of the individual markers in the group, it is sometimes more relevant to come back to the original data and assess the relevance of a group by how well it can contribute to discriminating the phenotypes investigated, typically by combining together its markers in a predictive model. This is particularly important when individual markers are not or barely predictive for the response of interest, while combinations of markers are.

For example, one may assume that only a subset of the features in a group g should be used to discriminate between the phenotypes, leading to the notion proposed by Lee et al. (2008) of Condition-Responsive Genes (CORG) in the context of gene expression analysis. The CORG of a group g is defined as the subset of genes in the group whose mean expression delivers optimal discriminative power for the disease phenotype, measured with a t-test statistic. A computational difficulty in this definition is that there is an exponential number $2^{|g|} - 1$ of subsets for a group $|g|$. Instead of finding the best subset, Lee et al. (2008) therefore propose a greedy search strategy starting from the gene most associated with a phenotype (using a t-test), and adding one by one other genes for as long as the association of the group activity (defined as the mean activity of the selected genes) and the phenotype increases.

An alternative to the selection of a subset of markers within a group, and

---

[1]The original score used in GSEA is the signal to noise ratio (SNR), defined as the difference in means of the two classes divided by the sum of the standard deviations of the two classes (Subramanian et al., 2005). Other scoring such as nonparametric statistics has also been proposed (Bayá et al., 2007).

the scoring of the group based on the association of the selected markers, is to directly assess the discriminatory power of each group by estimating the classification error of a classifier restricted to the markers in each group. In practice, any method for supervised classification may be used to learn a discriminative function from a subset of the markers (see **Section 6.2**), and the discriminatory power can be estimated by resampling and cross-validation strategies (see **Section 6.2.5**). For example, Pavlidis et al. (2002a) define the *learnability* of a gene set as the leave-one-out error or a 1-nearest neighbour classifier restricted to the genes in the set. We note that the definition of CORG may also be interpreted in this framework, where the predictive model is defined as the mean level of a subset of markers to be optimised.

6.4.4 Discriminative group modelling

All methods discussed in **Section 6.4.1**–**Section 6.4.3** score all groups $g \in \mathcal{G}$ independently from each other. It is then possible to rank the groups and identify the important groups by looking at the top of the list. Although this may be interesting to provide a biological interpretation of the discrimination between phenotypes or of a discriminative signature, it does not directly influence the predictive model estimated from individual markers, as discussed in **Section 6.2** and **Section 6.3**. In particular, no improvement in prediction performance can be expected from such analysis.

In order to impact the classifier by exploiting the prior knowledge encoded in the groups $\mathcal{G}$, one must go back to the predictive model construction step (see **Section 6.2**) and consider integrating the set of groups $\mathcal{G}$ in the inference process itself. For example, one can define new features to summarise each group, such as the mean activity of its members or of its CORG. This can lead to a decrease in the number of features, from p to G.

Alternatively, one may stay at the level of individual markers, but use the group information to drive the model inference towards models "coherent" with the group structure. For example, if the groups form a partition of the set of all markers, *i.e.* if all markers $m \in [1, p]$ are in one and exactly one group $g \in \mathcal{G}$, one can generalise the sparse linear discriminative methods **Equation 6.11** by the following *group-sparse* linear discrimination:

$$\hat{w} = \operatorname*{argmin}_{w \in \mathbb{R}^p} \frac{1}{n} \sum_{i=1}^{n} \ell(f_w(x_i), y_i) + \lambda \sum_{g \in \mathcal{G}} \| w_g \|_2 , \qquad (6.12)$$

where w_g denoted the $|g|$-dimension restriction of w to the features in group g. The penalty $\| w \|_{1,2} = \sum_{g \in \mathcal{G}} \| w_g \|_2$ is often referred to as the $\ell_{1,2}$ norm, or the *group lasso* penalty (Yuan and Lin, 2006). It boils down to the classical ℓ_1 norm where all groups are singletons, resulting in sparse models involving only a subset of markers when **Equation 6.12** is solved (see **Section 6.3.3**). For general groups $\mathcal{G}$, one can show that solving **Equation 6.12** again results in sparse models, but at the level of groups: the weights of individual markers within a group are shrunken to zero together, and the resulting selected markers with nonzero weights also form groups. **Equation 6.12** is therefore an embedded feature selection that forces the selection of markers by group, resulting directly in easily interpretable signatures.

This is, however, only true when the groups in $\mathcal{G}$ form a partition of all markers, which is rarely the case in practice where gene sets often can overlap. For example, many genes are likely to belong simultaneously to many functional groups. If we solve **Equation 6.12** for a set of groups $\mathcal{G}$ which is not a partition of $[1, p]$, then the markers will again be shrunken to zero by groups, but consequently the selected markers will not form groups anymore (Jenatton et al., 2011). In order to solve this issue, Jacob et al. (2009); Obozinski et al. (2011) proposed a variant called the *latent group lasso* which is equivalent to the group lasso (see **Equation 6.12**) when the groups form a partition, and leads to the selection of markers which form unions of groups even when the groups overlap.

6.5   Network-level analysis

Throughout **Section 6.4** we discussed several analytical approaches which can be followed to investigate a supervised inference problem with a large number p of features, when in addition a set of groups of features $\mathcal{G}$ is available as prior knowledge. We saw in particular how the knowledge of groups in $\mathcal{G}$ can help in the biological interpretation of difference between the conditions or responses considered, and potentially lead to more accurate predictors by reducing the statistical complexity of the inference problem.

In this section we investigate similar questions when, instead of groups of features, we want to exploit as prior knowledge a *graph of features* (see **Box 6.4** for graph formalism and terminology). This goal is particularly relevant in the context of systems biology, which describes many complex relationships between molecules and biological processes as networks, including for example Protein–Protein Interaction (PPI) networks, coexpression and regulatory networks, or signalling and metabolic pathways (see **Box 4.3**). It has for example been observed that genes associated with similar disorders tend to have physical interactions between their products, suggesting the presence of "hot spots" of proteins on the PPI network (Ideker and Sharan, 2008). Analysing cancer gene expression data in light of PPI network may therefore not only reduce the complexity and instability of the task by using a specific prior knowledge, but may also point out possible interacting complexes or pathways related to the network dynamics of the disease

Since such networks are increasingly deciphered experimentally or computationally, and easily available through various databases (see **Section 4.5.1**), it is not surprising that much work recently has been devoted to the question

---

❏ **BOX 6.4: Graph formalism and terminology**

A *graph* is a mathematical representation of a network, *i.e.* a set of objects (*nodes* or *vertices*) where some pairs of the objects are connected by links (*edges*). Formally, a graph is represented by pair $G = (V, E)$ where V is the set of vertices and E the set of edges. Each edge $e \in E$ has two *endpoints*, and is said to *connect* or *join* the two endpoints. A *directed graph* or *digraph* is a graph whose edges are ordered pairs of vertices, called the *head* and *tail* of the edge. It differs from an ordinary *undirected graph*, whose edges are unordered pairs of vertices. The *order* of a graph $|V|$ is the number of vertices, while the *size* of a graph $|E|$ is the number of edges. The *degree* of a vertex is the number of edges that connect to it.

In a directed graph, if a is the head of an edge with tail b, a is said to be a *direct predecessor* of b, and b a *direct successor* of b. A *path* is a sequence of vertices such that from each of its vertices there is an edge to the next vertex in the sequence. If a path leads from x to y, then y is said to be a *successor* of x and *reachable* from x, and x is said to be a *predecessor* of y.

---

of how these networks can help in the analysis of omics data, in particular in the context of predictive modelling. In this section, we review several strategies to perform such analysis, parallelising the organisation of the previous section which focused on the use of groups instead of networks. To fix notations, we assume as usual that each sample is characterised by p features (*e.g.* the expression level of p genes), and that we know in addition an undirected network with vertex set $\mathcal{V} = [1, p]$ (*i.e.* the vertices of the graph are the features) and a set of edges $\mathcal{E} \subset \mathcal{V} \times \mathcal{V}$ (see **Box 6.4**). Some networks like influence networks or gene regulation networks are naturally captured as a directed, valued graph, but they will not be considered here.

6.5.1   Network-level analysis of a signature

Following a differential or supervised inference analysis, it is common to obtain a signature containing a small set of features selected because they are different between the conditions investigated, or because they allow a good prediction of a response of interest. Just like detecting overrepresented groups in a signature allows to capture a biological meaning for the signature when a set of groups of features is available (see **Section 6.4.1**), it can be interesting to study a network-level interpretation of the signature when a network is given. If for example, many features of the signature are near each other on the graph, one would like to automatically detect and quantify this nonrandom localisation and express it in terms of subnetworks or pathways.

A common approach to perform such an analysis is just to navigate on the network using various software for network visualisation and analysis, such as Ingenuity's IPA, GeneGo, Ariadne Genomics' Pathway Studio. For example, Rhodes et al. (2005b) maps a set of genes of interest to the large PPI network from the Human Protein Reference Database (HPRD), in order to identify deregulated subnetworks and key players within our outside of the signature.

Once the elements of a signature $M \subset \mathcal{V}$ are mapped to a network, there exist different ways to automatise the network-level analysis of the signature. For example, Franke et al. (2006) proposed to automatically identify important vertices and prioritise them by computing for each vertex $v \in \mathcal{V}$ the "density" of vertices of the signature in the neighbourhood through, *e.g.* a kernel density estimator of the form

$$f(v) = \sum_{i \in M} e^{-\gamma d(v,i)} \,,$$

where $d(v,i)$ represents the shortest-path distance between vertices v and i on the network. Vertices can then be ranked by decreasing densities in order to focus on the ones surrounded by many signature elements. Many other formulations have also been proposed, such as Karni et al. (2009) who reformulates the search for "central" genes as a combinatorial optimisation problem that aims at finding a small number of nodes in the network which are within a small distance of all genes in the signature. Scott et al. (2005) propose to find a small set of vertices $\mathcal{A} \subset \mathcal{V}$ to add to the genes in the signature in order to obtain a connected subgraph which minimises:

$$D(\mathcal{A}) = \sum_{v \in \mathcal{A}} -\log(1 - p_v) \,, \tag{6.13}$$

where p_v is a p-value of vertex v to characterise how different it is between the two conditions. Intuitively, $D(\mathcal{A})$ is small when genes in $\mathcal{A}$ have small p-values, *i.e.* tend to be differentially expressed. The search for the best $\mathcal{A}$ is an instance of an NP-hard problem called *node-weighted Steiner tree* problem, for which no efficient algorithm is known. While the optimal subset $\mathcal{A}$ can be found when the size of the signature is small (typically less than 10), only an approximate solution can be found for larger signatures. Note that a variant of this approach is to try to connect all genes of the signature with as few additional nodes as possible, *i.e.* to minimise $D_1(\mathcal{A}) = |\mathcal{A}|$. Scott et al. (2005) however, claim that taking into account the additional nodes' p-values through (see **Equation 6.13**) is more promising to detect biologically relevant subnetworks such as regulatory subnetworks.

6.5.2   Finding differential subnetworks

Just like gene set analysis can be more sensitive than gene list analysis to detect small but consistent variations in gene expression (see **Section 6.4**), one may inspect globally which subnetwork seems to exhibit coordinated variations instead of first selecting a short list of interesting genes and mapping it on to the network. This idea was pioneered by Ideker et al. (2002) who assigned a statistical Z-score z_g to each gene g using expression data and searched for subnetworks that display a statistically significant amount of differential expression. This is formally expressed as trying to detect subnetworks $\mathcal{A}$ with large combined Z-scores defined by

$$z_{\mathcal{A}} = \frac{1}{\sqrt{|\mathcal{A}|}} \sum_{g \in \mathcal{A}} z_g. \qquad (6.14)$$

Finding the networks which maximise the combined Z-score is, however, an instance of the NP-hard problem called the *maximum weight connected subgraph problem*, and only approximative optimisation methods such as simulated annealing (Ideker et al., 2002) can be applied to find candidate subnetworks with large combined score. Such subnetworks may correspond to small sets of proteins participating in a common complex or metabolic pathway, for instance, and may contain proteins with no significant individual differential expression if they lie on a path that connects sets of differentially expressed genes.

This method has been used and extended by many authors (Nacu et al., 2007; Cabusora et al., 2005). For example, Rajagopalan and Agarwal (2005) propose a variant of the scoring function and of the optimisation method. Patil and Nielsen (2005) apply it to a metabolic network, scoring both vertices (when a differential expression score can be computed) and edges (when correlation between gene expressions is computed). Liu et al. (2007) use the method to identify differentially expressed subnetworks, and then test for association of these subnetworks with predefined gene groups in the so-called Gene Set Enrichment Analysis (GNEA). This allows them to find interesting groups in diabetes. Similarly, Sohler et al. (2004) proposes a greedy heuristic method (*significant area search*) to identify subnetworks that are significant according to specified p-values, where individual p-values are combined using Fisher's inverse χ^2 method. The algorithm starts with a set of seed genes according to a specified threshold, and performs a greedy expansion by including the most significant neighbouring gene in each step.

A simpler approach for finding significant differential subnetworks is implemented in the BiNoM Cytoscape plugin (Zinovyev et al., 2008). Starting from a ranked list of all genes, from the most to the least differentially expressed one, the genes are mapped one by one to the network used for the analysis. When the top q genes are mapped on the network, the largest connected component of the subgraph they induce is detected. The size $C(q) \leq q$ of this largest connected component is compared to the mean size $R(q)$ of the largest component of the subgraph of q randomly chosen genes on the network (with the same connectivity distribution as the q genes selected), to form a the score $S(q) = (C(q) - R(q))/q$. This score is constructed in such a way that it can be compared across various values of q, allowing to choose the number of differentially expressed genes q which optimise the significance of the largest connected component among them. **Figure 6.5** shows the result of this analysis on the Wang breast cancer dataset, when differential expression between basal-like cancers and other subtypes (see **Table 5.1**) is investigated. The most significant connected component on the HPRD network is obtained with the top 600 genes. It includes several protein complexes such as a cluster of overexpressed *MCM* proteins which play a role in DNA replication, or a set of keratins also overexpressed in basal-like tumours.

6.5.3   Finding discriminative subnetworks

The methods presented in **Section 6.5.1** and **Section 6.5.2** are useful to interpret at the network-level the result of a differential expression analysis. However, they do not directly capture the coordinate dysregulation within a pathway, since the network information is not used to score or rank the genes. Instead of first computing a score of differential expression for each gene, and subsequently searching subnetworks with large aggregated scores according to **Equation 6.14**, an alternative proposed by Chuang et al. (2007) is to define the activity $a_{i,\mathcal{A}}$ of a subnetwork $\mathcal{A}$ in a given sample or patient i by the aggregated expression levels of the corresponding genes in that patient. For example, the subnetwork expression can be defined as

$$a_{i,\mathcal{A}} = \frac{1}{\sqrt{|\mathcal{A}|}} \sum_{g \in \mathcal{A}} z_{i,g} \, ,$$

where $z_{i,g}$ is the expression level (raw level or normalised by transforming it to a Z-score) of gene g in sample i. In a second step, one can score subnetworks for the phenotype of interest by computing a statistical measure of association between the subnetwork activity and the response across samples. Typically, one can estimate the mutual information between them (see **Section 6.3.1**). The search for high-scoring subnetworks is, however, complicated because the score of a subnetwork is a nonlinear function of all its genes. Chuang et al. (2007) propose to search high-scoring subnetworks with a greedy approach which starts from small candidate subnetworks and iteratively adds new neighbouring genes which increase most the subnetwork score. Applying this method on two breast cancer datasets to detect prognostic markers, they find subnetworks enriched in functional annotations characteristic of the hallmarks of cancer, observed that the corresponding gene lists are more robust than lists obtained by single gene analysis across two different cohorts and result in better prediction accuracies.

The objective function proposed by Chuang et al. (2007) is combinatorial in nature, and the bottom-up greedy heuristic they propose may therefore seriously lack global awareness. This motivated Chowdhury and Koyutürk (2010) to reformulate the search for coordinately dysregulated subnetworks as a discrete optimisation problem, a variant of the set-cover problem, solved by state-of-the-art approximation algorithms. Expression data are first quantised into binary expression levels, and the goal is to find subnetworks made of genes that together *cover* all samples, *i.e.* complement each other to discriminate samples with different phenotypes. An alternative formulation is proposed

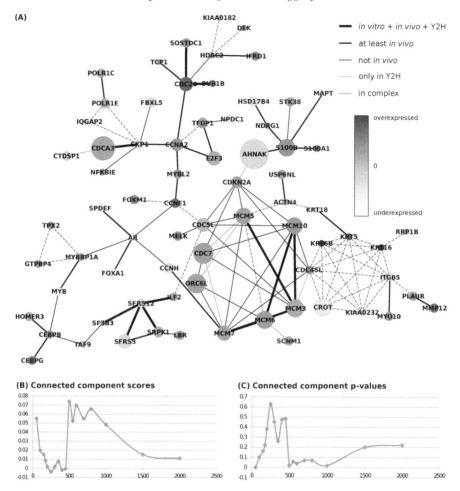

FIGURE 6.5 Detection of differentially expressed subnetworks with the BiNoM Cytoscape plugin. (A) The largest connected component of the network extracted from the PPI network of the HPRD database, among the 600 most differentially expressed genes between basal-like tumours and other subtypes in the Wang breast cancer dataset. The node sizes are proportional to the ratio between the local node connectivity and the global node connectivity (node connectivity specific to the network). The node colors visualise the level of under- or overexpression. (B) Dependence of the score of the largest connected network component formed by direct interactions between the most differentially expressed genes (note that using 600 genes gives a significantly high score value). (C) p-values for the score plotted in (B). (See colour insert.)

by Ulitsky et al. (2010), who define dysregulated pathways as subnetworks composed of products of genes that are dysregulated in a large fraction of samples and formalise it as an optimisation problem.

6.5.4 Network-driven predictive models

In addition to searching for important subnetworks, which may lead to new biological interpretation, one may also want to incorporate the known network in the predictive modelling step in the hope of improving the robustness and the performance of the model. Different ideas have been proposed for that purpose, which typically use the network to constrain the statistical inference procedures discussed in **Section 6.2** in order to infer predictive models "coherent" with the network.

For example, Rapaport et al. (2007) proposed linear discriminative methods where the classical Euclidean penalty in **Equation 6.5** is replaced by a penalty which enforces the weighs of connected vertices to be similar:

$$\hat{w} = \underset{w \in \mathbb{R}^p}{\operatorname{argmin}} \; \frac{1}{n} \sum_{i=1}^{n} \ell(f_w(x_i), y_i) + \lambda \sum_{i \sim j} (w_i - w_j)^2 \,, \qquad (6.15)$$

where $i \sim j$ means that vertices i and j are connected on the network. The rational behind the penalty $\Omega(w) = \sum_{i \sim j}(w_i - w_j)^2$ in **Equation 6.15** is to enforce weights which, once mapped to the network, do not vary too much along edges and therefore easily lead to the identification of regions with positive or negative weights, as shown on **Figure 6.6**. Interestingly **Equation 6.15** is equivalent to first smoothing the raw data on the graph, followed by a standard linear discrimination on the smoothed profiles, and can be extended to a variety of smoothing strategies based on Fourier analysis on the graph; we refer the interested reader to Rapaport et al. (2007) for further details and discussion.

We note that the solution of **Equation 6.15** is smooth on the graph, but is not sparse, in the sense that all vertices have nonzero weights and contribute to the discrimination. If one wants in addition to select a limited number vertices, while still maintaining the smoothness of the network, a simple solution is to combine the smoothing penalty with a sparsity inducing penalty such as the ℓ_1 norm to obtain, for example, the following problem:

$$\hat{w} = \underset{w \in \mathbb{R}^p}{\operatorname{argmin}} \; \frac{1}{n} \sum_{i=1}^{n} \ell(f_w(x_i), y_i) + \lambda \sum_{i \sim j} (w_i - w_j)^2 + \mu \sum_{i=1}^{p} |w_i| \,, \qquad (6.16)$$

where we now have two regularisation parameters λ and μ to adjust, which control respectively the smoothness and the sparsity of the solution.

Jacob et al. (2009) considered a weaker form of network constraint. Instead of constraining the weights of connected genes to be similar, as in **Equation 6.15**, they put no constraint on the gene weights but instead enforce the selection of a few genes which tend to be connected to each other. In other

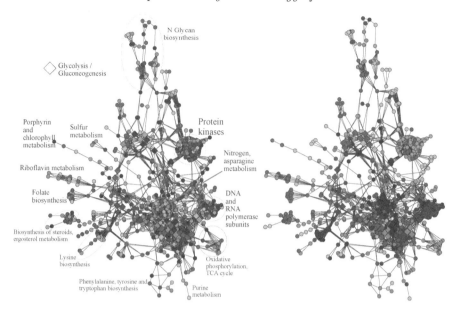

FIGURE 6.6 Network-driven linear discriminative model. This picture compares the weights of a linear classifier trained by a standard SVM (left) or a SVM using the network-driven penalty (see **Equation 6.15**) (right). The network is the metabolic network of the KEGG database. While interpreting the weights of the standard SVM at the functional level is uneasy, the network-driven SVM highlights networks areas with negative weights (*e.g.* kinases) or positive weights (*e.g.* TCA cycle). Image from Rapaport et al. (2007). (See colour insert.)

words, they enforce the selection of a gene signature which is formed from a limited number of connected components on the gene network used as prior knowledge. The solution proposed by Jacob et al. (2009) is to adapt the latent group lasso (see **Section 6.4.4**), which enforces the selection of features forming unions of predefined groups, to the case of networks by defining as groups all singletons or edges of the network. For example, **Figure 6.7** illustrates a gene signature trained to predict relapse in the Wang breast cancer dataset, using the PPI from the HPRD database as prior knowledge. We clearly see subnetworks pop out in this signature, suggesting the implication of functional networks including a subnetwork of 20 proteins containing 9 ribosomal proteins (*RPS4X*, *RPS6*, *etc.*) as well as the DNA repair genes *RAD50* and *RAD51*, and another subnetwork containing genes involved in cell cycle, such as the transcription factor *E2F1*, cyclins *CCNB2* and *CCNE2* or cell division cycle gene *CDC25B* (see **Chapter 7**).

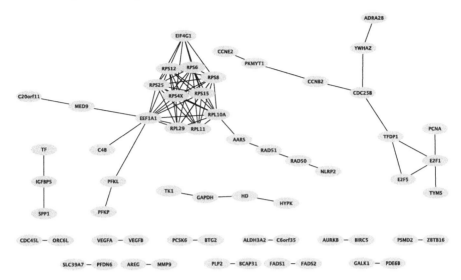

FIGURE 6.7 Network-driven feature selection. A prognostic signature estimated from the Wang breast cancer dataset, using the graph-driven feature selection
method of Jacob et al. (2009). Image courtesy of Anne-Claire Haury.

6.6 Integrative data analysis

We have so far mostly discussed methods to analyse and infer predictive
models from a single series of experiment, such as gene expression data collected on a cohort of samples with different prognoses. As high-throughput
genomic technologies such as **microarrays*** or **NGS*** (see **Chapter 3**) are
quickly becoming mature and widespread, it becomes possible to collect a
variety of measurements on each given sample, and to share this information
across different studies. Integrating such heterogeneous information is likely to
be key to increase the statistical power of inference methods, by increasing the
number of samples analysed jointly and combining evidence from different levels of biological observations. In this section, we discuss different approaches
to carry out this data integration in the context of predictive analysis and
modelling.

6.6.1 Combining heterogeneous data in predictive models

Let us first discuss the possibility to jointly analyse and integrate in a
unique predictive model the different views of a sample including genome,
transcriptome, epigenome *etc.* Assuming that each view $i = 1, \ldots, H$ char-
acterises the sample by p_i features stored in a vector $x^{(i)} \in \mathbb{R}^{p_i}$, the most

obvious method to integrate the various views is to *concatenate* them in a single vector x of dimension $p = \sum_{i=1}^{H} p_i$, and to run any feature selection and predictive modelling from the resulting p-dimensional vector.

When a linear model $f_w(x) = w^\top x$ is inferred on the concatenated vector $x \in \mathbb{R}^p$ (see **Section 6.2.3**), and if we decompose $w \in \mathbb{R}^p$ as a concatenation of $w^{(i)} \in \mathbb{R}^{p_i}$ for $i = 1, \ldots, H$, then a simple computation shows that the score $f_w(x)$ can be rewritten as a sum of scores obtained from individual views:

$$f_w(x) = w^\top x = \sum_{i=1}^{H} w^{(i)\top} x^{(i)} = \sum_{i=1}^{H} f_{w^{(i)}}\left(x^{(i)}\right).$$

In other words, concatenating vectors with linear models can be thought of as learning predictive models from each view, combined by addition to give jointly the best prediction. Implementing this strategy in linear discriminative methods discussed in **Section 6.2.3** amounts to replacing **Equation 6.5** by:

$$\hat{w} = \underset{w \in \mathbb{R}^p}{\operatorname{argmin}} \frac{1}{n} \sum_{i=1}^{n} \ell(f_w(x_i), y_i) + \lambda \sum_{i=1}^{H} \| w^{(i)} \|^2. \qquad (6.17)$$

This approach is popular in particular in the context of integrative analysis with kernel methods (see **Section 6.2.4**). Indeed, the inner product is additive under concatenation ($x_1^\top x_2 = \sum_{i=1}^{H} x_1^{(i)\top} x_2^{(i)}$), so the equivalent operation when a kernel K_i is defined for each view $i = 1, \ldots$ is to simply define a new integrated kernel as the *sum* of individual kernels:

$$K = \sum_{i=1}^{H} K_i,$$

and to run a kernel method with the integrated kernel K. The advantage of the kernel formulation instead of the direct concatenation is that it allows the combination of nonlinear functions of each view, or the integration of different prior knowledge for each view if a specific kernel is defined for each view (such as the network-driven analysis of gene expression discussed in **Section 6.5.4**). This strategy was found very efficient in the context of gene function prediction from heterogeneous characterisation of genes by Pavlidis et al. (2002b).

If many views are available, and/or if many kernels are defined for each given view, and one assumes that only a few views are relevant for the predictive problem to be solved, then a variant is to replace the linear discrimination algorithm of **Equation 6.17** by a group lasso penalty akin to **Equation 6.12**:

$$\hat{w} = \underset{w \in \mathbb{R}^p}{\operatorname{argmin}} \frac{1}{n} \sum_{i=1}^{n} \ell(f_w(x_i), y_i) + \lambda \sum_{i=1}^{H} \| w^{(i)} \|. \qquad (6.18)$$

Just as for the simple concatenation, this group lasso formulation can be extended to the combination of kernels, resulting in the so-called Multiple

Kernel Learning (MKL) algorithm (Lanckriet et al., 2004a; Bach et al., 2004). Lanckriet et al. (2004b) showed in particular the relevance of MKL for genomic data fusion.

6.6.2   Meta-analysis of multiple datasets

A second important question in data integration is the problem of integrating given information (such as gene expression data) over different studies, a problem usually referred to in statistics as **meta-analysis***. Hundreds of large-scale cancer profiling experiments have been carried out in different laboratories, and central repositories to collect these data have emerged (see **Section 4.7**). This wealth of information by far exceeds what any single laboratory could produce, and can be a useful resource for several purposes.

- It can serve to *validate* results found in one experiment on independent experiments. For example, Ramaswamy et al. (2003) identified on a cohort of 76 samples a gene-expression molecular signature that was differentially expressed in **metastatic*** tumours of diverse origins relative to **primary tumours***. They were then able to show that, on several other publicly available datasets spanning various solid cancer types, their signature was associated with clinical outcome and metastatic disease.

- Alternatively, jointly analysing thousands of samples profiled in tens or hundreds of experiments is a way to increase statistical power and robustness by increasing the number of samples analysed, and to investigate biological phenomena that may be present across different cancers of cell types. For example, Rhodes et al. (2004a) perform a meta-analysis of 40 published cancer microarray datasets, comprising more that 3700 samples, in order to identify a signature related to **neoplastic*** transformation and **tumour progression***. For breast cancer, Wirapati et al. (2008) performed a large meta-analysis of more than 2,800 tumours to validate and consolidate signatures to classify breast cancers in terms of *ER* signalling, *HER2* amplification and proliferation.

Although jointly analysing thousands of samples from tens or hundreds of datasets is tempting, it also raises several computational and statistical challenges. First, in spite of efforts to centralise and normalise data format (see **Section 4.5.2**), collecting data and annotations remains often a tedious process requiring manual data clean-up, reformatting, name standardisation and normalisation. Fortunately, several efforts such as Oncomine (Rhodes et al., 2007) and CleanEx (Praz et al., 2004), which collect and curate a large number of datasets, can greatly help data collection. Second, jointly analysing multiple datasets requires special care from a statistical point of view, in particular to handle variations between datasets (Hedges and Olkin, 1985). Directly combining raw data (after some form of normalisation within each dataset) is arguably the simplest option since it allows one to treat multiple datasets as a single one and to directly use methods and software for single-dataset analysis. However, it raises questions in terms of statistical validity, since samples are intrinsically stratified by the different studies which may reduce the statistical power of tests and sometimes leads to paradoxical conclusions (see **Box 6.5**).

❏ **BOX 6.5: Simpson's paradox**

The Simpson's paradox, also known as the Yule-Simpson effect, is a paradox in statistics and decision making in which a correlation present in several datasets is reversed when the datasets are combined. For example, in the following picture, there exists a clear positive correlation between x and y on each dataset (solid lines), but the correlation is reversed when both sets are analysed jointly (dashed line).

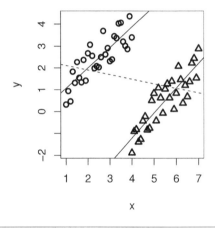

Alternatively, one may first carry out statistical analysis within each study, and *a posteriori* integrate the findings of each study. For example, Rhodes et al. (2004a) identify a limited number of genes differentially expressed in many studies, and then combine the different signatures by selecting genes present in many signatures. An obvious limitation of this approach is that a gene which is not strongly differentially expressed will be absent from the individual molecular signatures, and therefore from the final one. In other words, such *a posteriori* data integration suffers from a loss of power due to premature thresholding. A more interesting approach is certainly to jointly perform the statistical analysis on all studies simultaneously, in order to gain statistical power, without directly combining the raw data (Hedges and Olkin, 1985). For example, Fisher's combined probability test combines the p-values

$p_1, \ldots, p_K$ obtained in K different studies in a single statistic:

$$X^2 = -2 \sum_{i=1}^{K} \log p_i \,,$$

which follows a χ^2 distribution with $2K$ degrees of freedom when all the null hypotheses are true and the tests are independent. A closely related technique is the inverse normal method, which computes a Z-score for a meta-analysis by combining the Z-scores $Z_i = \Phi^{-1}(1 - p_i)$ of each study (where Φ is the standard normal cumulative distribution) with the formula:

$$Z = \frac{\sum_{i=1}^{K} Z_i}{\sqrt{K}} \,. \tag{6.19}$$

We refer the reader to Hedges and Olkin (1985) for more details on these and other statistical techniques for meta-analysis. For example, Wirapati et al. (2008) performs a meta-analysis of 18 breast cancer expression studies by combining Z-scores with **Equation 6.19**. This allows to jointly analyse 2,833 expression profiles and consolidate signatures associated with ER signaling, *HER2* amplification, and proliferation.

6.7 Conclusion

The heterogeneity of tumours discussed in **Chapter 5** has immediate bearings on cancer clinical management. Since different cancers have different prognoses and respond differently to different treatments, it is important to predict as precisely as possible the prognosis and drug responsiveness of each tumour in order to be able to propose the most adapted treatment to each patient. Although cancer subtypes defined by unsupervised clustering of tumours (see **Chapter 5**) represent a first step towards cancer stratification with prognostic and predictive value, more accurate prognosis and drug response prediction are usually obtained by supervised statistical learning techniques.

The learning tasks to be solved are, however, particularly challenging from a statistical viewpoint because the measured features by far outnumber the samples available to train the model. We saw that simple models such as NB sometimes compare favourably to more complex nonlinear models, suggesting that improvement in prediction accuracy will not come from more complex and realistic models, but from a better understanding of the trade-off between modelling issues, which require realistic models, and statistical issues, which call for simple models. We also saw that, despite its popularity, the approach which consists in constructing a molecular signature using feature selection techniques suffers from instability, and is not always justified in terms of model performance. Regularisation with prior knowledge offers a particularly promising avenue to integrate biological knowledge into the learning framework, allowing one to reduce the statistical estimation burden while focusing on biologically relevant models.

It is likely that the performance of prognostic and predictive models will increase as more data to train supervised models will be collected. In particular, current and future clinical trials to validate prognostic and predictive molecular signatures will generate large datasets that will progressively allow us to benefit more from complex machine learning strategies, paving the way to more accurate prediction and more interpretable models. In parallel, better molecular characterisation of tumours should enable defining subtypes with better homogeneity (see **Chapter 5**), and integrating this knowledge into the supervised inference paradigm should also be beneficial.

✎ **Exercises**

- When we evaluate the performance of a molecular predictor involving feature selection in cross-validation, why is it important to perform feature selection at each fold and not only once with all data?

- Measure the performance of molecular predictors as a function of the size of a signature, as in **Figure 6.4**, for different prediction problems from gene expression data (diagnosis, prognosis, prediction of drug response). When does feature selection improve the performance of the model?

- On the Wang breast cancer dataset, or the gene expression dataset of your choice, select differentially expressed subnetworks between samples with good and bad prognosis with the BiNoM Cytoscape plugin. Do you recover the most differentially expressed genes in the selected subnetworks?

⇨ **Key notes of Chapter 6**

- Molecular predictors have started to replace classical models based on clinicopathological parameters for prognosis and prediction of drug response.

- Molecular predictors can be learned by various linear or nonlinear methods for supervised statistical inference.

- Feature selection can be used to learn a molecular predictor based a small set of markers, called a molecular signature. The gain of performance resulting from the selection of a few markers is however not always observed, and the robustness of genes selected in signatures is often limited.

- Integrating prior knowledge such as gene sets or biological network is useful to interpret molecular predictors and increase their robustness and performance.

- Integrative data analysis such as meta-analysis and heterogeneous data fusion can be useful to increase the number of samples and the variety of data analysed.

Chapter 7

Mathematical modelling applied to cancer cell biology

Mathematical models are used to translate biological problems into a formal language, to apply this language for formal reasoning and to provide new insights on the problems. We propose, in this chapter, to study mathematical models of problems related to cancer cell biology. After explaining our motivation, goals and method of modelling biological systems using formal mathematical tools, we will present two examples of mathematical models of the cell cycle using both chemical kinetics and logical formalisms. From these models, we will extract motifs of feedback loops and study them independently.

7.1 Mathematical modelling

7.1.1 Why construct mathematical models?

Biological systems are complex by nature (Kitano, 2002a). This complexity arises not only from the enormous amount of elements that compose these systems and their variety, but also from the difficulty to define their interactions in a simple and linear way. These elements (or constituents, entities) can be proteins, genes or, at a different scale: cells, tissues, organs and even organisms.

The long-term goal of mathematical modelling is to construct a virtual object that would mimic this object's behaviour in real conditions, understand and predict the behaviour of perturbations. However, we claim that this virtual object does not need to contain all the possible pieces that compose the real object. Some experimental groups have already tackled this ambitious problem by trying to reconstruct synthetically an engineered cell capable of surviving, proliferating, *etc.* with a minimal amount of genes (Tomita et al., 1999). Similarly, Denis Noble, pioneer in systems biology, and colleagues have proposed in the early 1960s, the construction of the first virtual organ, the heart. Although the project seemed unfeasible at the time, they managed to simulate the impact of a defect at the organ scale. With this modelling experience, Noble argued *"if you tried to reconstruct everything - all of the molecules in all of the heart's cells - no current computer could cope."* This could easily be confirmed by the following numbers: there are between 20,000 and 25,000 genes in the human genome that would express from thousands to one million proteins. Considering the high number of interactions among these proteins, no computer would be able to treat this amount of data at present. But even if such a computer existed, would we be able to understand the functioning of the human body?

Even though it is impossible, as of today, to describe every protein and every interaction implicated in cellular events, we can still get some insight about particular cellular mechanisms, derive some hypotheses on how a signal is transmitted from one cellular compartment to another and predict some behaviours of the cell in diverse conditions with a systemic approach. Systems biology aims at sketching the main traits of the cell, *i.e.* the major transitions, the **signalling pathways***, or the interactions between key players involved in a biological process. The amount of details to include in the model along with the mathematical formalism used to describe a process should be led by the biological question. Very often, the most complex model is not the most instructive: a simple model may be as effective if not more, and its simplicity may facilitate its interpretation.

7.1.2   Why construct a mathematical model of cancer?

As previously mentioned, the high numbers of proteins and associated functions are not easily manageable. However, if, as of today, we are still limited by our technology on one side, we are overwhelmed by it on the other side. Indeed, our knowledge of the nature of the interactions between these constituents and their organisation in networks has also increased tremendously over the past decades, due to the emergence of high-throughput approaches (see **Chapter 3**) and our capability of analysing them (see **Chapter 5** and **Chapter 6**). As a result, when trying to understand the function of a gene, a protein or even a process and its real role in a process, it can no longer solely be based on intuitive reasoning.

Waddington (1957) assumed that cellular processes are based on complex networks of interacting genes and proteins. More recently, cancer was referred to as a *systems biology* or a *network* disease (Hornberg et al., 2006). Cancer can indeed be seen as a pathology of the processes that govern differentiation, proliferation, apoptosis, *etc.*, that is to say, a deregulation of these networks. As a result, modelling cancer starts with the study of possible deregulations of a normal cell cycle. The questions that can be answered with systemic approaches are of the following types: What are the cellular pathways involved in a pathology? How to use these pathways to improve predictions? What are the effects of a perturbation on a pathway? *etc.* Mathematical modelling provides tools to apprehend some of these issues.

7.1.3 What is a mathematical model?

In interdisciplinary fields, the choice and the meaning of words is very important. A model can have different definitions according to the background of the persons speaking. A general definition of a mathematical model could be an abstract representation of reality, but more particularly here, we choose to define it as follows.

A model, as we intend it in this chapter, is a set of biochemical laws, composed of variables and of parameter values that describe a biological process. It serves as a mean to formulate hypotheses; test the coherence of disseminated and uncorrelated published data; identify misunderstood zones and contradictory facts; propose a logical functioning of particular underlying biological processes; establish predicting facts concerning the importance of some players or the outcome of perturbations; anticipate ways to compensate, rescue or silence altered pathways; *etc.* A model needs to be comprehensive; the model needs to be capable of reproducing all experiments related to the components of the system. Finally, more importantly, a model needs to be falsifiable. In order to be a good thinking tool, the model has to be challenged by unexpected experiments for the purpose to be improved. To conclude, a model has to be regarded as a transient object that assists the biologists. It is in the nature of scientific models to become obsolete. Thus, a model is soon, and has to be, superseded by more refined and more up-to-date models.

7.1.4 Different forms and formalisms of a model

There are different forms of mathematical models: data-driven models vs. (prior) knowledge-based models. Data-driven models can be based on machine learning methods (see **Chapter 6**), where the structure of the model is inferred from the biological data. Knowledge-driven models are based on theoretical knowledge compiled and constructed by human expertise. Here, we concentrate on the knowledge-based model.

Mathematical models are a translation of cellular phenomena in formal terms. The models can be qualitative or quantitative, statistical or mechanistic, static or dynamical, discrete or continuous, deterministic or stochastic. There is no universal formalism that will answer all possible biological problems and some will be more appropriate to treat problems such as cases dealing with a low number of entities (cells, proteins, *etc.*), with qualitative or quantitative inputs, with or without cellular compartments, with a certain level of details, *etc.* All formalisms have advantages but also limitations. Without entering into too many details, some of the formalisms used in systems biology are described below. Note that the list is not exhaustive but highlights the diversity of formalisms that are available in the field (for further readings and references for these formalisms, read the detailed review of de Jong (2002) and a more recent mini-review of Machado et al. (2011)):

- The chemical kinetics approach describes the change of concentrations of genes, messenger RNA (mRNA), proteins or metabolites over time. However, the rates of synthesis, degradation, binding constants, or concentrations of proteins are not easy to deduce from experiments. To cope with these issues, some approximations of the chemical kinetics equations have been proposed (such as S-systems or Piecewise-Linear Differential Equations (PLDE) mentioned below).

- The Power-Law models or the S-systems approximate polynomial equations by simpler linear combination of powers of species concentrations.

- In contrast, Boolean modelling provides a highly qualitative approach with a coarse-grain description of the biological processes.

- If Petri nets relied on graphical representations to tackle discrete events at first, they are now more widely used for continuous or stochastic approaches.

- PLDEs combine a continuous description separated by discontinuous events illustrating switch-like behaviours of genes.

- To include stochasticity or to treat noncontinuous or nondeterministic cases, Stochastic Differential Equations (SDE) or Stochastic Master Equations (SME) are used.

- Directly applied from theoretical computer science, the Process Algebra (PA) languages bring up the issue of stochastic communication and competition for a particular event between agents or processes.

- The rule-based formalisms offer the possibility to describe relationships among entities in (close to) natural language and deal with combinatorial complexity (Kappa, *etc.*).

- To take space into account, diffusion-reaction based approaches using Partial Differential Equations (PDE) are best suited.

- Finally, Flux Balance Analysis (FBA) is a formalism that assumes steady state and **homeostasis***** and is often used to model metabolic networks.

Using chemical kinetics, diffusion-reaction approaches (based on Ordinary Differential Equations (ODE), Partial Differential Equations (PDE) respectively) or Boolean approaches to model biological processes is suitable when the major players and processes represent elementary physical entities (molecules, fluxes, chemical reactions). However, applying these modelling techniques when the players can themselves be complex objects with nontrivial behaviours is difficult. In this case, an agent-based approach can be used. In particular, it is appropriate to create multiscale models of cancer as discussed below.

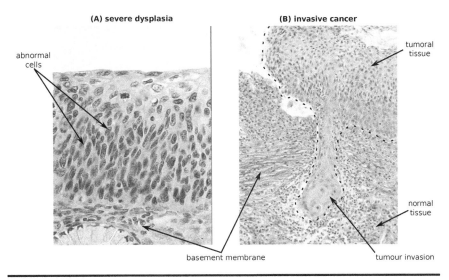

FIGURE 2.10 Tumour progression and invasion in cervix cancer.

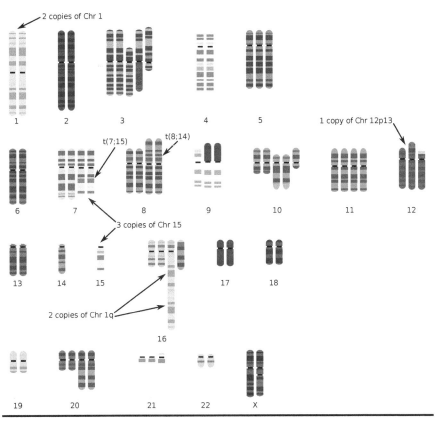

FIGURE 2.12 Karyotype of the T47D breast cancer cell line.

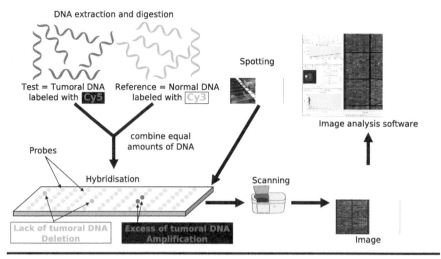

FIGURE 3.3 Array-CGH protocol.

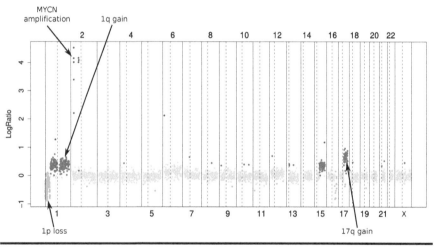

FIGURE 3.5 aCGH profile of the IMR32 neuroblastoma cell line.

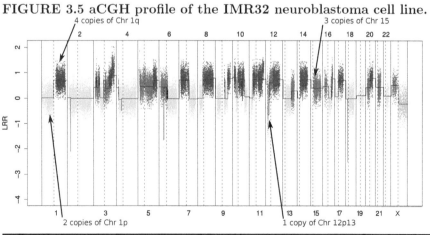

FIGURE 3.8 LRR and BAF profiles for the T47D breast cancer cell line.

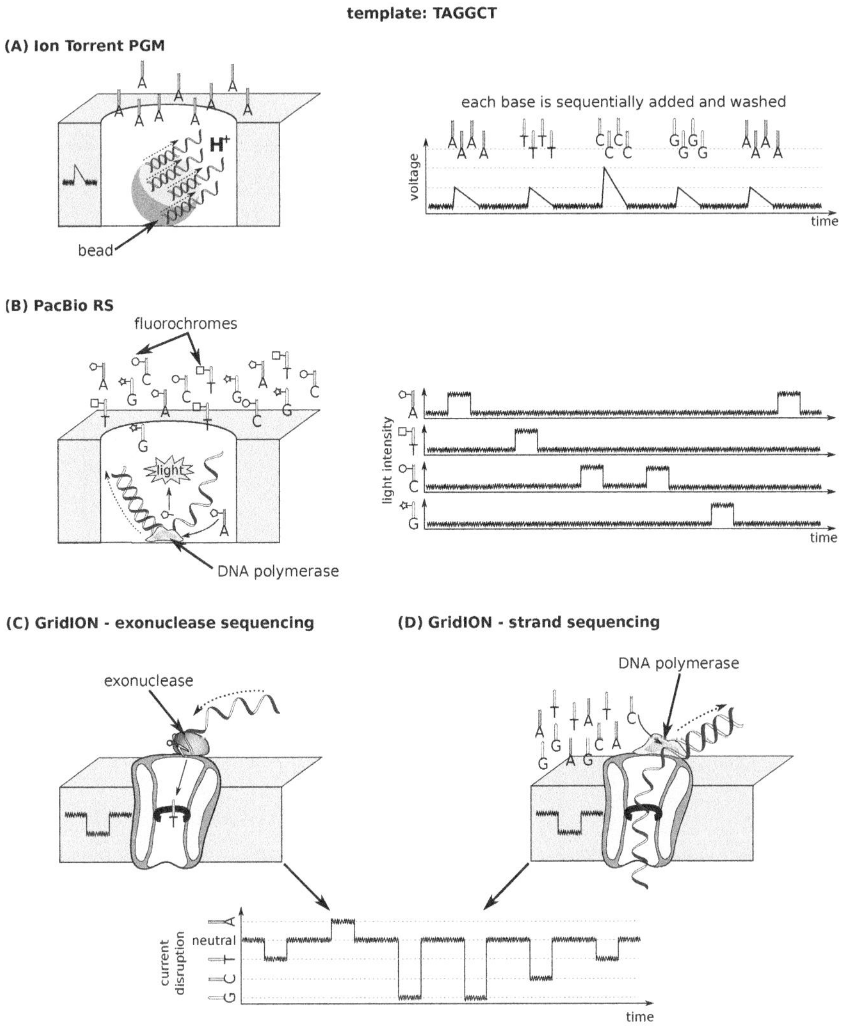

FIGURE 3.14 From second- to fourth-generation sequencing, illustration on **TAGGCT** template.

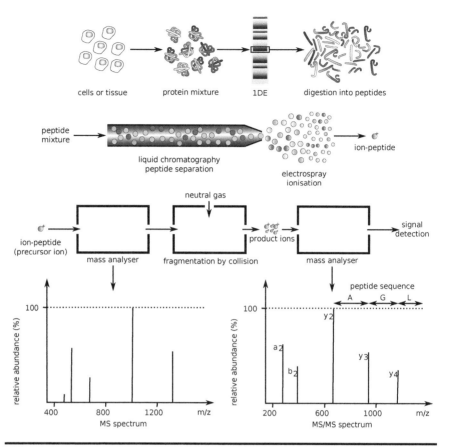

FIGURE 3.19 Mass spectrometry protocol.

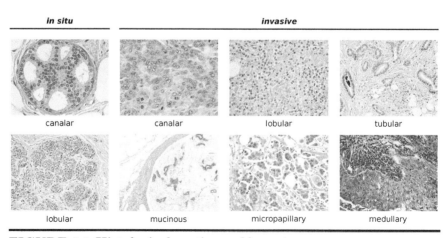

FIGURE 5.1 Histological sections of breast cancer.

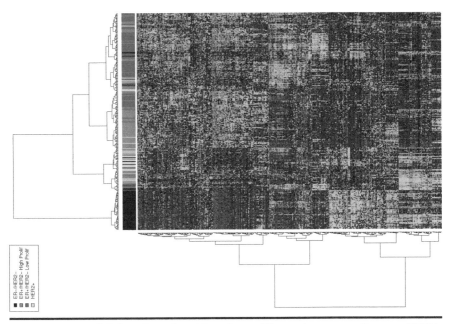

FIGURE 5.3 Molecular classification of breast cancer from mRNA expression profiles.

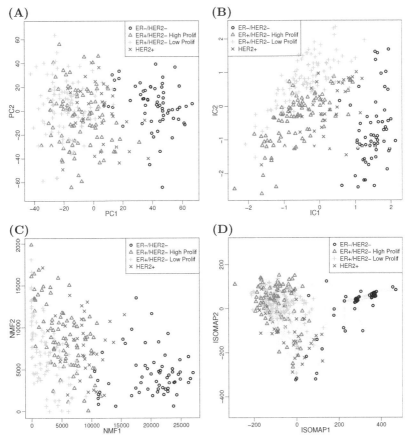

FIGURE 5.6 Breast cancer diversity in 2 dimensions.

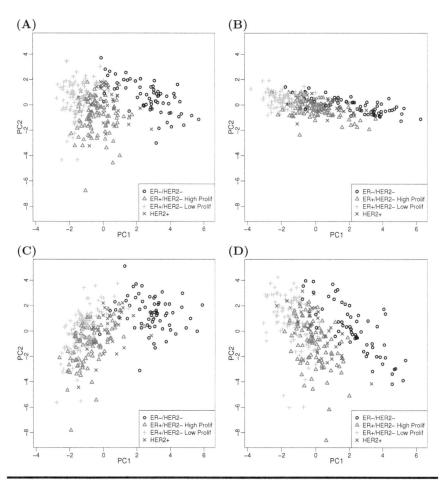

FIGURE 5.7 Multiple Factor Analysis on breast cancer data.

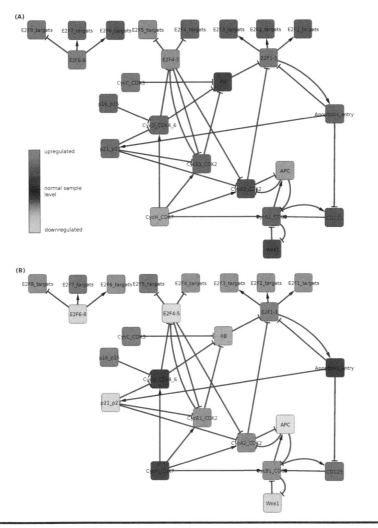

FIGURE 5.8 Modular map of RB pathway.

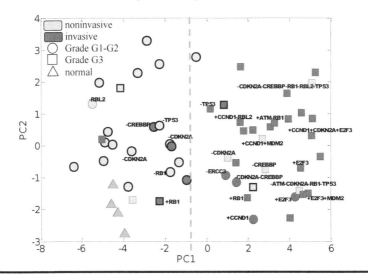

FIGURE 5.9 PCA of the modules activity.

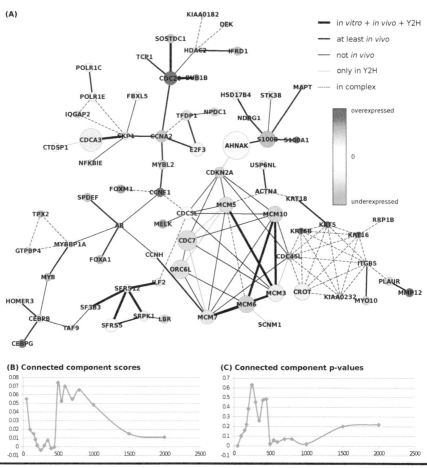

FIGURE 6.5 Detection of differentially expressed subnetworks with the BiNoM Cytoscape plugin.

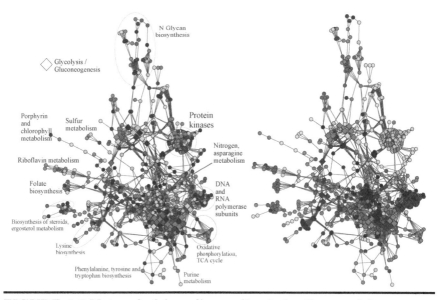

FIGURE 6.6 Network-driven linear discriminative model.

7.1.5   Agent-based and multiscale modelling in cancer

Agent-based modelling is a powerful simulation modelling technique that has seen a number of applications in the last few years, including cancer biology (Zhang et al., 2009). It allows one to encapsulate complex patterns of objects' behaviour in the form of rules (Bonabeau, 2002). Agents interact in a competitive and repetitive fashion. This approach relies on the power of computers to explore the dynamics usually out of reach in pure mathematical methods and takes its roots in *cellular automata modelling*.

In practical biological applications, an agent can be a biological species, an organism, a cell, an organelle or other subcellular structure. An agent can also be a protein or any other molecule when its behaviour can be described by a decision-making process rather than regular physico-chemical laws. The agents of different types, for instance cancer and stromal cells, are placed in space, with some properties, with a particular interaction topology, such as a grid, in some non agent environment. Agents can move, measure environmental properties, self-replicate and interact with other agents according to some state-dependent rules. As a result, the states of all agents can be changed and each agent is able to influence the environment.

Agent rules can be purely phenomenological. However, in more advanced agent-based techniques, agent rules can incorporate physical and chemical laws. For example, in an agent-based model of liver lobule response to intoxication by paracetamol, the rules governing cell behaviour were based on physical adhesive cellular properties and physical motion equations (Drasdo et al., 2011).

The main idea of agent-based modelling is the *emergence of complex system behaviours* from relatively simple internal agent rules and thus, suitable for modelling self-organisation phenomena. Evolutionary problems can also be treated in the agent-based modelling approach. A simple example of agent-based modelling (forest-fire model) will be considered in **Section 10.3**.

CellSys is a modular software tool that simulates growth and organisation processes in multi-cellular systems in two- and three dimensions (Hoehme and Drasdo, 2010). It implements an agent-based model that approximates cells as isotropic, elastic and adhesive objects. Cell migration is modelled by an equation of motion for each cell. The software includes many modules specifically tailored to support the simulation and analysis of virtual tissues including real-time three-dimensional visualisation and Virtual Reality Markup Language (VRML) support.

Due to its flexibility, agent-based modelling is frequently used in constructing multiscale models. *Multiscale modelling* takes into account multiple spatial, temporal or structural scales. Multiscale modelling is of particular importance in studying cancer. Let us imagine a comprehensive model of how ultraviolet radiation causes skin cancer. The model should take into account the effect of radiation in damaging DNA (in nano- and milliseconds time scale). As a consequence, signalling cascades are rewired and the cell cycle is deregulated (seconds and minutes time scale). This leads to phenotypical changes on macroscopic level (hours and days time scale). The cells form a tissue organised and remodelled in the course of **tumorigenesis*** (months and years time scale). Spreading metastases is taking place at the level of the organism and requires a description at a larger spatial scale.

In the description of tumorigenesis, at least two scales are considered: at the intracellular level, describing functioning of signalling cascades regulating cell death, cell survival and proliferation, and at the level of cell population at which changes in the functioning of signalling cascades lead to macroscopic tumour growth (examples of modelling tumour invasion can be found at **Section 8.1.3**). In practice, the types of models developed in the rest of this chapter could be seen as models to simulate the intracellular behaviour, and rules between cells (defined as agents) could be set to simulate the cell population behaviour. For reading more about applications of the multiscale modelling approach in cancer biology, see Deisboeck et al. (2011).

In this chapter, we propose to briefly treat two mathematical formalisms: chemical kinetics based on nonlinear ODEs approach and the Boolean logic approach.

7.1.6   Hallmarks of mathematical modelling

Mathematical models can provide the appropriate tools to interpret all sorts of experimental observations in a rigorous and systematic manner. More specifically, we propose four characteristics or hallmarks of a mathematical model (see **Figure 7.1**):

- A model achieves *formalisation of our knowledge*: it aims at describing biological phenomena in a formal and unambiguous way; it recapitulates (integrates all facts) and summarises (provides concise description) what is known about a biological process; it allows the identification and the listing of the key players and mechanisms involved in a process; it also provides a way to visualise biological data; *etc.*

- A model facilitates the *generation and comparison of hypotheses*: it enables analyses and interpretation of biological data; this leads in turn to the formulation of hypotheses on the network structure, on the plausible molecular interactions behind a process; *etc.*

- A model proposes *predictions*: it can predict results such as mutant phenotypes, response to drug treatments, identification of therapeutic intervention points, behaviours in particular cellular contexts; it can also test and rank predictions by performing *in silico* experiments; it can anticipate side-effects of drugs or perturbations; *etc.*

- A model enables *conceptualization*: it can be used as an abstract thinking object and it allows introduction of new concepts in biology: in

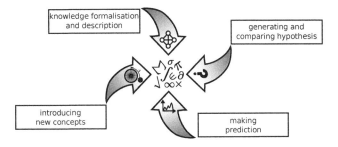

FIGURE 7.1 Hallmarks of mathematical modelling. A model can be: a descriptive object - *knowledge formalisation and description*; a hypothesis generator - *generating and comparing hypotheses*; a predictive tool - *making predictions*; and a conceptual tool - *introducing new concepts*.

particular it can give formal definition for a class of behaviours and cellular observations (*e.g.* any dynamical behaviour of individual motifs, see **Section 7.4**).

Note that, even though the two roles of hypothesis generator and predictive tool seem very similar, they fulfil different functions. The first one derives hypotheses based on how the pieces of the puzzle could fit together, whereas the latter allows to perturb the system and to derive predictions on how it would react and behave.

7.2   Mathematical modelling flowchart

In this section we present a multistep procedure that we use for building mathematical models to answer a biological question. The purpose is not to provide a universal recipe but rather to propose steps to follow when modelling a biological question. We will focus on knowledge-based models, the data-based approaches being the subject of **Chapter 5** and **Chapter 6**. Examples of network modelling will then follow.

1. *Definition of the biological problem*: A clear question or a list of questions is formulated. It is crucial to take the time to carefully write down the biological problem, clearly state the question, and delineate the purpose of the modelling. This step needs to be done in close collaboration with the modeller and the experimentalist.

2. *Gathering of information*: Any type of information related to the studied processes are gathered (articles, discussions with experts, *etc.*).

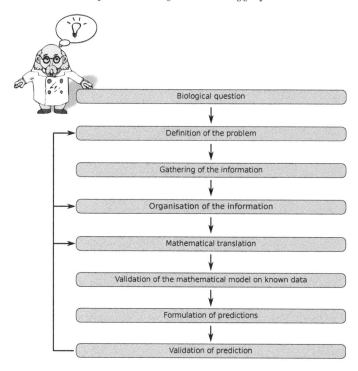

FIGURE 7.2   **Flowchart of mathematical modelling.**

3. *Organisation of the information*: The format used here is a diagram recapitulating the collected information; in systems biology the diagram is typically composed of nodes and edges (see **Box 4.4**). However, it can be a database or anything most suited to the modeller. Whatever format is used, each protein, entity, reaction, compartment, function, processes, both observed and expected outputs, *etc.* are annotated with comments and references.

4. *Mathematical translation*: Depending on the formalism used for modelling, the diagram is translated into a mathematical object. Each node of the diagram is assigned the dynamics that best describe its behaviour.

5. *Validation of the mathematical model*: The model (set of differential equations, set of Boolean rules, *etc.*) is often integrated numerically. The parameters of the model are tuned so as to fit the known physiology (wild type behaviours, normal cellular context, *etc.*). The model is then perturbed and the result of the simulations is confronted to the corresponding experimental result. The perturbations correspond to the conditions of mutants or drug treatments. The model should account for all experimental variations considered in the initial set up.

6. *Formulation of predictions*: Typically, phenotypes of alterations (mutants or drug treatments) that have not been performed yet can be proposed.

7. *Validation of predictions*: The hypotheses formulated by the model and validated *in silico* are to be tested in a wet lab. The results can also be verified in existing experimental data (gene expression microarrays, high-throughput sequencing, *etc.*)

8. *Feedback from experiment to model*: The model can then be refined, as long as it does not recapitulate satisfactorily the observations. Several iterations from experiment to model to experiment may be needed to achieve a good result. The feedback can be at different levels: the initial problem can be reformulated based on the results of the literature search, the information can be organised differently, the level of details adapted, the mathematical formalism chosen can evolve from discrete to continuous and vice versa, from deterministic to stochastic and vice versa. The model can be modified according to the experimental results.

7.3   Mathematical modelling of a generic cell cycle

7.3.1   Biology of the cell cycle

Cancer genes are genes whose **mutations*** or alterations could increase the risk of cancer development and promote tumorigenesis (see **Chapter 2**). Many of these cancer genes are involved in pathways related to the cell cycle. Cancer cells often fail to respond to external signals that would halt proliferation of normal cells, therefore, in cancer cells, cell cycle checkpoint mechanisms that should stop the cycle in abnormal situations are altered. For all these reasons, the study of a normal cell cycle seems to be a good starting point to the study of cancer cell cycle (Sherr, 1996).

The cell cycle is often described as an alternation between two phases: DNA replication (S phase) and Mitosis (M phases) separated by two gap phases (G1 and G2). These gaps ensure that one phase has been completed properly before the other one starts. The presence and activity of the Cyclin Dependent Kinases (CDK) and their cyclin partners determine at which stage of the cycle the cell lies. To ensure that everything is ready to advance in the cell cycle, there exist safe mechanisms on which the cell relies. The transitions from one phase to another are thus monitored by checkpoint controls. They supervise the good progression of the cell cycle and halt it when problems are encountered in order to give time to the cell to repair the damage. That way, proper genetic material can be passed on to the next generation. Since cancer is a disease of uncontrolled and excessive proliferation, some of these checkpoints malfunction in many different cancers.

The first mathematical models of the cell cycle were mainly developed for yeasts and frogs. As early as 1962, one of the first mathematical interpretations of cell cycle was published (Koch and Schaechter, 1962). Later, in the 1990s, groups like Tyson's, Novák's or Goldbeter's developed complex models of cell cycle mechanisms. Since then, articles related to cell cycle modelling have exploded (Csikász-Nagy, 2009). If modelling can help understand the various importance of some proteins and complexes and their temporal organisation, some questions remain: how is the cell cycle related to cancer and why is mathematical modelling needed?

One of the checkpoints mentioned above concerns the tumour suppressor gene *RB* (referred to as retinoblastoma or *RB1*). *RB* is altered in many cancers. It is a key regulator of the G1/S transition. Its activity is linked to external cellular signals. In a normal cell and early in the cycle, RB is fully active in its unphosphorylated form. In response to growth signals, a series of events leads to the association and activation of the *CDK4* and *CDK6/CCND1*, where CCND1 refers to CyclinD1, and the consequent phosphorylation of RB. This is the starting point for the progressive inactivation of RB and the passage through the G1/S transition. In 1974, Pardee proposed the theory of a restriction point (see **Box 2.2**). In many cancer cells, since the restriction point is altered through *RB* **mutation***, loss or deregulation, proliferation occurs independently of the presence or absence of mitogens. The cell behaves as if it were constantly receiving positive growth signals.

Many models have explored the restriction point from a theoretical point of view. Among them, Aguda and Tang (1999), studied the effect of varying the concentration of the genes that have an influence on the restriction point; Qu et al. (2003a), among others, characterised the molecular features of the restriction point; (Novak and Tyson, 2004) reproduced the experiments done by Zetterberg et al. (1995) that aimed at identifying the precise time of the restriction point in late G1; and finally, (Conradie et al., 2010) measured the importance of some cell cycle actors in the control of the positioning of the restriction point.

Let us study one of these models with both chemical kinetics and Boolean approaches (see **Box 7.3** for differences between these two formalisms).

7.3.2   Chemical kinetics formalism

When the biological question requires quantified answers and when the experimental data permit, a quantitative approach is recommended. One popular formalism used when modelling quantitative data is the use of chemical kinetics (Guldberg and Waage, 1864), based on systems of nonlinear ordinary differential equations (ODEs). Even though it is more detailed than the most qualitative formalisms, ODE models remain a rough abstraction of reality for which many assumptions are made: the cellular environment is considered to

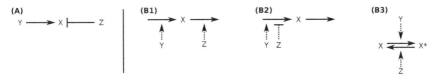

FIGURE 7.3 From an influence network to a reaction network. (A) Influence network: X is activated by Y and inhibited by Z. (B) Biochemical network interpretations of the influence network. Plain arrows are biochemical reactions and dashed arrows are influences of a gene on a reaction. (B1) Y mediates the synthesis of X and Z mediates the degradation of X. (B2) Y mediates the synthesis of X while Z inhibits it. (B3) X can appear in two forms, either active ($X*$) or inactive (X). Its activation is promoted by Y and its inactivation by Z.

be homogeneous, well mixed, with a high probability that molecules meet, and with fixed temperature and pressure.

In chemical kinetics, some well-established kinetic laws are used to formulate reaction rate equations, among them, the law of mass action, Hill or Michaelis-Menten kinetics. Each equation is deterministic and follows the rate of concentration of the proteins or their activity state over time.

Network representation of biological knowledge is often represented as an influence network in biological papers, where, for instance, a protein activates or inhibits another protein without a detailed description of the type of activation or inhibition. The most appropriate type of networks for ODE modelling being reaction networks, in this particular case, the influence networks should be translated into reaction networks.

Note that some works have been proposed on the translation of Boolean models to ODEs models (Wittmann et al., 2009) and that some methodological works are currently done on the automatic translation of reaction networks to influence networks. Both problems are not easily solvable.

Here are some examples of interpretations of an influence network as a reaction network, when the biochemical details are not known. Let us consider three variables X, Y and Z. If what is known from the literature is: X is activated by Y and inhibited by Z, then there are several ways to interpret it (see **Figure 7.3**).

In the case of mass action kinetics, the rate of a reaction (synthesis or degradation, for instance) is proportional to the product of the concentrations of its reactants. The rate of change of a variable X can be described as follows:

$$\frac{dX}{dt} = \dot{X} = k_1 \cdot Y - k_2 \cdot Z \cdot X, \tag{7.1}$$

where Y is promoting the accumulation or the activation of X at a rate k_1, and Z is mediating the depletion or the inactivation of X at a rate k_2 (see **Figure 7.3B1**).

If the synthesis is governed by highly nonlinear dynamics (as it is often the case), a Hill function controlled by the Hill parameters, K and n, can be used:

$$\frac{dX}{dt} = \dot{X} = \frac{k_1 \cdot Y^n}{K^n + Y^n} - k_2 \cdot Z \cdot X. \qquad (7.2)$$

There are many ways to write the differential equations that correspond to an influence network. We just gave the example of the case where Y promotes the activation of X and Z promotes the inactivation of X. However, the inhibition of Z depicted in the influence network (see **Figure 7.3A**) could also be interpreted as an inhibition of the synthesis of X (see **Figure 7.3B2**). In this case, we would write:

$$\frac{dX}{dt} = \dot{X} = \frac{k_1 \cdot Y}{k_{1'} + k_{1''} \cdot Z} - k_2 \cdot X. \qquad (7.3)$$

Finally, if both the activation and the inactivation of X are controlled by Michaelis Menten kinetics, the switch (the successive ON and OFF states of X) with these two coupled Michaelis-Menten terms is referred to as a Goldbeter-Koshland switch (Goldbeter and Koshland, 1984):

$$\frac{dX}{dt} = \dot{X} = \frac{k_1 \cdot Y \cdot (X_{tot} - X)}{K_1 + (X_{tot} - X)} - \frac{k_2 \cdot Z \cdot X}{K_2 + X}. \qquad (7.4)$$

where X_{tot} is a parameter for the total amount of X if we assume that the total amount of X in the cell is the same all the time, *i.e.* the mass of X is conserved. The time at which the switch occurs depends on the ratio of Y and Z and the sharpness of the transition depends on the values of K_1 and K_2 (the smaller the values, the stiffest the switch) (see **Figure 7.3B3**).

As quickly shown here, the translation from an influence network to a biochemical description is not unique. Note that all the three networks of **Figures 7.3B1** to **7.3B3** could be written using mass action, Hill or Michaelis-Menten kinetics. The nature of the biochemical reaction and the choice of the mathematical representation for this reaction will depend not only on the properties of the components catalysing the reaction (if they are kinases or phosphatases, transcription factors, stoichiometric inhibitors or part of the **proteasome*** complex, *etc.*) but also on the availability of the knowledge provided by the experimental results (inhibition by complexation, by degradation, *etc.*). Interpreting and transcribing the information found in the literature into a biochemical reaction network is not as straightforward as it seems.

The identification of the parameter values is an even more difficult task. Ideally, the parameters that control the speed or activity of the reactions would be directly derived from the literature, found in databases such as SABIO-RK or BRENDA or fit mathematically to experimental data by optimisation methods. If none of these is possible, then they should be chosen manually so as to fit the experimental data. Provided that the formats of the equations are already chosen (as presented in **Equations 7.1** to **7.4**) and are not going to be changed, this step is a difficult and tedious one. Even when the parameter values can be derived directly from the literature, one has to consider that all the experiments are performed in different conditions, on different cell lines with different technicians with different moods, different room temperature, *etc.* and may not be appropriate for a model of a different cell line, or a generic model. These values have to be handled carefully. Moreover, the number of parameters has to be accurate in order to avoid overfitting. Indeed, the number of these parameters and the choice of their values need to be constrained by a reasonable amount of biological data that they need to reproduce.

The first requirement that the model must meet is to reproduce the conditions of the system in a normal situation, when it is not — or very lightly — perturbed. Once the wild type model fits the observed phenotypes and behaves as expected, the model can be challenged by performing *in silico* experiments. Let us imagine that the gene Z is deleted from the cell. In this case, X will not be inhibited — or degraded. In order to simulate the deletion of the gene in the mathematical model, k_2, and only k_2, will be set to 0 (or if Z's concentration varies over time, its synthesis term will be set to 0 as well as its initial condition). The solution of the simulated perturbation will have to show that X remains active at all times when Y is present.

7.3.3   Cell cycle as an ODE model: Application to the restriction point

Novak and Tyson (2004) proposed a model of the dynamics and the role of the restriction point. The model accurately reproduces experimental results published by Zetterberg et al. (1995), validating the choices made on the topology of the network (see **Figure 7.4**) and on the dynamics of the mathematical model (see **Figure 7.5**).

The network includes several modules: growth factors (with early and delayed response genes), RB and its interaction with E2F, the antagonism between the cyclins/CDK complexes (CycE/CDK2 and CycA/CDK2) and their inhibitors (p27, referred to as KIP in the model), the antagonism between all the cyclins/CDK complexes (CycA/CDK2, and CycB/CDK1) and their degradation machinery (CDC20 and CDH1). The characterisation of the mathematical model uses mass action kinetics, Goldbeter-Koshland switches, and Hill terms (equations of the model).

The authors have constructed the core of the mammalian cell cycle from a model of the budding yeast cell cycle (Chen et al., 2000) arguing that even a basic molecular network with little biochemical details such as that of the budding yeast can explain the dynamics of the restriction point. The model is also able to reproduce the cells' behaviour after removal of growth factors and re-entry into the cell cycle after the growth factors are added back. Zetterberg et al. have measured the length of the cell cycle of individual cultivated mouse fibroblast cells. They have explored the effect of transient growth factor deprivation at different times throughout the cell cycle by treating the cell with cycloheximide, a drug that blocks protein synthesis, and then washing it away

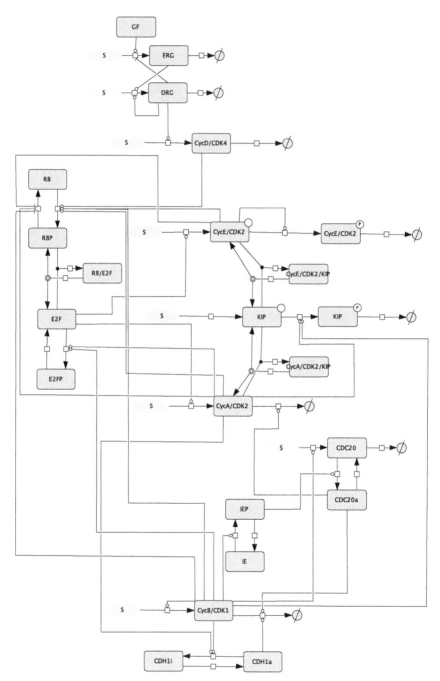

FIGURE 7.4 Network of Novak and Tyson cell cycle. Continuous model cycle of the restriction point developed. Figure adapted from Novak and Tyson (2004).

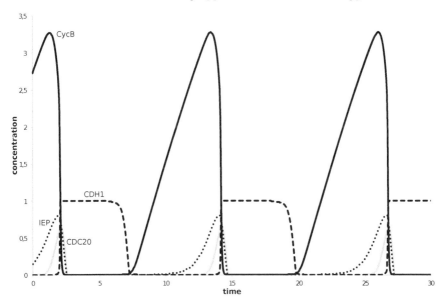

FIGURE 7.5 Simulations of Novak and Tyson cell cycle model. Simulation of the continuous model cycle of the restriction point developed by Novak and Tyson. The concentrations of CycB, CDC20, IEP, and CDH1 are plotted as a function of time.

after 1 hour. The different *in vitro* experiments show that (1) with constant growth factors, the division time is about 14 hours with 7 hours spent in G1; (2) when the cells are treated during the first 3 hours of the G1 phase, they divide with an 8–9 hour delay for the first division following the treatment and with no delay for the second division; (3) when the cells are treated after the fourth hour of the beginning of the G1 phase, no delay is observed in the first or the second division, setting the restriction point between the third and the fourth hour of the G1 phase, (4) and when the cells are treated late in the cycle, they show no delay at the first division but might show a delay in the second division. Some more recent computational analyses have been further performed on this model by Conradie et al. (2010). The model can be found and simulated on the Biomodels webpage with the ID: BIOMD0000000265.

A similar computational model of the G1 to S transition (Qu et al., 2003b) showed that, for this transition, bistability arises from a positive feedback loop involving CycE/CDK2 and E2F, whereas the oscillatory behaviour is governed by another positive feedback loop involving CycE/CDK2 and p27 but these oscillations are born only when p27 is phosphorylated at multiple sites (ultra sensitivity is required). The model is also available in the Biomodels database with the ID: BIOMD0000000110.

Once the model is capable of reproducing experimental observations, it becomes a tool to test other experiments *in silico* that have not been performed yet, and can suggest quantitative results such as length of delays or cell cycle times to diverse perturbations.

7.3.4 Boolean formalism

Boolean (logical) formalism offers a more qualitative approach to the dynamical analysis of a biological process that the ODE formalism. As a consequence, less precise information is needed for building a Boolean model. Because of their apparent simplicity, the extension and refinement of logical models are more easily manageable than the extension of an ODE model. The counterpart is that a qualitative model can only give qualitative, therefore limited, insights. The activity — rather than the concentration — of the biological species is associated to each node of the regulatory graph. Moreover, real time is not considered in this framework. Every step in the model corresponds to an event, but the time associated to an event is not explicitly given.

To construct a Boolean model, first, the biological information needs to be translated and summarised into a regulatory network where nodes are biological species and edges are influences (see **Box 7.1**). Then, the conditions for a node to turn ON or OFF are defined by a Boolean logic (see **Box 7.2**). Finally, the updating strategy needs to be chosen: synchronous, asynchronous or a mixture of both.

❑ **BOX 7.1: Boolean models**

They are composed of a set of variables and of logical update rules. In our context, biological knowledge is represented as a graph (see **Box 6.4**). We define two types of graphs:

Regulatory graphs are *influence* graphs. They are signed and directed graphs, composed of nodes and edges. **Nodes** are variables and can be genes, proteins, complexes, processes, small molecules, RNA, *etc.* **Edges** are directed arrows. They are relationships between two nodes and provide two types of information: they can be activating (in this case, the sign of the interaction is +), or inhibiting (in this case, the sign of the interaction is −).

State transition graphs consist of nodes and arcs. **Nodes** are states (composed of the value of each node). **Arcs** are transitions between these nodes.

More specifically, a Boolean regulatory graph connects, with logical rules (or logical functions), a set of discrete variables whose states depend on the state of the other variables. Each variable corresponds to a node in the graph. A logical rule is assigned to each node of the graph, defining how the different inputs (incoming arrows) combine to control its level of activation. From given initial conditions and at each discrete time step or event, one or more variables will change according to the chosen updating strategy. If the strategy is synchronous, then all the variables that can change are updated (in a deterministic fashion). If the strategy is asynchronous, then one of the variables that can change will be updated at a time (in a non-deterministic fashion). The different scenarios of all the possible changes are described in a graph called the *state transition graph* (see **Box 7.1**). The state transition graphs are usually different for the synchronous and the asynchronous cases.

There are two types of asymptotic solutions or attractors in a logical model: stable states or limit cycles. If the successor state of one state is itself, it is considered to be a stable steady state or point attractor. If a set of states is visited more than once in a repeated order, it is considered to be a limit cycle attractor.

❏ **BOX 7.2: Boolean logic**

It is a mathematical formalism for which a set of variables (a, b, c, *etc.*) can only take two values: TRUE or FALSE; 1 or 0; present or absent; active or inactive; expressed or not expressed. Each variable is updated according to the logical rule that defines its activity. The state of a variable is a function of its inputs or regulators. Some operands connecting the variables are defined. The ones that are mainly used in our study are the following:

- a AND b: is TRUE if only if the two variables are TRUE
- a OR b: is TRUE if at least one of the two variables is TRUE
- NOT a: is TRUE if only if the variable is FALSE

7.3.5 Cell cycle as a Boolean model: Application to the restriction point

With the purpose of translating the continuous model of the restriction point presented by Novak and Tyson into a Boolean model, Faure et al. (2006) transposed the biochemical reaction network into an influence network (see **Figure 7.6**). The authors not only translated the model into a different formalism but they also extended it by adding a E2 ubiquitin enzyme, UbcH10 responsible for the CDH1-dependent degradation of CycA. They then derived the logical rules.

To account for the sequential events of the cell cycle, they first proposed to separate fast and slow processes. By doing so, they were able to group

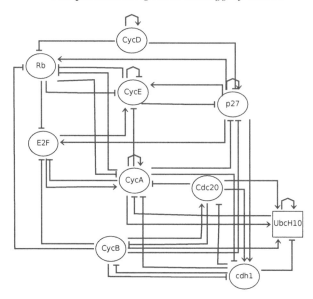

FIGURE 7.6 Boolean cell cycle of the restriction point developed by Faure et al.. Figure adapted from Faure et al. (2006).

the variables of the model into two classes: fast and slow. They referred to these classes as *priority* classes. Through thorough analysis of the obtained asymptotic solutions with the asynchronous updating strategy, they realised that some processes inside each of these classes were controlled by similar mechanisms and therefore could be updated synchronously. They further separated the fast and slow classes into two classes updated with synchronous and asynchronous strategies leading to four classes: fast synchronous, fast asynchronous, slow synchronous, fast asynchronous.

By defining these four classes and introducing some biological considerations in the updating choices, the authors were able to qualitatively reproduce the behaviour of both wild type and of mutants.

7.4 Decomposition of the generic cell cycle into motifs

Even the simplest model can reveal interesting aspects of a biological process. For the neophytes or the systems biologist skeptics, the use of mathematics to apprehend a biological problem can be seen, at first, as mysterious and somewhat magical. However, looking closely, there exists a link between the physiology and the result of mathematical simulations. For instance, ev-

❏ **BOX 7.3: Boolean vs. Chemical kinetics**

The differences of the Boolean and chemical kinetics approaches in terms of formalism, type of associated diagrams, input, and output are listed below:

Formalism	Diagram	Input	Output
Boolean	influence network	Set of logical rules	State transition graph with stable state solutions
		Set of initial conditions	Trajectories in the state transition space
Chemical kinetics	reaction network	Set of chemical reactions, set of kinetic laws, set of parameter values, set of initial conditions	Time series (concentration of variables)

ery time an oscillatory behaviour is observed, such as in the cell cycle or in the circadian rhythms, the mathematician may look for a functional negative feedback loop in the network coupled or not with a positive feedback loop that could explain the process. Similarly, an abrupt transition will most certainly have a positive feedback or a feed-forward loop hidden somewhere in the network. Theoretical works have explored some of these network motifs (Alon, 2007b; Tyson et al., 2003; Csikász-Nagy et al., 2009). We propose to study two motifs with both the ODE and the Boolean approaches. The motifs are extracted from the models of Novak and Tyson and of Faure et al. presented previously. However, for each purpose, the kinetics are simplified and the parameters slightly modified.

7.4.1 Positive feedback loop

To illustrate the positive feedback loop dynamics, we propose a two-node network illustrating the transition from M phase to G1 phase. In M phase, the cyclin-dependent kinase complex CycB/CDK1 is active and maintains CDH1 inactive through phosphorylation. After a series of activation, CDH1 is dephosphorylated and turned on. CDH1 is an ancillary protein that brings the CycB/CDK1 complex to the Anaphase Promoting Complex (APC) for degradation. When the level of CycB drops, the cell exits mitosis.

For simplicity, several hypotheses have been made on the networks and on the dynamics:

1. *Hiding implicit molecules*: the complexes Cyc*/CDK* is often represented by Cyc* (CycD, CycE, CycA or CycB) only while the CDK* (CDK1, CDK2, CDK4 or CDK6) partner is implicitly present. It is absolutely required for the activation of the complex. The reader has to assume in the rest of the chapter that Cyc* refers to the complex Cyc*/CDK*;

2. *Simplifying complex biochemical events* (with a compact and phenomenological mathematical formulation): in the continuous model, the activation and inactivation of CDH1 are considered to be nonlinear abrupt events. To model the switch between phosphorylated CDH1i (i for inactive) and unphosphorylated CDH1a (a for active) forms, the function representing a Goldbeter-Koshland switch is used;

3. *Synchronicity of the population*: the dynamics is that of one cell that would mimic the behaviour of a synchronised population of cells, the obvious stochasticity of cells is not taken in account here.

The corresponding equations of the positive feedback loop involving CycB (recall CycB stands for CycB/CDK1) and CDH1 (see **Figure 7.7A**) are written as follows:

$$\frac{dCycB}{dt} = k_1 - (k_{2a} + k_2 \cdot CDH1a) \cdot CycB \tag{7.5}$$

$$\frac{dCDH1a}{dt} = \frac{k_3 \cdot CDH1i}{J_3 + CDH1i} - \frac{k_4 \cdot CycB \cdot CDH1a}{J_4 + CDH1a} \tag{7.6}$$

with the corresponding parameters used in the simulation of **Figure 7.7B** and **C**: $k_1 = 0.15$, $k_{2a} = 0.2$, $k_2 = 1$, $k_3 = 0.2$, $k_4 = 1$, $J_3 = 0.05$, $J_4 = 0.05$ and $CDH1a + CDH1i = 1$.

As expressed in **Equations 7.5** and **7.6**, CDH1a activates and inactivates abruptly. For CycB, both activation and inactivation are chosen to be governed by mass action kinetics: it is synthesised at a constant rate (k_1) and inactivated by a constant term (k_{2a} for background degradation and k_2 for inactivation by CDH1a).

Depending on the initial conditions: CycB=1, CDH1a=0 or CycB=0, CDH1a=1, the system shows two possible outputs, either a M arrest where CycB remains active all the time because CDH1a cannot activate (see **Figure 7.7B**), or a G1 state where CDH1a is active and keeps CycB from activating (see **Figure 7.7C**).

A change of behaviour can be observed by changing parameter values. For instance, one can simulate exit from mitosis or the passage from M to G1 by varying the value of k_1, the parameter controlling the synthesis of CycB. The initial condition needs to be set to a G1-like state. Therefore, we let CycB=0, and CDH1=1. We follow the variation of k_1 as a function of the activity of CycB (see **Figure 7.7D**). The system exhibits what is referred to as a hysteresis: for specific starting conditions and for a certain range of

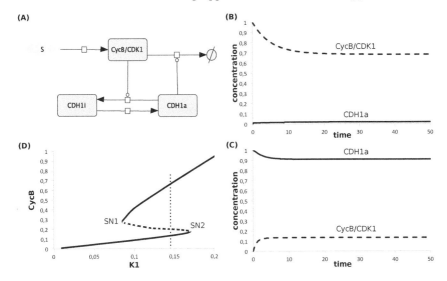

FIGURE 7.7 ODE model of the positive feedback loop. (A) CycB is synthesised (and implicitly here forms a complex with CDK1). Note that the variable CycB/CDK1 taken from Novák and Tyson's diagram corresponds to the variable CycB of **Equation 7.5**). CycB then phosphorylates and inactivates CDH1 (CDH1i). When CDH1 is in its unphosphorylated form (CDH1a), it is able to promote the degradation of CycB. (B) Continuous simulation of the nonlinear differential equations with initial conditions CycB = 1 and CDH1a = 0 and for $k_1 = 0.15$. (C) Continuous simulation of the nonlinear differential equations with initial conditions CycB = 0 and CDH1a = 1 and for $k_1 = 0.15$. (D) Bifurcation diagram of the positive feedback loop with SN1 = 0.085 and SN2 = 0.17.

k_1 values, there exist two different coexistent stable solutions that depend on the system history. The upper and the lower branches are stable, the middle branch is unstable.

Let us start with CycB equal to 0 for low values of k_1. CycB is inactive (lower branch of the hysteresis). As k_1 increases, it reaches a threshold value (SN1 = 0.085, where SN stands for Saddle Node bifurcation point). The qualitative stability of the steady state does not change but two stable solutions now coexist. The two stable states surround a middle unstable state (dashed line). Because of where the system started from, CycB remains inactive. As k_1 increases, it reaches the second threshold value (SN2 = 0.17). At this value, the stability of the solution is lost and the system jumps to the higher stable branch where the CycB turns on. For k_1 above the threshold value of SN2, CycB is considered to be activated and as a result starts to phosphorylate and inactivate CDH1a. CycB switch is rather abrupt. If at this point, k_1 is decreased, the system remains on the higher branch until SN1 threshold value is reached again and CycB can start to activate *de novo*. The resulting graph

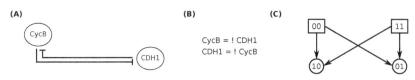

FIGURE 7.8 Boolean model of the positive feedback loop involving CycB and CDH1. (A) Boolean positive feedback loop involving CycB and CDH1 (CDH1a). (B) Logical rules associated to each node CycB and CDH1. (C) Transition graph of the Boolean model.

is called a bifurcation diagram (see **Box 7.4**) which follows the stability of the differential system as a key parameter is changed.

The reaction network showed in **Figure 7.7** can be simplified into an influence network. All biochemical reactions of the graph are interpreted as activations or inhibitions based on the biological meaning of these interactions. Thus, in the reaction network, CycB phosphorylates and inactivates CDH1. In the influence network, it is translated into: CycB inhibiting CDH1. Similarly, CDH1 is mediating the degradation of CycB in the reaction network, therefore, in the influence network, CDH1 inhibits CycB (see **Figure 7.8A**).

The logical rules (see **Figure 7.8B**) are derived from the influence network and depend on the sign of the influences (positive for activation, negative for inhibition). A feedback loop is positive if the product of the signs of all the influences is positive. The logical rules determine the state of the nodes at the following event. In asynchronous update strategy, only one node can be updated at a time. Therefore if the initial condition is [CycB, CDH1] = [0,0], then either CycB or CDH1 can turn ON. If the initial condition is [CycB, CDH1] = [1,1], then either CycB or CDH1 can turn OFF. If the initial condition is [CycB, CDH1] = [0,1], then the system is stable and if [CycB, CDH1] = [1,0], the system is also stable.

The construction of the asynchronous transition graph (see **Figure 7.8C**) shows the existence of two solutions or stable steady states: the first one, [CycB, CDH1] = [0,1] which is equivalent to a G1 arrest where CDH1 keeps CycB from accumulating, and the second one, [CycB, CDH1] = [1,0] which is equivalent to the M phase state in which CycB level remains high.

7.4.2   Negative feedback loop

To illustrate the negative feedback loop, we consider a three-node loop extracted from Novak and Tyson's model (see **Figure 7.9A**). The loop involves CycB (noted CycB/CDK1 in the diagram) and its regulation in the M phase. When CycB is synthesised (and implicitly forms a complex with CDK1), the cell enters the M phase. CycB is known to switch on its own degradation pathway creating a possible functional negative feedback loop. Many studies have shown that in a continuous framework, there need to be at least three species

---

❏ **BOX 7.4: Bifurcation**

bifurcation diagram is a diagram that shows solutions of the set of differential equations as a parameter is varied. It highlights the points at which the stability changes qualitatively (Kuznetsov, 2004). The solutions of the dynamical system can be steady states, stable or unstable, and limit cycles.

bifurcation point is a sudden qualitative change in the solution of the dynamical system for a small change in the parameter value. These qualitative changes can be: a change of stability (from unstable to stable and vice versa), the apparition or disappearance of a solution (limit cycle oscillations), *etc.*

---

in order to give rise to oscillations in a negative feedback context (Ferrell et al., 2011). Between the activation of CycB and the activation of CDC20, which is involved in the degradation of CycB, Novak and Tyson have proposed a delay that could be imposed by the activation of a nonidentified intermediary enzyme, IE, that would need to be phosphorylated to be activated, IEP (see **Figure 7.9A**). The three-node network forms a negative feedback loop.

The ODEs are written with the hypothesis that CDC20 is governed by a Goldbeter-Koshland switch whereas both CycB and IEP follow mass action kinetics. IEP's kinetics has been simplified from the initial model of Novak and Tyson in which it was described as a Goldbeter-Koshland switch as well. Note that other hypotheses in the choice of the format of the equations might be appropriate as well. The translation of a reaction network into chemical kinetics framework is not unique.

The corresponding equations are written as follows:

$$\frac{dCycB}{dt} = k_1 - (k_{2a} + k_2 \cdot CDC20a) \cdot CycB \tag{7.7}$$

$$\frac{dCDC20a}{dt} = \frac{k_{a20} \cdot IEP \cdot (CDC20_{tot} - CDC20a)}{J_{a20} + (CDC20_{tot} - CDC20a)} \\ - \frac{k_{i20} \cdot CDC20a}{J_{i20} + CDC20a} \tag{7.8}$$

$$\frac{dIEP}{dt} = k_{1IP} \cdot CycB - k_{2IP} \cdot IEP \tag{7.9}$$

with the corresponding parameters used in the simulation of **Figure 7.9B**: $k_{a20} = 1$, $k_{i20} = 0.7$, $J_{a20} = 0.01$, $J_{i20} = 0.01$, $CDC20_{tot} = 1$, $k_1 = 0.15$, $k_{2a} = 0.01$, $k_2 = 1.5$, $k_{1IP} = 1$, $k_{2IP} = 0.3$ and initial conditions $CDC20a = 0$, $IEP = 0$, $CycB = 0$.

For the chosen initial conditions of **Equations 7.7** to **7.9**, the system

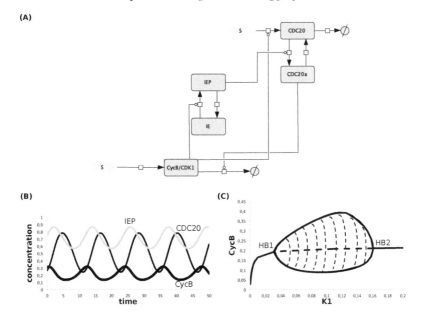

FIGURE 7.9 Continuous model of the negative feedback loop. (A) Reaction network of a negative feedback loop involving CycB (noted CycB/CDK1 in the diagram), IE, and CDC20. (B) The 3 proteins exhibit sustained oscillations with different amplitudes but with the same period (simulations with XPPAUT). (C) the bifurcation diagram shows the qualitative change of CycB as a function of the parameter k_1, governing the synthesis of CycB. HB1 and HB2 correspond to the Hopf bifurcation points. Plain lines correspond to stable steady states and stable limit cycles, and dashed lines to unstable steady states.

shows one cycle before reaching a stable limit cycle oscillatory regime (see **Figure 7.9B**). CycB is the first to rise (increasing concentration), followed closely by IEP and then by CDC20a. When CDC20a reaches a critical value, CycB starts to get degraded and inactivates (decreasing concentration). As a consequence, IEP inactivates as well, leading to a more abrupt drop in CDC20a activity. CDC20a being off, CycB has the possibility to rise again. And another cycle starts.

We choose k_1 as the varying parameter to explore the solutions of our system of ODEs (see **Figure 7.9C**). For low values of k_1, the system cannot oscillate. It is stuck in a stable steady state. As k_1 reaches a critical value (HB1 = 0.0318, where HB stands for Hopf Bifurcation point), the stable state becomes unstable and oscillations are born. The upper and lower values of the amplitude of these oscillations are indicated on the graph. For values close to the bifurcation point, the amplitudes are small and the system is sensitive to perturbations. Indeed, for a small change in k_1, it can easily move back to the stable state. As we get away from the bifurcation point, the amplitudes become

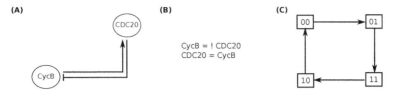

FIGURE 7.10 Boolean model of the negative feedback loop involving CycB and CDC20. (A) Influence network representing the negative feedback loop. (B) Logical rules governing the negative feedback loop. (C) State transition graph showing the solutions of the system.

more pronounced to finally die at the second bifurcation point (HB2 = 0.1608). The unstable state becomes stable again. With the bifurcation diagram (see **Box 7.4**), we have been able to prove the existence of oscillations and to delimitate the range of possible values for k_1 for which the system would exhibit this oscillatory regime.

When the reaction network of the negative feedback loop is translated into an influence network (see **Figure 7.10A**), the intermediary enzyme IE is no longer needed. The three-node reaction network becomes a two-node influence network. To each node is associated a logical rule (see **Figure 7.10B**). The transition graph shows no stable steady state (see **Figure 7.10C**). From any initial condition, the system is caught in the cycle solution, where each state is visited in an ordered suite: [00, 01, 11, 10].

7.4.3 Positive and negative feedback loops

If we combine both positive and negative feedback loops into one single model and if we choose the appropriate parameter values, the saddle node and the Hopf bifurcations (see **Figure 7.11A**) are maintained in the full model. With another set of parameters though, all bifurcation points may not have been conserved.

The positive and negative feedback loops share one component: CycB. The differential equations are then combined into one complete model from the positive and negative feedback models previously translated. The equations are the following:

$$\frac{dCycB}{dt} = k_1 - (k_{2a} + k_{2p} \cdot CDC20a + k_{2pp} \cdot CDH1a) \cdot CycB \quad (7.10)$$

$$\frac{dCDH1a}{dt} = \frac{k_3 \cdot CDH1i}{J_3 + CDH1i} - \frac{k_4 \cdot CycB \cdot CDH1a}{J_4 + CDH1a} \quad (7.11)$$

$$\frac{dCDC20a}{dt} = \frac{k_{a20} \cdot IEP \cdot (CDC20_{tot} - CDC20a)}{J_{a20} + (CDC20_{tot} - CDC20a)}$$
$$- \frac{k_{i20} \cdot CDC20a}{J_{i20} + CDC20a} \quad (7.12)$$

$$\frac{dIEP}{dt} = k_{1IP} \cdot CycB - k_{2IP} \cdot IEP \quad (7.13)$$

$$\frac{dCDH1a}{dt} = \frac{k_3 \cdot CDH1i}{J_3 + CDH1i} - \frac{k_4 \cdot CycB \cdot CDH1a}{J_4 + CDH1a} \quad (7.14)$$

with the corresponding parameters used in the simulation of **Figure 7.11B**: $k_1 = 0.15$, $k_{2a} = 0.02$, $k_{2p} = 1$, $k_{2pp} = 0.65$, $k_3 = 0.2$, $k_4 = 1$, $J_3 = 0.01$, $J_4 = 0.01$, $k_{a20} = 1$, $k_{i20} = 0.7$, $J_{a20} = 0.01$, $J_{i20} = 0.01$, $CDC20_{tot} = 1$, $k_{1IP} = 1$, $k_{2IP} = 0.3$, $CDH1a + CDH1i = 1$ and the initial conditions $CDC20a = 0$, $IEP = 0$, $CDH1 = 1$, $CycB = 0$.

All simulations of the ODE models of this chapter, including **Equations 7.10** to **7.14**, are computed using XPPAUT. The system oscillates in an ordered manner and with the same period (see **Figure 7.11B**). CDH1 is turned off as CycB starts to be synthesised. As CycB rises, IEP, after reaching a threshold, activates CDC20 which starts to turn off CycB. When CDH1 rises again because CycB is starting to decrease, it degrades CycB enough to lead to exit of mitosis and entry in G1 phase.

With this simple model, the separation between the four phases of the cell cycle is not as obvious as it is in Novak and Tyson's description of the cell cycle. As more components and biochemical details are added, the description becomes more and more refined and the phases better separated.

The bifurcation diagram shows that to maintain the oscillatory behaviour, k_1 must be between HB1 and HB2 values (see **Figure 7.11C**). The bifurcation diagram confirms the existence of a hysteresis combined with a region of oscillations. Adding the positive feedback to an oscillating system lengthens the cell cycle time by introducing a G1 phase. The region for hysteresis is very small (k_1 between SN1 and SN2). In this model, for this specific set of parameters, the transition G1 to S is not prevalent.

The combined positive and negative feedback loop model is translated into Boolean formalism with CycB, CDH1, and CDC20 (see **Figure 7.12A** and **B**). The state transition graph contains 8 states (2^3) leading to one stable steady state (see **Figure 7.12C**) corresponding to a G1 arrest: [CycB, CDC20, CDH1] = [0,0,1]. There exists also a cycle attractor which involves CycB and CDC20 only, for which the state of CDH1 is not changing (equal to 0): [000, 010, 110, 100].

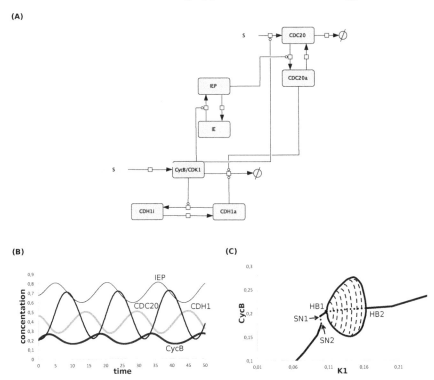

FIGURE 7.11 Continuous model of the combined positive and negative feedback loops. (A) Reaction network representing the positive and the negative feedback loops. (B) Continuous simulation and bifurcation diagram of the feedback loops using XPPAUT. (C) The bifurcation diagram shows the qualitative change of CycB as a function of the parameter k_1 with SN1 = 0.097, SN2 = 0.099, HB1 = 0.107, HB2 = 0.1624.

To study network motifs more in depth, some thorough reviews and detailed articles from a biology, mathematics and physics point of view are available: Alon (2007b), Tyson et al. (2003), Prill et al. (2005), Ferrell et al. (2011), Kholodenko (2000), Kashtan et al. (2004), Alm and Arkin (2003), Thomas (1981), *etc.*

7.5 Conclusion

Mathematical models provide a systematic tool to investigate cell behaviours in various contexts and propose mechanisms under the form of a

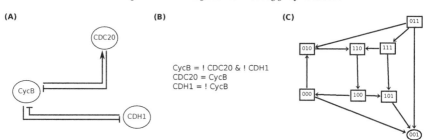

FIGURE 7.12 Boolean model of the combined positive and negative feedback loops. (A) Influence network representing the positive and the negative feedback loops. (B) Logical rules governing the model. (C) State transition graph.

network that recapitulates experimental observations. They translate biological knowledge into mathematical terms. They aim at shedding some light on some observed contradictions and paradoxes and summarising what is known about particular biological processes. With a mathematical model, hypotheses can be tested *in silico* before performing experiments in the wet lab. In the next chapter, we will see how mathematical models are used to understand the different hallmarks of cancer.

✎ Exercises

- Using GINsim (or another software for Boolean modelling), build the combined positive and negative feedback loop model and visualise the state transition graph. What happens to the stability and the state transition graph when CDH1 is deleted? when CDC20 is deleted?

- In a thorough study of Feed Forward Loop (FFL) (Csikász-Nagy et al., 2009) presented the features of coherent and incoherent feed forward loops. For 3 genes, A, B and C, a coherent loop links A to C with two paths, one direct and another one indirect, going through B. The signs of both paths ending in C are the same. For example: A activates C and A activates B which activates C. In an incoherent loop, the signs of both paths ending in C are opposite. For example: A activates C and A activates B which inhibits C. This design might present some advantages in certain cases. Both paths may be able to account for different time scales of concurrent or compensating signals.

  Problem: In a Boolean framework, create a model of an incoherent feed forward loop including p27, an inhibitor of CycE. CycE and p27 are both activated by E2F. E2F is considered as an input here. Build the state transition graph corresponding to the model. What are the expected two stable states?

- In the ODE model including both positive and negative feedback loops (**Section 7.4.3**), simulate the temporal simulations and the bifurcation diagram if the effect of CDC20 degradation on CycB, k_2 was reduced from 1.5 to 0.15 with the same initial conditions of the model.

⇨ Key notes of Chapter 7

- Mathematical models can be a descriptive object, a hypothesis generator, a predictive tool and a conceptual tool.

- The biological questions determine which mathematical formalism is the most appropriate to answer them.

- A complex model can be decomposed into feedback loops: positive, negative, feed forward, *etc.*

- Positive feedback loops can give rise to bistability.

- Negative feedback loops can give rise to oscillations.

- Saddle node bifurcations are associated to bistable behaviours.

- Hopf bifurcations are associated to oscillatory behaviours.

Chapter 8

Mathematical modelling of cancer hallmarks

As mentioned before, in 2000 Hanahan and Weinberg published a seminal review on hallmarks of cancer (also presented **Chapter 2**) in which they proposed six features for cancer formation:

1. Sustaining proliferative signalling
2. Evading growth suppressors
3. Activating invasion and **metastasis**[*]
4. Enabling replicative immortality
5. Inducing angiogenesis
6. Resisting cell death

According to *Merriam-Webster* dictionary, a hallmark is a *distinguishing trait, characteristic or feature.* Therefore, a hallmark of cancer is what defines a cancerous cell (see **Chapter 1**). In 2010, Lazebnik questioned to some extent the term *hallmarks* used by Hanahan and Weinberg in the sense that benign and malignant cells share a lot of these hallmarks. According to him, among the six hallmarks listed in the review of 2000, only one is really specific to malignant tumours, *i.e.* invasiveness. The other five share common features of benign cases. The question is then: Are these hallmarks appropriate to characterise cancer as we intend it? Cancer is a combination of deregulations of more than one feature, one can wonder how many of these cellular functions need to be affected to cause malignancy? Is one hallmark safe? Are two or three hallmarks sufficient to cause cancer? If so, which ones? Most of the mathematical models on cancer concentrate on an individual hallmark but some investigate the roles and the dynamics of these hallmarks put together. Among them, Abbott et al. (2006) implement the six hallmarks in a model based on a cellular automaton.

In 2011, Hanahan and Weinberg revised the list of hallmarks by adding four more traits, two enabling characteristics and two emerging hallmarks:

7. Genome instability and **mutation**[*]
8. Tumour-promoting inflammation
9. Reprogramming energy metabolism
10. Evading immune destruction

In this chapter, we propose to review the current status of knowledge around some of these hallmarks from a mathematical perspective and focus on some specific aspects of each hallmark and characteristic as an example.

8.1   Modelling the hallmarks of cancer

Mathematical models related to the eight individual hallmarks and the two enabling characteristics of Hanahan and Weinberg (2011) along with the questions they raise are reviewed here. Note that, as of today, some of these hallmarks have not been studied enough from a mathematical point of view to provide a review for it.

8.1.1   Sustaining proliferative signals

Extracellular signals are passed on to the inside of the cell via the activation of a complex network which ultimately leads to cell responses such as the promotion or the inhibition of cell proliferation. Both the growth activation and inhibition are monitored by **signalling pathways*** initiating from an external signal to diverse checkpoints of the cell cycle.

The questions that motivated the mathematical modelling of these pathways are of different types. Among them:

1. How are the initial signals passed on to the cell? What is the molecular mechanism involved in the cascade activation? What are the key components of the cascade? What are the critical parameters that control the activation of the cascade?

2. How is the signal amplified? How is the noise filtered? What is the shape of the signal? What is the role of feedbacks in the propagation of the signal?

3. What deregulations of the signalling cascade lead to cancer?

It is known that cancer cells drive their own proliferative signals and do not rely on external stimuli to activate growth. In normal conditions, growth signals (growth factors, mitogens, stress, heat shock, *etc.*) from the outside of the cell are transmitted to the inside of the cell through the Mitogen-Activated Protein Kinase (MAPK) cascade. The MAPK cascade is playing a role in gene expression, cell cycle machinery, survival, apoptosis and differentiation. Many mutations of the MAPK pathway have been associated with cancer (Dhillon et al., 2007).

There are six groups of MAPKs: the extracellular signal-regulated kinases, *ERK1* (and *ERK2*), *ERK5*, *ERK3* (and *ERK4*), *ERK7* (or *ERK8*), *JNK* and *p38*. Each cascade has the same format: there are three kinases that are sequentially activated through double phosphorylations (at serine and threonine sites). At the top of the cascade, ligands bind to receptor tyrosine kinases that lead to dimerisation and autophosphorylation of the receptors which then bind to adapters (such as *GRB2*). The Mitogen-Activated Protein Kinase Kinase Kinase (MAPKKK) are activated by phosphorylation or by RAS, a GTP protein, itself activated by external stimuli. Consequently Mitogen-Activated Protein Kinase Kinase (MAPKK) and MAPK are activated, very often in the presence of **scaffold proteins.**[*] These successive activations lead to the synthesis of genes including transcription factors. According to the transcription factors that are transcribed, cells will enter the cell cycle, start apoptotic events, differentiate, *etc.*

The constant activation of this pathway is often observed in cancers (Downward, 2003). The constitutive growth signal might come from different sources:

Constitutive ligand supply: The constant production of growth factor ligands can result from the deregulation of paracrine (from one cell to its neighbours within the same tissue but through diffusion) or autocrine (within each cell) loops. There exist other ways for growth factors to abnormally send signals to other cells: through the deregulation of juxtacrine loop (from one cell to adjacent cells after secretion) and of endocrine loop (from one cell to distant cells transported through the bloodstream).

High levels of receptors: The amount of receptors is such that they are not limited by ligand supply. The high concentration of receptor proteins at the cell level can be the result of a mutational defect that leads to constant production of the receptor or a structural alteration of the receptor itself. Mutations leading to overexpression or upregulation of the EGF receptor, a member of the ErbB family identified as an **oncogene**[*], are often found in most **carcinomas**[*] and overexpression of *HER2* is often associated to breast cancer.

Constitutive activation of components of the pathway: Mutations of any actor of the MAPK pathway such as *RAS* or *BRAF*, downstream of the receptor activation, can overpass any incoming signals.

Amplification[*] **of transcription factors:** At the bottom of the MAPK cascade, transcription factors such as *MYC* are activated. In many cancers, they are found to be amplified and therefore independent of the MAPK activation.

The biochemical description of the pathway has raised a lot of interest and many mathematical models, both descriptive and quantitative, have proposed mechanistic explanations of the pathway functioning (Kholodenko, 2002; Schoeberl et al., 2002; Levchenko et al., 2000; Chen et al., 2009). More particularly, the activation of this pathway has been associated with a bistable, ultrasensitive and irreversible switch in frogs (Huang and Ferrell, 1996) and most probably in mammals as well. Mathematical models have shown that the MAPK pathway is controlled by a hysteresis (Bhalla and Iyengar, 1999) (see **Section 7.4.1**), and that the ultrasensitivity arises from the multiple phosphorylations on the receptor sites. All these features of the switch allow the filtering of noise and the activation of the pathway only when the signal is frank.

As presented in a review on modelling of signalling pathways and more particularly that of the MAPK cascade (Klipp and Liebermeister, 2006), mathematical models are used to model different aspects of the signalling pathway: the relative amount of phosphatases and kinases (Hornberg et al., 2005; Bhalla and Iyengar, 1999), signal amplification (Heinrich et al., 2002; Shibata and Fujimoto, 2005; Mayawala et al., 2004), the effect of feedback loops (Kholodenko, 2000; Huang and Ferrell, 1996), the effect of scaffolding (Levchenko et al., 2000), and the importance of crosstalks between the MAPK cascade and other pathways (Schwartz and Baron, 1999). Heinrich et al. (2002) have proposed a model that explored three key aspects of the cascade: the amplitude of the signal, the time of the signal, and the duration of the signal. According to the value and the kinetics associated to these variables, the output signal will be different.

Note that in the cell cycle model presented in **Section 7.3**, in response to growth factor activation, the MAPK cascade leads to the synthesis of CyclinD1, the activation of the complex CyclinD1/CDK4 (referred to as CycD/CDK4 in the model) and the inactivation of RB.

8.1.2   Evading growth suppressors

It is the role of tumour suppressor genes to negatively regulate cell proliferation. Among these tumour suppressor genes, *RB* and its role in the cell cycle has been mentioned in **Chapter 2**. Another important tumour suppressor gene is *TP53*. *TP53* plays a significant role in cell death and more particularly the regulation of apoptosis in response to DNA damage (see **Section 8.1.5**).

Similar to the MAPK kinase, there exist external negative signals that activate inhibitors of the cell cycle when growth needs to be stopped. In normal conditions, these signals are often ON when the cells have entered a post-mitotic state, senescent or differentiated state. The same types of questions that were posed for the growth activation pathways can be addressed for the growth inhibitory pathways.

The transforming growth factors (TGFβ) pathway is one of these signalling pathways that ultimately leads to the transcription of cell cycle inhibitors such as *p15INK4b* and *p21CIP1*, blocking the cycle at G1/S or G2/M transitions. It is also largely involved in loss of contact inhibition and cell evasion (see **Section 8.1.4**)

Depending on the cell conditions, TGFβ is both a **tumour suppressor**

gene[*] and a growth promoting gene. The TGFβ family members are cytokines involved in anti-proliferation and evasion mechanisms. Depending on the cell type, TGFβ can be involved in other pro-cancerous mechanisms. Indeed, it can promote proliferation, apoptosis, angiogenesis, cell motility, differentiation and survival.

TGF ligands bind to Type I and Type II receptors at the cell surface. The newly-formed receptor unit phosphorylates cytoplasmic regulators, the R-Smads. Once phosphorylated, the R-Smads recruit other Smad partners and translocate to the nucleus, where genes are transcribed in a cell-type dependent manner.

Some of the mathematical models developed around TGFβ concentrate on the regulation at the receptor level (Vilar et al., 2006), others focus on Smad phosphorylation (Clarke et al., 2006), or on the nuclear transport and the transient activating signal coming from a negative feedback loop (Melke et al., 2006). Some studies have been proposed on the different roles of TGFβ such as its role in cancer and more particularly in cell motility (Wang et al., 2007). More comprehensive models include aspects of TGFβ signalling (both mechanistic and quantitative) that offer a more precise view of the possible deregulations of this cascade in cancers (Zi and Klipp, 2007).

8.1.3 Activating invasion and metastasis and the role of tumour microenvironment

Tumour expansion into adjacent tissues and the appearance of metastasis are two important steps in **tumorigenesis**[*] converting a tumour into malignant disease threatening the life of the organism. In this sense, the hallmark of cancer connected to invasion is one of the most important from a clinical perspective. *Tissue invasion by tumour* is usually associated with dissemination of cancer cells into adjacent tissues while *metastasising* is associated with *remote* colonisation of distant tissues. A **metastatic**[*] colony is a result of a continuous process starting from the early growth of the **primary tumour**[*], and detachment of invasive tumour cells from the primary tumour leading to the colonisation of other organs.

Genetic changes causing an imbalance of growth regulation lead to uncontrolled proliferation necessary for both primary tumour and metastasis expansion. However, unrestrained growth does not, by itself, cause invasion and metastasis. This phenotype may require additional genetic changes. Thus, tumorigenicity and metastatic potential have both overlapping and separate features. For example, it was demonstrated that in addition to loss of growth control, an imbalanced regulation of motility and proteolysis appears to be required for invasion and metastasis (Liotta and Kohn, 2003). Therefore, interaction of cancer cells with surrounding *tumour microenvironment* is of extreme importance for understanding the mechanisms of invasion.

One of the most important molecular mechanisms involved in tissue invasion and metastasis is **Epithelial-to-Mesenchymal Transition (EMT)**[*].

It is an important process during embryonic development, cancer dissemination and wound healing. In each of these events, cells are required to migrate from one location to another. In order to achieve this, the cell must acquire the capacity to infiltrate surrounding tissue and to ultimately migrate to a distinct site. An overt feature of this process is a change of cell morphology from a sedentary type to one that facilitates migration in the **Extracellular Matrix (ECM)*** and settlement in an area involved in tissue growth and repair (or cancer metastasis). EMT is characterised by loss of cell adhesion, repression of E-cadherin, and increased cell mobility.

To establish a successful colony, a cancer cell must successfully bypass a number of obstacles: leave the primary tumour, enter the lymphatic and blood circulation, survive within the blood circulation, overcome host defences, extravasate and grow as a vascularised metastatic colony. At each step, there are big chances for a cancer cell to die. As a result, a very small percentage ($< 0.01\%$) of Circulating Tumour Cells (CTC) ultimately initiates successful metastatic colonies. Interestingly, it was shown recently that not only tumoral but also normal cells are capable to travel through the body and fix themselves in distant organs (Podsypanina et al., 2008).

All these experimental facts posed a number of theoretical questions on the process of metastasising and evolution of metastatic potential in tumours. Below we list some of the most important of them:

1. Can metastatic potential of a cancer cell be a subject of Darwinian natural selection, and in what sense?
   Selection of a cancer cell for fitness with respect to use of nutrients and survival inside a tumour is easy to understand. What kind of selection could drive a cancer cell to evolve to metastatic phenotype is not yet clear.

2. How early is the metastatic potential acquired by cancer cells in tumorigenesis?

3. How is metastatic potential related to the tumour size?
   It is known that the metastatic potential depends on the tumour size in some cancers (*i.e.* breast adenocarcinoma) while in other cancers (*i.e.* small cell lung cancer) even small tumours can metastasise very early.

4. Is metastatic potential a relatively rare trait in the population of cells inside a primary tumour?
   A related question is: do distant metastatic colonies possess the same genetic alterations leading to cancer as in the primary tumour?

5. How to explain tissue preference for seeding metastasis in particular cancers?
   For example, prostate cancer usually metastasises to the bones and colon cancer or uveal **melanoma*** have a tendency to metastasise to the liver.

Answering these questions is extremely important in developing successful strategies for treating malignant tumours. There are several aspects here. For example, in clinical practice, many stage I solid tumours are treated according to the worst-case scenario, namely that the tumour will metastasise. In reality, however, only $10 - 20\%$ of people with cancer will die from disseminated cancer. However, if the metastatic potential appears very late in **tumour progression**[*] (after the treatment) then the lethality **prognosis**[*] for the treatment outcome might be very unreliable based on the biopsy taken *before the treatment*. If the metastatic colonies are genetically very different from the dominant clones of the primary tumour then it might make any targeted therapy very questionable: the biopsy of the primary tumour might not be able to give any information on what are the driving tumorigenic mutations in metastasis.

Unfortunately, most of the important questions in this field do not have yet a clear answer. Nowell (1976) proposed the *clonal selection hypothesis* where the mutator phenotype results from sequential rounds of clonal selection. Accordingly to this hypothesis, within the primary tumour, there are one or several tumour cell sub-populations able to complete the metastatic process (Fidler and Kripke, 1977). Some later hypotheses on the process of appearance of metastasis underlined several specific possible scenarios which nevertheless can be considered as *special cases* in the process of clonal selection (Talmadge, 2007). Hypothesis of *parallel evolution* states that the seeds of metastasis can be disseminated relatively early in tumour development such that to the moment of clinical tumour detection the primary tumour and metastasis can already have followed significantly different evolutionary history. *Dynamic heterogeneity* hypothesis states that the cells with metastatic phenotype can be continuously generated and lost within the primary tumour, and that the metastatic phenotype is a transient phenomenon. It suggests that *the frequency* of such transitions plays a determinate role in metastasising. The dynamic heterogeneity model has also been suggested to incorporate reversible EMT. *Clonal dominance* theory suggests that metastatic phenotype is advantageous in clonal competition such that most of the cells in the primary tumour possess it to the moment of metastasis appearance. Finally, the **Cancer Stem Cell (CSC)**[*] hypothesis states that only a small fraction of stem-like cancer cells are able to spread in the organism and seed themselves at distant sites.

Until so far, mathematical modelling has contributed to clarifying the mechanisms of tumour invasion and metastasis in several relatively narrow aspects. Firstly, machine learning techniques were applied to derive prognostic molecular signatures of metastasis-free survival for cancer patients before they are treated. More recent approaches aimed at using the current knowledge on biological networks such as Protein–Protein Interactions (PPI) in ameliorating these signatures. These methods were described in more detail in **Chapter 6**.

Mechanistic modelling of tumour invasion and metastasis is based on using spatial and multiscale modelling techniques and describes interaction between tumoral cells and tumoral microenvironment which can include several types of normal cells (of the stroma or immune system) and various structures such as ECM. Proliferation and death of tumoral cells coupled with their diffusion through tissue is modelled by using diffusion-reaction equations, or by using agent-based techniques with explicit rules on movement and transformation of cancer cells in some (usually discrete) space. Multiscaleness of modelling in this case is manifested in embedding more detailed intra-cellular models of biological networks in the set of formal rules governing the cellular behaviour.

One of the simplest models of tumour invasion in a passive diffusion-like fashion for describing invasion of a brain by **glioma**[*] was suggested by Swanson et al. (2003). The invasion is modelled by a standard diffusion equation including the term responsible for local cell proliferation. In the case of glioma, even this simple description is able to realistically model the tumour expansion, taking into account the brain shape, effect of surgical **resection**[*] and action of **chemotherapy**[*].

More elaborated models of tumour invasion include taking into account active transformation of the surrounding environment by tumoral cells. In this way, Anderson et al. (2006) developed a hybrid discrete-continuous model of tumour invasion. The model represents a mix of cellular automata and continuous modelling approaches, in which four components are considered:

1. The cancer cells, in the form of cellular automata, which are able to move in a two dimensional space

2. Nutrients necessary for a cancer cell to survive represented by local oxygen concentration

3. ECM represented by the concentration of matrix macromolecules

4. The concentration of the Matrix Degrading Enzyme (MDE)

The model considers a possibility of clonal heterogeneity and natural selection by introducing mutational process in the rules governing the cancer cell behaviour. The most interesting prediction from this model was that the harsh microenvironment (characterised by **hypoxia**[*], heterogeneous ECM) can provide the conditions to select for more invasive clones and reduce the clonal heterogeneity. This happens because those cells which acquired partial independence in the tumour microenvironment might receive a selective advantage in harsh conditions. The hypothesis that the tumour microenvironment in certain conditions might entrain cancer cells for invasion and metastasis has important clinical consequences. In particular, the model predicts that invasive tumour properties are reversible under appropriate microenvironment conditions and suggests that the therapy aimed at cancer-microenvironment interactions may be more successful than making the microenvironment harsher (*e.g.* chemotherapy or anti-angiogenic therapy).

One important aspect of interaction between tumour cells and tumoral microenvironment and normal stroma consists in changing the acidity of the environment by cancer cells. Recently, this *acid-mediated tumour invasion* model received a lot of attention from the point of view of mathematical modelling. In Martin et al. (2010) a mathematical model of this phenomenon was suggested extending an older acid-invasion reaction-diffusion model created by Gatenby and Gawlinski (1996). The model includes five processes: (1) tumour cells produce excess acid which diffuses into surrounding tissue; (2) acidification of the environment causes death of untransformed normal cells; (3) death of normal cells produces potential space into which the tumour cells may proliferate; (4) in order to invade, the tumour cells must not only kill the normal cells by environment acidification but also degrade the ECM; (5) the ECM is remodelled by active Matrix MetalloProteinases (MMP) which are formed only at the interface between tumour and normal cells. In a way, the model states that for invasion, tumour cells must cooperate with normal cells. As a result, two interesting predictions come from this modelling. First, the model predicts that there is an optimal level of environment acidification for invasion of cancer cells. This level represents a trade-off between excessive killing of normal cells (and, as a result, failing to produce sufficient amounts of MMPs) and insufficient killing of normal cells (losing competition for space). Second, the model predicts that very aggressive cancers can be encapsulated by creating a gap between the tumour and normal stroma, a feature observed by clinicians in some cancers. Both predictions have potential for developing cancer treatments by preventing tumour invasion, by increasing local acidity of the environment in advanced cancers and decreasing local acidity in early carcinomas.

The multiscale mathematical model developed in Ramis-Conde et al. (2009) represents in greater detail the process of cell passage through vessel walls. The model utilises a multiscale approach, modelling interaction between β-catenins and N- and VE-cadherins and disruption of the bonds between endothelial cells by tumoral cells travelling in the blood stream. The intra-cellular mechanisms in this model are represented through the standard reaction network approach and the inter-cellular interactions are modelled using a biophysical approach taking into account cell shapes. The model is able to reproduce many experimental observations on the quantitative parameters of extra- and **intravasation**[*].

The collection of above mentioned models represents the first attempt to formally describe the complex and multistep process of tumour invasion and metastasising. Further efforts will be needed to create new models representing the missing steps of invasion (such as overcoming host defences by metastatic cells travelling in the body) and integrate existing models in a single model of tumour invasion. This future model recapitulating many scenarios of how tumours invade the host organism will have a strong impact on developing treatment strategies aimed at preventing the appearance of metastasis, hence, saving patients' lives. Also efforts should be made for closing gaps in understanding and formalising the process of tumour invasion. For example, currently there is no mathematical model to describe the molecular process of EMT which apparently plays a crucial role in equipping cancer cells with environment independence and motility. Together with this, *in silico* reconstruction of the tumoral microenvironment and the mechanisms of its functioning and determining the cancer cell fates is another outmost important direction in this field.

8.1.4   Inducing angiogenesis

Angiogenesis is the process by which cells grow new blood vessels from already formed vessels to supply nutrients and oxygen to the tissues. Angiogenesis is involved in wound healing and reproduction but also in tumorigenesis, cardiovascular diseases, diabetic ulcers, *etc.* In healthy cells, angiogenesis is controlled by a balance of pro-angiogenic vs. anti-angiogenic factors. In tumour cells, this balance is perturbed and an abnormal proliferation of blood vessels is observed.

Most of the models are built so as to test possible anti-angiogenic therapies. More particularly, mathematical models have provided insights in the understanding of the different processes of angiogenesis, in the reproduction of the dynamics of angiogenesis and in exploring means to re-establish the lost angiogenic balance by exploring possible targets for angiogenesis therapies.

There are major steps in angiogenesis presented schematically here. The activation of quiescent endothelial cells lining the cell walls is triggered by angiogenic signals. In response to these stimuli, the endothelial cells become motile by losing cell-to-cell contact and start developing a capillary sprout. This is made possible by the activation of the MMPs that degrade both the basement membrane of the vessel and the close ECM surrounding the sprout. The MMPs allow the endothelial cells to grow and to start sprouting in the direction of the angiogenic stimulus. The growing and dividing endothelial cells merge and form a hollow tube in their centre, the lumen. The maturation of vessels insures that pericytes and smooth muscle cells surround and stabilise the tube formed by endothelial cells towards the tumour. New branches and loops form and eventually the tumour can receive the necessary nutrients and oxygen to survive and spread.

The stimuli of angiogenesis are essentially composed of growth factors. They trigger intracellular signalling pathways that control biological responses such as proliferation, death, differentiation, senescence or survival. In tumours, angiogenesis is induced by excessive secretion of these growth factors. The main growth factors that play a role in angiogenesis are the following:

- The fibroblast growth factors (FGF family), which, once active by autophosphorylation and dimerisation, activate cascades of events and eventually lead to biological responses that favour growth of endothelial cells, smooth muscle cells and fibroblasts.

- The vascular endothelial growth factors (VEGF), which drive the formation of new capillaries. When cells become oxygen-deficient, they start to express VEGF which stimulates endothelial cells.

- The transforming growth factors (TGFβ), which can either stimulate proliferation in some cell types or block it in other cells.

- The epidermal growth factors (EGF), which activate cascades that lead to proliferation, differentiation or apoptosis.

- The angiopoietins and Tie receptors, where the Tie receptors, Tie1 and Tie2 are expressed in angiogenic cells and the angiopoietins Ang1 and Ang2 participate in the formation of mature blood vessels and act as ligands to the Tie receptors, and more particularly to the Tie2 receptors.

- The platelet-derived growth factors (PDGF).

- The insulin-like growth factor (IGF).

So far, most of anti-angiogenic therapies have concentrated on VEGF and on how to inhibit it, being the main signalling molecule to angiogenesis. Among the existing drugs, the angiogenesis inhibitor bevacizumac has been used since 2004 to treat colon cancer and **glioblastoma*** with chemotherapy treatment. Bevacizumac binds and inhibits VEGF (Shih and Lindley, 2006). It has been proven to slow down tumour progression when used in combination with chemotherapy but not to cure these cancers. Other inhibitors such as sorafenib and sunitinib block the angiogenesis signalling pathway at different levels. The first one acts on hepatocellular carcinoma and kidney cancer and the latter on both kidney cancer, gastrointestinal stromal tumours and neuroendocrine tumours.

The formalisms used for modelling angiogenesis are essentially of two types: continuum and discrete (see **Section 7.1.4**). Continuum models explore the distribution in time and space of various variables (cell density, species, *etc.*). They are described as continuous variables governed by reaction-diffusion equations in one-space dimension (Balding and McElwain, 1985; Orme and Chaplain, 1996), two-space dimension or/and three-space dimension (Orme and Chaplain, 1997; Anderson and Chaplain, 1998; McDougall et al., 2006). Most of the models use Partial Differential Equations (PDE), stochastic or not (Merks and Glazier, 2006), and some models use nonlinear Ordinary Differential Equations (ODE) to model tumour growth (Arakelyan et al., 2002). Discrete models are of several types: Boolean models where model variables are components of the signal transduction pathways activated during angiogenesis (Bauer et al., 2010) and the second one where behaviour and interactions of cell components of angiogenesis are described as lattice models such as agent-based models (Stokes and Lauffenburger, 1991; Alarcon et al., 2003; Owen et al., 2009). Both types of discrete models are often simulated stochastically. Some mathematical models have included both types of formalisms to tackle different features of angiogenesis and with a multiscale perspective (Macklin et al., 2009; Perfahl et al., 2011; Wu et al., 2009). Very detailed reviews on mathematical modelling of angiogenesis describe the different formalisms, the purpose of the models and their findings (Peirce, 2008; Mantzaris et al., 2004).

On a more biological point of view, existing models have focused on particular biological aspects of angiogenesis such as initiation of capillary growth in response to angiogenic factors or initiation of buds from primary vessel (Levine et al., 2000; Orme and Chaplain, 1996), interaction between endothelial cells and the ECM (Anderson and Chaplain, 1998), cell movement and interaction with the environment (Anderson and Chaplain, 1998; Stokes and Lauffenburger, 1991), sprout branching with the addition of cell stochasticity (Anderson and Chaplain, 1998), therapeutic discovery and combined use of target components (Arakelyan et al., 2002; Mac Gabhann and Popel, 2006), *etc.*

8.1.5   Resisting cell death

Resisting cell death or escaping apoptosis is one of the most frequent hallmarks in cancer. The main questions that mathematical models of this hallmark have tried to answer focus on the biochemical details of apoptosis: what is the machinery capable of activating different modalities of cell death and what are the crosstalks between these modalities? As it is the case for angiogenesis, cell death is controlled by a balance between pro-apoptotic versus anti-apoptotic factors. More generally, **homeostasis*** is maintained by the equilibrium between death and proliferation of cells. The quantitative aspect in modelling cell death is therefore important. It permits one to tackle questions such as: what molecular factors contribute to this balance? Is the decision to enter cell death irreversible, and if not, when is the cell committed to die?

Over the past years, mathematical modelling of apoptosis has concentrated on both the activation of mitochondrial-dependent apoptosis in response to DNA damage via *TP53*, referred to as the intrinsic pathway, and on the receptor-triggered apoptosis via death receptors such as Fas or TNFR, referred to as the extrinsic pathway.

The intrinsic apoptotic pathway is initiated by TP53 in response to stimuli such as DNA damage. The dynamics of TP53 activation has raised interest. The focus was on how the cell knows when DNA cannot be repaired and how apoptosis is turned on, or also how TP53 dynamics and activity are linked to initiation of apoptosis. Periodic pulses of TP53 with varying amplitudes have been observed by several experimental groups (Lahav et al., 2004; Lev Bar-Or et al., 2000) and studied by theoretical groups (Ciliberto et al., 2005; Geva-Zatorsky et al., 2006; Ma et al., 2005). These TP53 oscillations may be born from the negative feedback involving MDM2 and TP53 but how this behaviour really triggers apoptosis remains obscure. This part of the apoptotic pathway needs more investigation both from an experimental and a theoretical point of view.

The extrinsic apoptotic pathway is initiated at the level of the receptors. Many studies have been proposed on the formation of the Death-Inducing Signalling Complex (DISC) and its implication in cell fate decision between death and survival (Bentele et al., 2004; Lavrik et al., 2007). The activation of the receptors can lead to cell death, cell growth and differentiation depending on the concentration of crucial players such as *cFLIP* (Han et al., 2008). Aguda and Algar (2003) proposed early in the history of cell fate modelling a very qualitative model, which introduced some feedback loops in the decision process and insisted on the need for nonlinearity to either enter the cell cycle or to provoke apoptotic death. Since then, more models of cell fate decision have been proposed (Gaudet et al., 2005; Lavrik et al., 2007; Philippi et al., 2009; Calzone et al., 2010) but they mainly concentrate on early decisions.

Fussenegger et al. (2000) were the first to present a model that contained both pathways in a great amount of detail. With a small model, Bhalla and Iyengar (1999) were able to show some emerging properties of the system. Among these emerging properties, the all-or-none response observed in caspase activation (Eissing et al., 2004) and in *BAX/BCL2* interaction (Cui et al., 2008) is brought into relief in models of no more than 4 and 5 variables respectively. Both bistability and irreversibility are shown at several levels and appear to be a recurrent feature in apoptotic models.

There exist also two types of apoptosis: Type I and Type II. Type I apoptosis can be found in some cell types. It is the program that concerns the positive feedback involving early caspases such as caspase-8 (*CASP8*) and the effector caspases such as caspase-3 (*CASP3*). This apoptotic program has been thoroughly studied by Eissing et al. (2004), as well as the perturbation of this feedback by specific caspase inhibitors (Stucki and Simon, 2005). It has been proven that the fight between caspases and their inhibitors can either destabilise or enhance the bistable behaviour (Choi et al., 2007; Legewie et al., 2006). Other bistable switches are involved in the type II cells. Type II cells concern events regulating mitochondria permeabilisation (Choi et al., 2007; Legewie et al., 2006) and consequential release of essential components leading to activation of effector caspases (Eissing et al., 2007; Nakabayashi and Sasaki, 2006; Rehm et al., 2009). Mathematical models have also shown that the amount of inhibitors could suppress the apoptotic phenotype. For instance, *BCL2* seems to have the capability to block type-II or intrinsic apoptosis (Hua et al., 2005; O'Connor et al., 2006). In a very detailed model of the extrinsic apoptosis, the switch-like behaviour brought about by the positive feedback implicating *CASP8* and *CASP3* and the role of the permeabilisation of the mitochondria membrane were deeply studied (Albeck et al., 2008). Similarly, the role of the apoptotic inhibitor *cFLIP* in the regulation of type-I cell death has been demonstrated by Bentele et al. (2004).

Most of the models of the apoptosis pathway use a continuous framework based on ODEs in order to follow the rate of change of protein concentration. Rehm et al. (2009) proposed the first spatio-temporal model of mitochondria outer membrane permeabilisation using PDEs. Some other discrete formalisms have also proved their validity and brought their contribution in the analysis of apoptotic mechanisms such as Boolean (Tournier and Chaves, 2009) and Petri Nets modelling (Heiner et al., 2004; Li et al., 2007). The review of models describing apoptosis and cell fate decision is not exhaustive but illustrates the growing interest, over the past 10 years, of understanding the mechanisms of cell death.

8.1.6   Genomic instability and mutation

Genomic instability, sometimes called *genetic instability* or *genome instability* or *genome plasticity* is believed to be characteristic of the majority of tumours. Most generally it can be defined as the inability of a cell to reproduce intact genome, its exact sequence and separation in chromosomes, from one cell generation to an other. This is a rather quantitative than qualitative notion since the normal process of DNA replication is never guaranteed free from errors. At average the rate of mutation with probability 10^{-7} per gene per cell division is considered as physiological (Komarova, 2005). Significantly higher mutation rates could be considered as genomic instability. To avoid confusion, *the process of genomic instability* should not be mixed up with the *observation* of a modified genome. Observations of genome abnormalities do not automatically lead to the conclusion about the presence of genomic instability (for example, tetraploid genomes can be stably reproduced in human cell lines). However, the presence of a large number of independent genomic abnormalities usually evidences existence of a period of genomic instability at some moment of the genome evolution.

Known mechanistic causes of genomic instability can be classified by the scale of the resulting genome modifications. A genome modification of the largest scale is the *whole genome duplication* which can lead, for example, to frequent tetraploidisation. Genomic instability at the scale of chromosomes is called Chromosomal Instability (CIN) and is manifested by loss and amplification of whole chromosomes or their large parts and inter- and intra-chromosome **translocations***. Mechanistic causes leading to both whole genome duplication and CIN remain obscure; however, some possible mechanisms leading to defective mitosis have been established. At smaller scale, genomic instability is manifested, for example, by Microsatellite Instability (MIN), associated with defective Mismatch Repair (MMR) or Homologous Recombination (HR) mechanisms. Replication errors or problems with Base Excision Repair (BER) can lead to base substitutions and micro-insertions and micro-deletions. In this section we will denote all instabilities leading to large-scale genome rearrangements as CIN, and all types of instabilities leading to smaller-scale genome modifications as MIN.

Increased rate of mutations can be induced by external factors leading to DNA damage, such as UV light exposure or toxic stress. If disappearance of the external source of DNA damage leads to genome stabilisation then this is usually distinguished from genomic instability; however, this distinction is rather subtle. A physiologically normal cell should be able to preserve its genome by inducing various DNA repair mechanisms or apoptosis. However, all these mechanisms can be leaky, especially at high rates of DNA damage, therefore, the fact of increased mutation rate might not necessarily indicate the existence of defective cell cycle checkpoints or DNA repair. Hence, genomic instability and the presence of DNA damaging factors (such as radio- or chemotherapy) create a nontrivial interplay which can be a subject of mathematical modelling.

At any scale, genomic instability is associated with genetic or epigenetic defects in proteins with the function of genome *caretakers*. These genes are usually considered as a special class of tumour suppressor genes with *BRCA1*, *BRCA2*, *ATM*, and *MLH1* being the most studied examples. Mutations in caretaker genes can be recessive like for the classical tumour suppressor genes or dominant-negative mutations (thus, requiring only single allele mutated for compromising the gene's caretaker function). Historically, the function of many caretakers was discovered from familial genetic studies.

The phenomenon of genomic instability being practically omnipresent in cancer cells led to formulation of two major theoretical questions:

1. Is genomic instability required for tumorigenesis or it is rather caused by tumorigenesis?

2. How early does genomic instability appear in tumorigenesis?

Despite many experimental and theoretical studies, both questions remain pending and hotly debated topics in current literature.

The multistep theory of tumorigenesis claims that most of the cancers require many (more than two) genetic changes for a single cancerous cell to appear. Taken simplistically, this leads to a conceptual problem, since the normal physiological mutation rate does not explain accumulation of sufficient number of genetic transformations *in one cell* during a typical human lifespan (70 years). Hence, the mutation rate should be increased at some point, and this is the essence of the *mutator phenotype* hypothesis suggested by Loeb et al. (1974), which states that mutator (read caretaker) mutations play a crucial role in tumorigenesis by accelerating the acquisition of oncogenic mutations. In its stronger formulation, the mutator hypothesis also postulates that a mutator phenotype should appear relatively early in the sequence of accumulated oncogenic mutations.

Since then this simple suggestion has been criticised along two lines of argument. The first objection states that genomic instability in most circumstances is deleterious for cancer cells and the cell clones possessing genomic instability should be rapidly eliminated (*negative clonal selection theory*). The second objection is known as *clonal expansion* theory, stating that each oncogenic mutation leads to expansion of premalignant lineages, creating by this a larger cell pool possessing the malignant mutation. Thus, the requirement of appearance of two independent mutations in *the same* single cell is relieved.

Recent sequencing of a cohort of cancer genomes has shown the absence of frequent (or total absence of) mutations in the known caretaker genes (Negrini et al., 2010). With some reservations, this suggests the *oncogene-induced DNA replication stress* model. According to this model, the malignant cell first should acquire a mutation in an oncogene leading to increased cell proliferation. After this, the activated oncogenes induce genomic instability through DNA replication stress which in particular affects specific genomic sites, called common fragile sites. This genomic instability leads to cell cycle arrest and massive apoptosis unless in one of the damaged cells the function of *TP53* is compromised, which results in cells evading cell death and senescence. This model is also different from the classical mutator hypothesis.

The last argument against the original mutator hypothesis is that appearance of genomic instability can serve to *reduce replication cost* rather than to increase mutability. The cells avoiding cell cycle delays connected with maintaining and repairing the genome can have selective advantages simply because they do not spend their resources on caring about genome integrity. Of course, these cells will be characterised by higher mortality; however, some of them will be able to occasionally survive.

Resolving these controversies crucially depend on the quantitative parameters of cancer cell evolution. Therefore, mathematical modelling traditionally plays a very important role in interpreting experimental observations and its role will only increase in the future.

The mathematical formalism largely used in this kind of study is based on the classical discrete or continuous Markov chain modelling. The possible scenarios of cancer cell evolution are usually systematised using so-called *mutation-selection networks* (Komarova, 2005). In these networks a state of a cell is characterised by probabilities of transition into other states (including transition in itself). For example, the simplest example of a mutation-selection network is shown by **Figure 8.1A**. In this network, three genetically different types of cells are presented: a wild type cell (state A), a cell with a mutation in one copy of a tumour suppressor gene (state B), and a cell with mutations in both copies of a tumour suppressor gene (state C). Transitions between these states represent the probability to obtain a single mutation u, and the probability to obtain two mutated copies of a tumour suppressor gene in the same cell from one mutated allele state. Using this diagram, differential equations for continuous Markov chain modelling can be written down, and the probabilities of tumour initiation can be computed. The model can reproduce two major scenarios: 1) acquisition of malignant transformation through fixation of a mutation in one allele and then a transformation to the state with two mutated alleles and 2) a *mutation tunnelling scenario* when a double-mutation phenotype appears more rapidly than a single-mutation phenotype is fixed in the population. For understanding details of this approach, the reader should refer to the excellent book written by Komarova (2005).

More complex example of a mutation-selection network is shown in **Figure 8.1B**. In this network, a simple Knudson two-hit model is extended by a possibility of genomic instability. On this diagram, the letter signifies the type of genomic instability phenotype (X for the state free of genomic instability, Y for the state with the CIN instability and Z for the state with the MIN instability), and the number signifies the number of mutated alleles in the tumour suppressor gene (0 for the wild type gene, 1 for one allele mutated, 2 for two alleles mutated). Thus, instead of obtaining a single-copy mutation of a tumour suppressor gene, cells can obtain the mutator mutation, *i.e.* a mutation leading to genomic instability, either of CIN type (one step below from the wild type state) or of MIN type (two-steps *mutations* above from the wild type). In this model, it is assumed that for obtaining the CIN phenotype, only a single mutation in a caretaker gene is required (negative-dominant mutation), and to obtain the MIN phenotype, two mutations in both caretaker alleles are required. When a cell obtains the CIN mutator mutation, the probability of obtaining the first and the second mutation in the tumour suppressor gene, is increased manifold. The model describes many possible scenarios of coming from the wild type state X_0 to one of the malignancy initiation states (X_2, Y_2 or Z_2). One possible scenario is depicted on the diagram by a series of dashed arrows. In this scenario, mutation in one tumour suppressor allele is fixed in the cell population, after which one of the cells acquires chromosomic instability by a mutator mutation. This leads to a relatively fast inactivation of the second allele.

A more complicated selection-mutation network (see **Figure 8.1C**) describes the appearance of the genomic instability in the presence of DNA damage. There are five states on this diagram depicting (1) a normal cell, (2) a damaged cell, (3) a cell in the state of DNA reparation and cell cycle arrest, (4) cell acquired a tumorigenic mutation and (5) the cell following programmed cell death (denoted apoptosis on this diagram, but it can be any other type of cell death). The model that can be constructed from this diagram describes the balance in how the damaged cell follows one of the three cell fates: being repaired, going into apoptosis, and acquiring the mutated tumorigenic phenotype. The model can take into account the relative efficiency of DNA repair, efficiency of apoptosis and the lethal consequences of tumorigenic mutations. Thus, the effect of three types of mutations can be considered in this model: those leading to tumorigenic transformation, those affecting DNA repair and those affecting apoptosis.

Using the mutation-selection network formalism for systematic modelling of **sporadic*** and hereditary cancer initiation with colorectal cancer as a prototype example, the following theoretical predictions were obtained (we do not aim here to give the exhaustive list of predictions) (Komarova, 2005, 2004):

1. Smaller-scale instabilities (such as MIN) and larger-scale instabilities (such as CIN) might have different roles in tumorigenesis. CIN cells are more likely to produce nonviable offspring than MIN. At the same time, it may be possible that CIN is easier to trigger. Mathematical analysis shows that if inactivation of MIN genes (genetic or epigenetic) occurs at sufficiently fast rate, around 10^{-6} per cell division, then at least in the case of colon cancer, MIN can precede mutation in the principal tumour suppressor gene APC. To decide whether CIN can be also advantageous for accelerating tumour development, several crucial parameters should

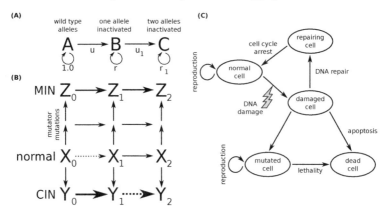

FIGURE 8.1 Examples of mutation-selection diagrams used to compute probabilities of cancer cell initiation for the simplest Knudson two-hit model of tumour suppressor inactivation. (A) Two-hit tumour inactivation in the presence of two types of genomic instability, MIN and CIN (B) Cell transformation in the presence of DNA damage, DNA repair and apoptosis. (C) Each node of the network represents a sub-population of cells in a certain state, edges represent transition probabilities. (B) In the network higher probabilities are indicated with thicker arrows and one particular scenario is shown by dashed arrows. In this scenario, mutation in one tumour suppressor allele is fixed in the cell population, after which one of the cells acquires chromosomic instability by a mutator mutation. This leads to a relatively fast inactivation of the second allele. Adapted from Komarova (2005).

be specified such as the number of dominant CIN genes in the human genome, the rate of CIN gene inactivation and the cost of having CIN (because of increased lethality). Absence of reliable knowledge on these parameters currently prevents us from having definite conclusions on whether CIN plays a triggering role in colon cancer.

2. The fitness of cancer cells (*i.e.* probability to divide) is optimised if the rate of chromosome loss is of the order of $10^{-3} - 10^{-2}$. This turned out to be a robust result which coincides with experimental values obtained for colon cancer cell lines (Lengauer et al., 1998). This rate can be a subject of natural selection in cancer cell colonies, hence it is predicted that it will be a characteristics for most of the cancers.

3. CIN mutation does not arise simply because it allows a faster accumulation of tumorigenic mutations; instead, CIN *must* arise because of alternative reasons such as environmental factors, oxidative stress or tumorigenesis (for example, inactivation of a tumour suppressor gene).

4. If the tumour arises because of activation of oncogenes rather than inactivation of a tumour suppressor then chromosomal instability is likely to be detrimental to the cancer. It happens simply because to turn on an

oncogene, a small scale mutation is needed whereas a chromosomal loss event can lead to inactivation of a functioning oncogene. The situation can become very complex if both oncogenes and tumour suppressors play a role in tumorigenesis.

5. Genetic instability in cancer increases (by accelerating the mutation rate) and decreases (by making many cells nonviable) the rate of cancer progression. As cancer progresses, the balance of these two selective pressures changes, and selection can switch cancer cells from *unstable* to *stable* regimes. Mathematical modelling confirms that this might be the most efficient *strategy* for cancer cell growth in terms of optimal control theory (Komarova et al., 2008).

It should be understood that the predictions obtained from all of the models depicted above depend very strongly on the assumptions made for the quantitative values of probabilities of state transitions. The most important among these probabilities are the rate at which key mutations are acquired and the rate of clonal expansion. Currently we have rather vague ideas about their possible numerical values. These probabilities integrate very complex biochemical mechanisms which are not presented on the diagrams explicitly. In this sense, the presented models are rather phenomenological descriptions that need to be detailed in the future.

Systems biology of cancer in its current state poses new challenges for modelling genomic instability and its role in cancer. Sequencing cancer genomes provides an unprecedented amount of data which can in principle shed light on the processes of cancer genome evolution, implementations of CIN and MIN mechanisms and the role of DNA repair defects in tumorigenesis. The final goal of systems biology in this case will be to unravel the mechanistic details of the most typical scenarios of acquisition of genomic changes leading to tumorigenesis and explain the observed patterns of genome modifications in concrete cancer types and subtypes. This remains a challenge for the next decade of mathematical modelling of tumorigenesis.

8.1.7 Tumour-promoting inflammation

As early as 1863, Rudolf Virchow made a connection between inflammation and cancer by noticing the presence of leucocytes in **neoplastic**[*] tissues. He suggested that cancer originates at sites of chronic inflammation. In the 2000s, our understanding of the inflammatory microenvironment of malignant tissues has supported Virchow's hypothesis, and the links between cancer and inflammation are starting to have implications for prevention and treatment.

It appears that the inflammatory response of a body to the presence of **neoplasia**[*] is similar to the processes of wound healing since both create environments in which excessive cell proliferation is taking place. Metaphorically speaking, cancer was called a *"wound that do not heal."* Also it was suggested that if genetic damage is the *"match that lights the fire"* of cancer, some types of inflammation may provide the *"fuel that feeds the flames"* (Balkwill and Mantovani, 2001).

Connection of tumorigenesis to inflammation is happening through two major routes that can be called *extrinsic* and *intrinsic*. Roughly speaking, inflammation can induce cancer (extrinsic route) and cancer can induce inflammation (intrinsic route).

The existence of the extrinsic route is proven by the fact that cellular inflammatory conditions connected to presence of pathogens or chronic diseases greatly increase cancer risk. Thus, the presence of *Helicobacter pylori* in the stomach, papilloma virus, hepatitis virus, and various autoimmune diseases connected to chronic inflammation and **prostatitis*** are important risk factors for cancer in the organs where the inflammation is taking place. However, it remains unclear whether chronic inflammation *per se* is sufficient for tumorigenesis. There is some emerging evidence that chronic inflammation contributes to genetic destabilisation of cancer cells through either inducing direct damage on DNA by elevated amount of reactive oxygen species or through interfering with DNA repair pathways and cell cycle checkpoints.

The intrinsic route of inflammation-induced tumorigenesis is that many known oncogenes have a crosstalk with pro-inflammatory pathways (such as angiogenic switch, recruitment of inflammation-associated cells such as leucocytes and macrophages). Due to this pre-existing crosstalk, activation of cell growth automatically leads to construction of an inflammatory microenvironment at neoplasia sites.

Among a large number of genes involved in inflammation one can distinguish several most important players which are the prime movers of the inflammatory response. Among them, *NFkB* and *STAT3* transcription factors and *IL1B*, *IL6*, *TNF* pro-inflammatory cytokines play the most important role and their participation in tumorigenesis can be unequivocally demonstrated. Moreover, the *NFkB* transcription factor is homologous to the retroviral oncoprotein v-Rel.

For historical reasons, the process of induction of the *NFkB* transcription factor by tumour necrosis factor (*TNF*) together with the interplay between *NFkB* and apoptotic pathways were the subjects of intensive mathematical modelling. As a matter of fact, the *NFkB* pathway is one of the most mathematically modelled with more than 30 mathematical models devoted to various aspects of this signalling (Cheong et al., 2008).

NFkB transcription factor was first characterised in the lab of Nobel Prize laureate David Baltimore in 1986. Later it was found that *NFkB* is encoded in genome by a family of 5 genes which can be subdivided in two classes: class I - *NFkB1* and *NFkB2* and class II - *RELA*, *RELB*, *REL*. The products of these genes perform their function in the form of homodimers (such as *NFkB1*, *RELA* dimers) or heterodimers (such as p60:p65 dimer).

The core of *NFkB* signalling, the subject of mathematical modelling, consists in a negative feedback architecture: *NFkB* is held in the cytoplasm by the inhibitor of its transcriptional activity called *IkB*. Bound to *IkB*, the *NFkB* dimers cannot penetrate the nucleus. Inducers of *NFkB* activity destroy *IkB* and releases *NFkB* which can then go to the nucleus and activate a number of its transcription targets. Among these targets there is *IkB* itself, thus *NFkB* shuts down its own activity with some delay which might give rise to damped oscillatory behaviour of the *NFkB* signalling (periodic shuttling of *NFkB* in and out of the nucleus) actually observed in experiments. Real implementation of this self-inhibitory schema is complicated by the presence of multiple dimers of *NFkB*, existence of three forms of the *IkB* inhibitor: IkBα, IkBβ and IkBϵ, presence of other feedback loops through, for example, A20 protein, existence of *noncanonical NFkB* pathways.

The earliest attempt to capture the dynamics of *NFkB* signalling with mathematical equations aimed at explaining how both*NFkB* transport into the nucleus and *IkB* rate of association/dissociation could keep most of *NFkB* in an inactive state in resting cells (Carlotti et al., 2000). However, this numerical model was of limited potential for experimental validation. Hoffmann et al. (2002) published the first predictive mathematical model of *NFkB*. The main focus of this modelling was on specific roles of *IkB* (*NFkB* inhibitor) isoforms IkBα, IkBβ and IkBϵ. It was known that mice deficient in any of these isoforms had distinct phenotypes. Thus, the model contained three inhibitor isoforms, one single and most predominate representative of *NFkB* family, namely the p65:p50 dimer and *IkB* kinase (IKK) as a molecule inducing *NFkB* signalling, and described their synthesis/degradation and association/disassociation.

Exploration of the model with computational simulations resulted in two major insights. First, it described how differential functions of the *IkB* isoforms could give rise to strikingly different *NFkB* dynamics in genetically reduced cells. The role of IkBα, whose expression is induced by *NFkB*, was to provide a negative feedback. This was aptly demonstrated by pronounced oscillations in *NFkB* activity in cells lacking the other isoforms. The role of IkBβ and IkBϵ was to dampen these oscillations. When all three isoforms were present, the *NFkB* response was biphasic, with an initial *NFkB* activity rising and falling within approximatively 1h, followed by a late activation phase characterised by a steady intermediate level of activity. Second, the *temporal dose-response* characteristics of *NFkB* signalling module were explored by simulating *NFkB* response duration for different stimulus durations. The model predicted that the module would generate the initial phase of 60 min of *NFkB* activity even with much shorter stimuli, while only for longer lasting stimuli (more than 1h) did the responses have durations proportional to the input duration.

Further mathematical modelling of the *NFkB* pathway went in several directions with several theoretical and practical questions. First, the initial study by Hoffmann already raised important conceptual questions: Can the temporal activity of *NFkB* target genes be programmed by the dynamics of the input signal? How complex can this programming be? This programming can be performed by such properties of the input signal as amplitude, rate of increase, duration, rate of decrease, or frequency. *NFkB* modelling became a useful concrete example for validating these possibilities. For example, by experiments and mathematical modelling it has been shown that different inflammatory stimuli generate distinct temporal profiles of the activity of the central node kinase IKK or transcription factor *NFkB* and that temporal regulation plays a key role in determining which subset of target genes are activated (Werner et al., 2005).

Another direction in which the mathematical modelling was able to help is understanding the possible role of additional feedbacks, present in the *NFkB* circuitry. Thus, in Kearns et al. (2006), with a mathematical model, it was shown that IkBε can provide delayed negative feedback in antiphasis with IkBα, which can shutdown *NFkB* signalling more quickly. Extracellular feedback through autocrine signalling was also investigated with use of computational modelling (Cheong et al., 2008). Additional negative feedback through the A20 protein was the subject of the study by Lipniacki et al. (2004). This model predicted that A20-mediated negative feedback is sufficient to produce the sharply peaked IKK activity profile resulting from persistent TNF*alpha* stimulation. However, this was not possible to confirm experimentally. In fact, mathematical modelling identified the importance of regulation of IKK for *NFkB* signalling (Cheong et al., 2008), but its detailed understanding remains an open direction.

A series of works were devoted to the study of a possible physiological role of *NFkB* oscillations in individual cells and in the average of sets of individual cells. These oscillations are largely hidden in wild type cells by the effects of IkBβ and IkBε (Hoffmann et al., 2002), and oscillations do not seem to alter gene expression programs when compared to the wild type biphasic response, raising doubts about their functional significance. Nevertheless, several theoretical studies investigated parametric conditions on the existence of oscillations and their properties. Thus, the *NFkB* pathway was studied with respect to sensitivity of *NFkB* oscillations to the variation of individual pathway parameters (Ihekwaba et al., 2005). Different aspects of *NFkB* oscillations, such as the timing and amplitude of peaks and low points are sensitive to different parameters in the original model, as measured by sensitivity coefficients. Some parameters are predicted to be broadly important for nearly all aspects of oscillations, and they all relate to reactions involving IkBα. The *NFkB* model served as an illustration for introducing new ideas on the different types of biological robustness (see **Chapter 9** and Gorban and Radulescu, 2007).

Several attempts were made in order to develop a consensus combined model of *NFkB* with little success to our knowledge so far. One of the obstacles is incompatibility of various models on the level of their wiring, parameterisation and the level of complexity. In Radulescu et al. (2008), the most detailed model of *NFkB* signalling was suggested containing 39 chemical species and 65 reactions, and the way to systematically reduce its complexity was suggested. This allowed one to compare the most detailed model to several existing models in a uniform fashion.

Despite the extensive mathematical modelling of the *NFkB* signalling pathway, it remains difficult to obtain significant insights into the role of *NFkB* signalling in tumorigenesis by modelling this pathway alone. More understanding should come from studying its connection with other cancer pathways such as apoptosis, MAPK pathway, hypoxia, angiogenesis, genetic instability and chemo/radio-resistance. Large efforts have to be undertaken in order to mathematically describe the specific role of *NFkB* signalling in various cancers and tissues, in the context of its interplay with other major signalling pathways. Having this motivation, Oda and Kitano (2006) charted the comprehensive map of molecular interactions in the toll-like receptor signalling network. This comprehensive map describes intensive crosstalk between *NFkB* and MAPK pathways, both regulated upstream by MyD88-dependent signalling as well as multiple feedback and feed-forward controls. Further efforts in this direction will be necessary.

8.1.8 Reprogramming energy metabolism

One of the first identified biochemical hallmarks of tumour cells was the shift in glucose metabolism from oxidative phosphorylation to aerobic glycolysis, referred to as the *Warburg effect.*

In **oncology***, the Warburg effect is the observation that most cancer cells predominantly produce energy by glycolysis followed by lactic acid fermentation in the cytosol, and not by glycolysis followed by oxidation of pyruvate in mitochondria like most normal cells do. The latter process is aerobic (uses oxygen). Malignant rapidly-growing tumour cells typically have glycolytic rates that are up to 200 times higher than those of their normal tissues of origin. This occurs even if oxygen is present and despite the fact that production of ATP via glycolysis is about 15 times less efficient through fermentation than via oxidation of glucose.

Otto Warburg believed that this change in metabolism is the fundamental cause of cancer, a claim now known as the *Warburg hypothesis*, which was not supported by the cancer research community during the 20th century. Today it is known that much of this metabolic conversion is controlled by specific transcriptional programs activated in response to mutations of tumour suppressor genes and oncogenes. Further study of mitogenic signalling pathways has revealed a number of essential and conserved cellular functions that couple the cell growth machinery to glucose and lipid metabolism, thereby coupling proliferation of cells and organisms to the nutrient status of their environment.

It remains unclear, though, what the selective advantages of the Warburg effect are. It was suggested that switching to anaerobic glycolysis can increase acidification of the tumoral environment and actively promote death of the normal cells of the stroma. Another hypothesis is that disabling the functioning of mitochondria by cancer cells (hence, switching off use of oxidative phosphorylation) serves for protection against activation of apoptotic programs.

Despite the fact that the Warburg effect seems to be a rather secondary event in tumorigenesis, there is a growing belief that the difference between normal and cancer cells in terms of the type of their basic energy metabolism can be the *Achilles' heel* of cancer cells. Then a relevant question for developing specific anticancer therapies is: *can one utilise the Warburg effect in order to selectively kill cancer cells?* One drug, dichloroacetate, based on this idea, is currently tested in clinical trials. Its action goes through restoring the original normal cellular metabolism in cancer cells, and thus promoting their self-destruction (Bonnet et al., 2007).

To improve the treatments that are based on the difference in energy metabolism, a particular effort needs to be made in understanding the connection between mitochondrial metabolism, programs of cell death, survival and other vital cellular functions. One way would be to suggest interventions that would make the cells with glycolytic energy production nonviable. The objectives in systems biology approaches related to metabolism can be summarised as follows:

- Collecting information on metabolic networks from the literature and assembling large reaction networks describing mechanisms of energy metabolism regulation and the connection to other cellular functions.

  Ideally, these representations should be in a standard computer-readable format (such as SBML, SBGN, see **Chapter 4**) amenable to formal analysis. Several large-scale efforts are already being made in this direction. Homo sapiens RECON 1 is *"a comprehensive literature-based genome-scale metabolic reconstruction that accounts for the functions of 2004 proteins and 2,766 metabolites participating in 3,311 reactions"* (Duarte et al., 2007). This network reconstruction was transformed into an *in silico* model of human metabolism and validated through the simulation of 288 known metabolic functions found in a variety of tissues and cell types. The Edingburgh human metabolic network reconstruction is made by integrating genome annotation information from different databases and metabolic reaction information from the literature (Ma et al., 2007). This network describes functions of nearly 3,000 metabolic reactions which are organised into about 70 human-specific metabolic pathways.

- Reconstructing metabolic networks using reverse engineering techniques and from available high-throughput experiments such as gene expression data, protein expression data, [13]**C-based metabolic flux data**[*] or high-performance liquid chromatography.

  Currently, these reconstructions are mainly limited to model organisms such as *S. cerevisiae* but human-related reconstructions of metabolic networks are on their way (Oberhardt et al., 2009).

- Analysing these networks from the point of view of their structure and dynamical behaviour.

  The aim of these analyses is to predict the most sensitive control parameters to intervene to the cellular energy metabolism. Metabolism modelling methods of Metabolic Control Analysis (MCA) and Flux Balance Analysis (FBA) are widely -used approaches (Fell, 1997; Palsson, 2006). The advantage of the FBA approach is that it requires knowledge of the *stoichiometric matrix* of the metabolic network rather than knowledge of the kinetic parameters.

A simple mathematical model of Warburg effect was suggested by Cloutier (2010). It includes a representation of glycolysis, a representation of mitochondrial oxidation of pyruvate, phosphocreatine buffering, exchanges of glucose, lactose and oxygen with the blood flow and production of ATP. The model is capable of reproducing the effect of damaging mitochondrial function leading to reorganisation of metabolites and fluxes, and identifies the most critical parameters that control the energy metabolism. One important conclusion derived from this model is that the sensitivities of metabolism parameters are very different in cancer and normal cells. This conclusion can orient the future developments of specific anticancer drugs.

Another metabolic network containing 60 metabolites participating in 80 metabolic reactions was constructed including the central pathways involved in Warburg effect: glycolysis, Tricarboxylic Acid Cycle (TCA) cycle, pentose phosphate, glutaminolysis and oxidative phosphorylation (Resendis-Antonio et al., 2010). The FBA approach was applied to model the growth curves of **HeLa cells***. An objective function was suggested for quantifying the optimal conditions for cancer cell growth. This function contains a linear combination of concentrations of lactate, ATP, ribose 5-phosphate, oxaloacetate and citrate, *i.e.* the components playing a central role in the production of energy and the most important cellular building blocks. The model was validated by experimental data and allowed the identification of some key enzymes controlling cancer cell growth.

A multiscale spatio-temporal approach to modelling competition between cells with normal function of mitochondria and cells switched to anaerobic glycolysis was developed in Astanin and Preziosi (2009). The model describes phenomenologically the irreversible transition of tumours from normal to glycolytic metabolism in the conditions of hypoxia. The model includes two populations of cells with different types of metabolism, extracellular liquid and ECM.

8.1.9   Other hallmarks

We have provided a set of mini-reviews related to mathematical modelling of some aspects of most of the hallmarks of cancer. As of today, two hallmarks (enabling replicative immortality and evading immune destruction) received less attention from the modelling point of view. There are only a few papers in which some aspects of these hallmarks are considered.

The question of telomere shortening and its connection to senescence and apoptosis were the subject of mathematical modelling in a series of papers

(Arino et al., 1995; Arkus, 2005; Rodriguez-Brenes and Peskin, 2010). In Rodriguez-Brenes and Peskin (2010), the model accounts for two processes: telomere length regulation for telomerase positive cells and senescence in somatic cells. The model can predict the length distribution for telomerase positive cells and describes both the time evolution of telomere length and the life span of cell lines if the levels of *TRF2* and the telomerase expression are known. The effect of a drug inhibiting telomerase, pentacyclic acridinium salt, RHPS4, at different stages of the cell cycle was analysed using simple continuous model of the cell cycle (Hirt et al., 2012).

The process of how a tumour evades destruction by the innate immune response was mathematically treated by de Pillis et al. (2005). In this work, an ODE-based mathematical model describes tumour-immune interactions, focusing on the role of natural killer (NK) and CD8+ T-cells in tumour surveillance. The model was parameterised from published mouse data and human studies.

For further reading on modelling hallmarks of cancer, we suggest (Auffray et al., 2011).

8.2   Discussion

8.2.1   Mathematical models of cancers

The amount of models mentioned in this chapter reveals the important effort made in the field of cancer systems biology. However, the final goal of cancer biology is to be able to realistically reproduce specific behaviour of individual types of cancers, and create mathematical models of tumours rather than generic descriptions of individual pathways involved in tumorigenesis. These models should describe specific interplay between the pathways most implicated in the tumour development, predict how the functioning of these pathways affects cell fates in the context of a particular tissue or tissue compartment, and predict the dynamics of tumour growth, including the invasion scenarios with their probabilities. One particular aspect of this modelling is to predict how a specific tumour would react to anticancer treatment. It is briefly discussed further in **Section 8.2.3**.

Here, we mention several efforts made in trying to grasp the whole behaviour of particular cancers with their specificities and in modelling individual cancers. There are several driver questions:

1. How to predict the dynamics of a tumour growth and its invasion into surrounding and distant tissues?
   This is one of the oldest direction in mathematical modelling of cancer (Byrne, 2010) with some recent contribution from machine learning and statistical modelling of high-throughput data mentioned in **Chapter 6**.

2. What are the specific networks involved in a tumorigenesis of a particular disease? This question received a lot of attention during the last decade with the appearance of high-throughput datasets that allow the identification of these networks.

3. How to combine the behaviour of intracellular pathways and the macroscopic cell behaviour in a multiscale mathematical model?
   This question remains a big challenge in systems biology of cancer for which several large-scale projects are already launched (see **Section 7.1.4**)

4. How to properly quantify the incidence curves of a particular cancer with respect to certain risk factors such as age?
   This question which makes rather a subject of cancer epidemiology will be very briefly discussed in **Section 9.8**.

We already mentioned spatial models of glioma invasion in the brain (see **Section 8.1.3**) reviewed in Swanson et al. (2003). These models take into account real configuration of a patient's brain from Magnetic Resonance Imaging (MRI) images to predict the rate of invasion and reaction on tumour resection or chemotherapy.

One of the most modelled individual cancers is the colorectal cancer, where somatic evolution was modelled (Fearon and Vogelstein, 1990). It was suggested that there exists a necessary set of events leading to the appearance of malignant growth in colon cancer. The authors suggested that mutations in at least four to five genes (*APC, KRAS, TP53* and others) are necessary for triggering tumour growth. Although there is a preferred sequence of genetic events, the exact sequence does not play a dominant role; the total accumulation of changes rather than their ordering is responsible for determining the tumour's fate. Since then, this model has been a basis for multiple mathematical models in which a particular stress was made on the role of genomic instability in cancer progression (Michor et al., 2005; Komarova, 2005) or on the mutated cell dynamics in the colon crypt together with their birth and death (van Leeuwen et al., 2006).

Growth of malignant cell populations in the breast has been a subject of mathematical modelling for a long time. The story of modelling breast cancer started from the question of realistic description of tumour size dynamics. One of the first mathematical models in cancer research had this aim and considered the growth of avascular multi-cellular tumour spheroids. These models were further developed taking into account various factors such as tumour angiogenesis (Byrne, 2010). Fluid mechanics approaches taking into account cellular adhesion properties were applied to model the initial stages of ductal carcinoma in situ (Franks et al., 2003).

There already exist large-scale European projects aiming at developing multiscale models of individual cancers. One of the most visible projects is CancerSys (http://www.ifado.de/cancersys/). Its goal is to establish a multiscale model for two major signalling pathways involved in the formation of hepatocellular carcinoma, the beta-catenin and RAS signalling pathways. The impact of these pathways on proliferation, tissue organisation and formation of hepatocellular carcinomas will be studied in the course of this project.

Network models of individual cancers started to appear recently. To give two examples, for Ewing's **sarcoma***, a network of specific molecular interactions was produced from identifying pathways activated downstream of the *EWS/FLI1* oncogenic transcription factor and from literature mining (Baumuratova et al., 2010). In (Pujana et al., 2007), the authors constructed a breast cancer-related network by taking four known breast cancer-associated genes: *BRCA1*, *BRCA2*, *ATM*, and *CHEK2*. The proteins were linked using data on phenotypic similarity, coexpression and genetic or physical interactions among orthologs of the proteins in other species. The network was extended from the four disease genes which implicated additional factors important for breast cancer progression.

8.2.2   Interdependency of the cancer hallmarks

It has been discussed throughout the chapters of this book that cancer is not the result of single mutations but rather a complex set of alterations in a particular order that trigger phenotypical disturbances. Some of the major actors of these alterations and the pathways leading to each hallmark or phenotype are represented as an influence network in **Figure 8.2**. The network is not complete but shows how intertwined the pathways governing the hallmarks of cancer are. As shown in the network, some data about certain of these pathways are still lacking and little experimental facts are reported on the influence that some genes may have on the various alterations of these pathways.

In their seminal paper, Hanahan and Weinberg (2011) noted that some of these hallmarks have influences on other hallmarks. As an example, telomeres (involved in enabling replicative immortality hallmark) have been revealed recently to be involved in the WNT pathway, to promote cell proliferation, to reduce apoptosis, and play a role in DNA damage repair. The interaction between the *RB* pathway (evading growth suppressors) and the *TP53* pathway (evading growth suppressor and resisting cell death) has been also known for a long time. Moreover, crosstalk has been brought into relief in diverse cancers: ER (oestrogen receptor) and HER2 (human epidermal growth factor receptor 2) through the MAPK pathway or TGFβ (Todorovic-Rakovic, 2005) in breast cancers, *etc.*

As reported in Berasain et al. (2009), evidence of crosstalk has been found in normal and cancer cells at the receptor level. The activation of EGFR can be linked to the activation of other receptors through transactivation (a process by which a protein or a ligand enhances the expression of other genes) such as the ADAM family members of metalloproteases involved in angiogenesis, the G-Protein Coupled Receptors (GPCR), the cytokine receptors, the integrins, the Receptor Tyrosine Kinases (RTK), or through physical interaction with Platelet-Derived Growth Factor Receptors (PDGFR) and IGF1R.

In modelling the hallmarks of cancer, there are two main tendencies: the modelling of individual hallmark pathway in detail and isolated from the influence of other pathways, and the modelling of the crosswalks between these signalling pathways. We have proposed reviews on some of the individual models. Some groups have concentrated on the studies on the crosstalks. Abbott et al. (2006) have proposed in an agent-based model to study the interactions between the initial six cancer hallmarks of Hanahan and Weinberg (2000) and showed the importance of sequences of events that lead to cancer. It becomes obvious that understanding the deregulations observed in cancers from the perspective of their signalling hallmark pathways requires one to consider their interplay. Crosstalks are indeed inevitable as most of these cancer pathways share a considerable number of genes. Levchenko et al. (2000) have identified the role of scaffolds, both experimentally and theoretically, as a possible way to reduce crosswalk between pathways.

From a different perspective, Cui et al. (2007) have provided a map of human cancer signalling, putting the emphasis on genes that are altered genetically and epigenetically. Putting together different types of diagrams from various databases including BioCarta and Cancer Cell Map, they have built a large network of 1,634 nodes and 5,089 links. Mapping datasets onto this map, they have showed that *"cancer mutated genes are enriched in positive signalling regulatory loops, whereas the cancer-associated methylated genes are enriched in negative signalling regulatory loops,"* knowing that the members of these loops are most probably involved in more than one single pathway.

Along with the multiscale aspect, modelling interdependencies and crosstalk of the signalling pathways will most certainly be one of the main challenges of cancer systems biology for the next decades.

8.2.3   Modelling and therapies

Functioning of cancer-related pathways and their crosstalk have been a popular subject of mathematical modelling in cancer systems biology over the past years and are of particular interest to improve therapeutical efficacy.

Models of cancer therapies can be roughly classified in two types. The first type of models predicts a very general effect of a therapy on tumoral growth. By their construction, these models are more phenomenological models. The second type of models predicts the action of concrete therapeutic drugs on cellular processes (such as angiogenesis or inflammation). These models are more mechanistic as opposed to the first type.

Modelling the therapeutic effect of a drug or its combination allows resolving such an important problem in cancer therapy as *optimisation of the treatment protocol*. This is of utmost importance for application of chemotherapy, for example, because this type of therapy is connected with heavy toxic

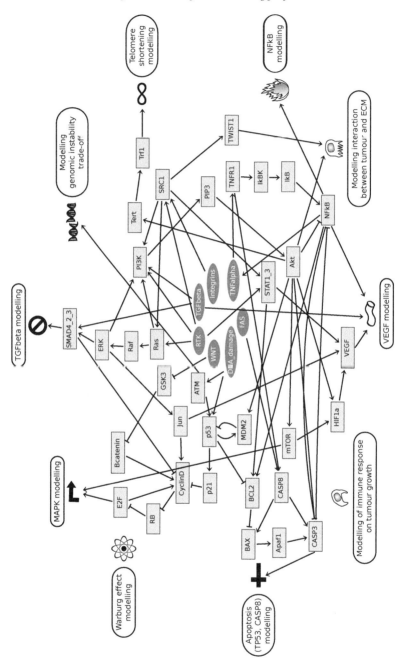

FIGURE 8.2 Pathways involved in hallmarks of cancer. In the centre of the map, some factors triggering the signalling pathways are gathered. The pathways lead to the 8 hallmarks and the 2 enabling characteristics. The most common types of models are mentioned for each of these hallmarks and characteristics.

stress on the patient's organism. Reducing this stress without decreasing the therapeutic effect should improve survival.

One of the first theoretical approaches developed to help optimise cancer treatment using traditional chemotherapy was based on the mathematical *optimal control theory*. The main metaphor of this theory is the following. The tumorigenesis process is characterised by some parameters such as the number of tumoral cells and the number normal cells (for example, the number of **blasts*** and healthy erythrocytes in blood). Left without treatment (*i.e.* external control), tumorigenesis develops by itself following some intrinsic laws such as exponential growth, or other laws such as the Gompertzian law (see **Box 8.1**) being one of the most popular descriptions. When treated, the growth law starts to be modified by external controls and the initial unperturbed tumorigenesis trajectory starts to deviate from its intrinsically determined route. The task of the optimal control (treatment) is to keep the trajectory of tumorigenesis in a certain region of its **phase space*** or bring it to some final stable point (for example, to the region where there are zero tumour cells) in an optimal way. Historically, optimal control theory was first applied to solve the problems of providing maximum height for a rocket by switching on and off its engines in an optimal program. The corresponding mathematical theory used to solve this problem is called the *Pontryagin's Maximum Principle*.

❑ **BOX 8.1: Gompertzian law**

According to the Gompertzian law, the cancer cell population grows as

$$N(t) = N_\infty e^{-be^{-ct}},$$

where $N(t)$ is the number of tumoral cells at time t, b, and c are some positive numbers and N_∞ is the target size of the cancer cell population.

Typically, treatment is described by a time-dependent vector of one or several therapeutic interventions $u(t) = \{u_1(t), ..., u_2(t)\}$. For example, if a single-drug chemotherapy is applied to the tumour in a periodic fashion (*e.g.*, once in a day for a couple of hours) then $u(t)$ contains a single component which is a periodic step-wise function of time with the corresponding period. The task of designing the optimal therapy is to suggest such a treatment plan $u(t)$ that will maximise or minimise an objective function. For example, Swan and Vincent (1977) published the first paper on the effect of one single drug which minimised the number of tumoral cells in multiple myeloma. Following this article, more complicated scenarios were used, consisting of using several drugs of different efficiency and specificity, describing the effect of surgery (removal of a certain fraction of cancer cells) and using more complex objective functions. Thus, in different papers, it was suggested to use as variants of objective functions: the cumulative drug level, the maximum intermediate drug level, the lower limit on the normal cell population, the maximum host survival time, and the maximum level of drug toxicity, avoiding drug resistance (Shi et al., 2011). The *treatment cost*, which might be very different in the first line (initial treatment) and of the second line (after resistance appearance), or certain preferences in the form of the treatment plan $u(t)$ (*e.g.* in a periodic fashion) can also be included in the objective function for optimisation (Shi et al., 2011).

So far, the impact of these studies on clinical practice has been very limited, though several papers become rather noticeable in the clinical domain. Among them, the work of Goldie and Coldman (1979) suggested a treatment plan that would avoid the appearance of drug resistant cells. This is of utmost importance since the resistance problem is the main cause of treatment failure in cancer (see also discussion of cancer robustness in **Chapter 9**). Day (1986) extended this model and suggested the *worst drug first rule* for the multidrug chemotherapies using drugs of various efficiency and activity spectrum. This result was generalised by Katouli and Komarova (2011). They took into account cross-resistance and developed both optimal timing and drug order. Some of these suggestions have already been introduced in small-scale clinical trials.

One particular aspect of optimisation of treatment is related to use of *chronotherapeutics* in cancer, *i.e.* administering drugs according to the Circadian Timing System (CTS). CTS represents cellular clocks coordinated by a hypothalamic pacemaker that controls cellular proliferation and drug metabolism. Therefore, it was suggested that taking into account circadian rhythms (of 24 hours periodicity) can enhance anticancer therapy and reduce the toxic load on the patient's organism, since the toxic drugs can be administered specifically in periods of active cancer cell division. Indeed, this strategy was shown to be able to reduce 2 to 10-fold the extent of toxicity of 40 anticancer drugs in mice or rats. These results can be extended to some degree to human patients (Levi et al., 2011). To optimise the administration of drugs in cancer chronotherapeutics, various models of cell cycle entrained by circadian clocks can be used, starting from the simplest automaton model describing durations of individual cell cycle phases to the detailed model of the mammalian cell cycle coupled with circadian clocks through expression of *WEE1* gene (Altinok et al., 2007, 2009). For example, administration of 5-fluoro-uracile DNA damaging agent can be incorporated in the simplest automaton model by assuming that cells exposed to 5-fluoro-uracile while in S phase have an enhanced propensity to quit the cycle at the next G2-M transition (Levi et al., 2011, 2008).

Another important application of systems biology approach in improving anticancer treatment is the creation of the whole-body physiologically-based pharmacokinetic-pharmacodynamic (PK-PD) models for anticancer drugs. Several attempts were already made in this direction (Karlsson et al., 2005), attracting a lot of interest from the pharmaceutical companies.

The effect of therapies, other than chemotherapy, was studied using mathematical modelling by defining a *default* model of tumorigenesis in both the absence and presence of therapy. Although these studies do not necessarily have the goal to optimise the described therapy, most of them have this potential. The effect of *radiotherapy* for early breast cancer was investigated with a mathematical model of growth and invasion of a solid tumour into breast tissue (Enderling et al., 2006). *Hormone therapy* in the form of androgen deprivation for treating advanced prostate cancer was treated mathematically in Tanaka et al. (2010). Mathematical modelling of *trastuzumab targeted therapy* on chronic myeloid leukaemia was reviewed by Abbott and Michor (2006). Response to trastuzumab in *receptor tyrosine kinase inhibitor therapies* was modelled by Faratian et al. (2009) predicting the role of PTEN protein in developing resistance. Finally, the effect of *tumour therapy with oncolytic viruses* on intra-tumour heterogeneity and, hence, cancer evolution and robustness was formally studied by Karev et al. (2006). This list is by no means comprehensive.

Several multiscale models taking into account both molecular mechanisms of anticancer drug action and the resulting net effect on the tumour growth by cell population modelling appeared during the last decade in the literature. For example, in Ribba et al. (2006) this type of approach was applied to study the effect of irradiation therapy on colorectal cancer and the dependence of sensitivity to irradiation on cell cycle phase.

⇨ **Key notes of Chapter 8**

- Cancer systems biology aims at realistically reproducing specific behaviour of individual types of cancers, and create mathematical models of tumours rather than generic descriptions of individual pathways involved in tumorigenesis.

- There are two approaches to modelling cancer hallmarks: individual hallmark description and interplay between hallmarks.

- Mathematical models of cancer are of particular interest to improve therapeutical efficacy.

- The signalling pathways involved in cancer are interconnected, and modelling crosstalk between them remains a challenge.

Chapter 9

Cancer robustness: Facts and hypotheses

Living organisms are complex systems characterised by **emergent properties***. One of such ubiquitous emergent properties is *robustness* (Kitano, 2004a). In a very general and intuitive form, robustness means that *living organisms are capable to continue performing their functions despite significant variations both in the environment and their physiological organisations including their genetic backgrounds.* This simple and intuitive view, however, needs to be made more precise and constructive. We should decipher what is exactly meant by variation and by capability to perform functions as well as to understand how such a property as robustness happened to be a common feature of living organisms.

From the point of view of mathematical modelling and computational systems biology, we could postulate the necessity of the robustness property and make it a design principle in mathematical modelling (for example, see Alon, 2007a). This would allow us to distinguish mathematical models possessing such a property and to claim that, among equally good models, a model characterised by robust properties is more likely to better correspond to the biological reality. To achieve this, one needs to introduce a definition of a biological robustness in mathematical terms and to provide tools to measure it. This nontrivial problem attracted a lot of attention during recent years and many attempts to define and classify various types of robustness were made.

In the cancer field the question of biological robustness becomes crucially relevant. There are two important questions: *How do tumour cells manage to appear and invade normal tissues?* and *Why is cancer so difficult to treat?* Both questions are related to the problem of evolution of tumours, functioning of biochemical networks, resistance to existing treatment strategies and, most importantly, tumour **relapse*** after an initially-successful treatment.

Cancer as a disease is robust because it has a significant chance to appear during the organism's lifespan and it is able to withstand therapeutic interventions including surgical tumour removal and application of various treatments. Posed in this way, the question of cancer robustness is transformed from a theoretical to a practical issue. If one is able to manage cancer robustness, for example, by finding well-targeted perturbations at some identified *fragility points*, then it will be directly related to defining new treatment strategies.

Arguably, this is the central objective of the systems biology of cancer, hence, it deserves to be discussed in this book in some detail.

The theory of biological and cancer robustness is still in its infancy. In this chapter, we provide a very general review of relevant existing ideas in this rapidly evolving field. This chapter is devoted to the description of general mechanisms leading to robustness in biological systems in general and robustness of cancer in particular, and to the use of the notion of robustness in cancer treatment strategies. In **Chapter 10**, the mathematical principles of biological and cancer robustness will be reviewed.

It is important to have in mind two levels of biological organisation in any discussion about cancer robustness. First, it is the *robustness of cancer cells*, *i.e.* the intracellular molecular mechanisms making cancer cells robust. Second, it is the *robustness of cancer as a disease* which is connected to evolutionary processes taking place in cancer cell populations. In this chapter both levels will be discussed.

Making a comprehensive review on the theoretical and experimental study of the biological robustness would lead to the writing of several separate books. Some of these books are already written (for example, see Wagner, 2005), but most of the knowledge in this field is dispersed in thousands of publications (PubMed contains 25 records with *robustness* within their title for the period of 1980–1989, 108 records for the period of 1990–1999, and 740 records for the period of 2000-2009, while the total number of PubMed records only doubled in 2000-2009 with respect to 1980–1989). The goal here is not to summarise the content of these books and articles but rather to formulate the meaningful questions in any constructive discussion about biological robustness. Some of the answers to the questions can be found in the above cited books, some of the key elements will be mentioned here, without pretending to be exhaustive. **Sections 9.1–9.5** and **9.8** are devoted to the general questions about robustness of living organisms while **Sections 9.6** and **9.7** are more specific to the cancer research field.

9.1   Biological systems are robust

We say that biological systems are robust; however, this statement needs to be refined. The first guiding principle is that together with the word *robust* we should always be ready to give an answer to two questions: **what** property is robust? and **with respect to what** perturbations is the property robust? Some well-documented examples of biological robustness are provided below (Wagner, 2005):

- The DNA and RNA sequences are robust to replication errors.

- The physicochemical properties of amino acids in a protein and the function of the protein are robust to point **mutations**[*] in codons.

- The structure of RNA molecules is robust to changes of individual RNA nucleotides.

- The three-dimensional structure and function of proteins are robust to changes in the amino acid sequence.

- The structure and function of proteins are robust to recombination, that is, swapping of contiguous stretches of amino acids among proteins.

- The expression pattern of a gene is robust to drastic changes in the regulatory regions of the genes caused by mutations.

- The flux of matter through a metabolic pathway and the outputs of the pathway are robust to drastic changes in enzyme activity.

- The flux of matter through a metabolic network and cell growth are robust to drastic decreases and increases in flux through individual chemical reactions, and complete elimination of such reactions.

- The expression pattern of *Drosophila melanogaster* segment's polarity genes, and gene expression patterns in other gene regulatory networks are robust to changes in regulatory interactions among network genes, in gene copy number, and in the expression patterns of genes *upstream* of the network.

- The developmental pathways that form phenotypic characters, and the characters themselves are robust to variations in the genes of these pathways.

- The organism's body plan is robust to both minor and massive changes in embryonic development, which are ultimately caused by genetic changes.

- The cell populations are robust to mutations appearing in individual members of the population.

- The cancer cell populations are often robust to conventional cancer treatments.

Note that we say that these are examples of biological robustness despite the fact that it is known that some perturbations can drastically change the properties of the systems presented in this list. Hence, the claim about robustness usually concerns *randomly occurring* rather than *targeted* perturbations.

Robustness is a property which is applied at many levels of living matter organisation and it can be seen as a ubiquitous multiscale organising principle. For our purpose to study of robustness in cancer, we will mainly concentrate on the robustness properties related to functioning of intracellular biological networks of molecular interactions and cell populations, leaving other robustness manifestations out of the scope of this chapter.

Why are living systems robust? If a living system were not robust (*i.e.* fragile) and not able to successfully withstand environmental and genetic perturbations, then it would quickly disappear in the history of life. Intuitively and somehow tautologically, robust systems should exist longer, because they are robust. As Wagner (2005) said *"A system will spend most of its time in a robust state, precisely because such a state is more robust. Put differently, if the world is rife with robust systems, it is because fragile systems are fleeting."*

Another consideration concerns our limits in cognition of things. All science is based on reproducibility of experiments, hence, the experimental system should, at least in some aspects, be robust with respect to the inevitable variability of experimental conditions. It would be very difficult to study systems which are hypersensitive to many factors. For a successful experiment, one should be able to prepare the conditions of that experiment in such a way that the system would be insensitive to almost all variabilities, in order to neglect them, and sensitive to a few targeted perturbations for which the effect is studied. Questions related to robustness and reproducibility of experiments, experimental design and data processing issues were considered in **Chapter 4**.

Thus, there is a certain bias in considering robust systems; they are selected both by nature and experimentalists. This simple consideration is completely relevant but as usual it explains only part of the complex reality. Principal complications come from trying to apprehend the following important questions:

- **What is persistent: A quantity, a state, a process, a structure or a function?** For example, some system parameters might even be drifting and despite that, the system is perfectly able to perform its function. For example, cancer cells can proliferate and change their genomes continuously or as a result of some large-scale events such as mitotic catastrophe (see **Chapter 2**). Another important example is the following: functioning of many kinases and receptors in cell signalling rely on certain conformational changes of the kinase molecules or complexes. These changes are often driven by thermal fluctuations, hence, one can say that for robust functioning of a kinase, a certain fragility of its structure with respect to thermal fluctuations is needed.

- **Do we consider robustness of an individual biological entity or a group of entities?** For example, lethal fragility of individual cells to certain perturbations, such as signals of the programmed cell death, might increase robustness of cell populations to mutation invasions in the long run.

- **What is the time scale of robustness for a given property?** Since we claim that a property persists in time then we should define precisely the time scale of this persistence. No property can persist forever. Some properties are considered as robust only in the sense of an average behaviour for a large period of time.

- **What is the biological context of a robust entity?** This question is crucial in considering living systems, since no living systems exist in isolation. A typical problem can be formulated as follows: under which conditions is a gene considered as essential for viability?

Considering these questions, one can realise that robustness is inevitably context-dependent. Changing the context might make robust properties fragile and *vice versa*.

Another source of complication comes from the fact that all living systems are replicating and inherit parental genetic programs. In a simple view, robustness of an organism as well as its fitness are phenotypic properties. These properties are inherited from the parental genomes and also influenced by the individual history of the phenotype development. An organism more robust to the changes in the environment will have better chances to survive, hence, higher fitness. However, one should also consider robustness with respect to modifications of the organism's genetic material (mutations), which makes the notion of robustness different from that of fitness. Natural selection acts on fitness differences: an organism characterised by better fitness has more chances to survive. However, two organisms with the same fitness but different robustness to mutations will have identical chances to survive (their genetic difference will be neutral), because mutations can affect only their offspring but not themselves. Thus, differences in robustness to mutations are selected indirectly and only when mutations occur. This simple consideration shows that cancer robustness is impossible to consider without referring to all complex questions of natural selection, population genetics and evolution of cell populations. These questions will be briefly treated in this chapter.

In different aspects and concrete applications robustness has multiple synonyms, such as *stability, persistence, resistance, permanence, insensitivity, survival, reliability, tolerance,* **homeostasis***, *buffering, resilience, canalisation, etc.* Giving exact and consistent definitions for all these words would be a daunting task, which we will try to avoid. Instead, we will talk here about robustness, defining every time what we mean by that.

The types of perturbations that a living system withstands can be classified as:

- **Varying environment** without violating the integrity of the system.

- **Perturbations changing the internal organisation of the system, causing internal damage** without changing the genetic program which will be passed to new generations.

- **Changes in the genetic program that is, mutations**. For this chapter it is very important to say that in our current understanding of **tumorigenesis***, cancer is mainly associated with this type of perturbation. In a way, cancer appears when robustness of normal cell population with respect to invasion of certain mutant cells is violated.

This classification is inspired by the definition of the robustness given in the beginning of this chapter. While being the starting point in many practical studies, this classification, however, becomes very quickly too naive. In considering changing environment, it is often difficult to neglect the feedback of the biological entity on the environment. Adaptation to environmental stress is often accompanied by internal reorganisation of a cell or a tissue. The processes of mutagenesis can be triggered by environmental conditions, *etc.*

Another important question to be asked is: How do biological systems achieve robustness? This is a complex issue and it would be better to decompose it into simpler questions such as:

- **Is robustness of a biological entity rather continuous or rather discrete or rather binary quantity?** Does it make sense to talk about *more robust* entity? In other words, can we talk about *ameliorating* robustness by small gradual changes? If low robustness means lethality then any gradual introduction of stabilising mechanisms is difficult, and they should appear *at once*, a scenario, which requires a careful explanation.

- **Can robustness be a subject of natural selection?** This is a nontrivial question being a subject of hot and long debates. The answer to this question depends crucially on what is meant by robustness.

- **Can robustness be considered as an independently selected property or is it coupled with other functions?** In many situations, robustness is coupled with other properties selected by biological evolution which might create either antagonism (leading to trade-off solutions) or synergy between this function and robustness (leading to *robustness for free*, Wagner, 2005).

- **What are the concrete mechanisms for achieving robustness?** Robust properties are ubiquitously found in most of the biological phenomena. The problem is that it would be difficult to distinguish the most important ones from the auxiliary ones. This field is full of questions like what is more important, natural selection or self-organisation, architecture of biochemical networks or distribution of kinetic rate parameters, structural stability or fine-tuned control? It is very unlikely that these dilemmas will be resolved soon due to lack of knowledge. However, constructive discussions on these topics improve our understanding of the general principles ruling the designs adopted by the biological systems during the history of their evolution.

9.2 Neutral space and neutral evolution

Wagner (2005) formulated the main principle of emergence of biological robustness as *"the same biological function can be implemented in billions of different ways."* In other words, this principle is equivalent to the existence of a large **neutral space***(see **Box 9.1**).

❏ **BOX 9.1: Neutral space and neutral evolution**

Neutral space is a collection of equivalent implementations of the same biological function. The notion of the neutral space appeared most naturally in studies of sequence-structure or structure-function relations of biological molecules but can be extended to many other types of problems related to evolution and robustness. Neutral molecular evolution theory states that the vast majority of evolutionary changes at the molecular level are caused by neutral mutations, *i.e.* those that do not change fitness (Kimura, 1983). The theory of neutral evolution postulates the existence of vast neutral spaces and claims that the absolute majority of systems' genetic changes is neutral with respect to natural selection. Hence the system is continuously drifting in a large region of equivalent configurations.

To illustrate the notion, let us give several examples of neutral spaces. The classical and the most studied one is the set of all RNA sequences leading to an identical RNA structure (Wagner, 2005; Reidys et al., 1997). It was mathematically rigorously shown that an RNA structure, which is robust against random mutations, typically has the biggest associated set of sequences that is mapped in the same structure. The most natural representation of a neutral space here is a graph containing multiple connected components. Another example is the set of all possible enzyme concentrations leading to the same optimal (or, nearly optimal) growth of a bacteria in a given environment. In this case, the neutral space can be represented as a surface or a manifold (with possibly a nontrivial structure) embedded in a multidimensional space of enzyme concentrations. A third example is a set of all possible genome modifications leading to the clinically identical tumorigenic cellular phenotype. This is the most complex example since the genotype-phenotype mapping here is extremely complicated, mediated by gene expression and various epigenetic mechanisms which are poorly characterised. The exact sequence of events leading to a genotype can be essential for having a particular tumour phenotype, *etc.* It is not even completely clear what would be an appropriate abstract mathematical representation of this type of neutral space. All these complications make any mechanistic and explicit study of this kind of neutral space difficult for the moment as opposed to the two previous ones.

Intuitively it can be claimed that a large neutral space should be a characteristics of a robust system's state. Such large neutral spaces have more chance to be met in the process of an evolutionary *blind grope*, as it was expressed by Wagner (2005). When such a neutral region is found, the state can randomly drift inside it without affecting the related function. However, different positions of a system inside the neutral space can be equivalent in terms of a function or a fitness, but, nevertheless are not equivalent in terms of robustness. Those regions of the neutral space which are at the maximal distance from the neutral space border can be assumed to be more robust, because more modifications of a system will be required to jump out of the neutral space. Hence, one can ask the following question: even if any point in the neutral space is equivalent from the point of view of a function or fitness, can natural selection drive the biological system to that region of a neutral space which is characterised by the largest robustness? In other words, do we expect to find a biological system or mechanism closer to the centre of the neutral space, or rather closer to its border? Evolution of robustness will be briefly discussed in **Section 9.5**.

Random drift inside the neutral space of genetic programs can be characterised by the Kimura's theory of neutral evolution (Kimura, 1983). The theory of neutral evolution clarified several important features of the neutral drift in populations. Firstly, a connection between neutrality and population size is established: the neutral evolution theory states that the natural selection does not notice relative changes in the fitness which are much smaller than $1/4N_e$ where N_e is an effective size of evolving population (it is a number reflecting not only the number of individuals in the population but also various features of their life style and history). Secondly, fixed neutral mutations (*i.e.* mutations attained a frequency one in the whole population) appear on average each $1/\mu$ generations, where μ is the rate of mutations per allele per individual per generation. The time in which such a neutral mutation will be fixed is inversely proportional to the effective population size $1/N_e$. As a consequence, the population is likely to be genetically heterogeneous (polymorphic) with respect to the alleles characterised by $N_e\mu \ll 1$ condition, and homogeneous (monomorphic) for the alleles for which $N_e\mu \gg 1$.

A warning should be made when the notion of neutral space is used to illustrate the principles of robustness and evolution. First, most of the conclusions for the properties of neutral spaces are drawn from the studies of mutations of genetic sequences and their impact on the structure of biological molecules. Thus, it might be dangerous to extrapolate on more complex situations. The properties of the neutral space are often assumed to be simple (such as connectivity, smoothness, ability of introducing a natural measure, *etc.*) which is not guaranteed in complex cases. Since the neutral spaces are typically multidimensional regions of high-dimensional spaces, their properties can be counterintuitive and lead to misinterpretation (as shown in the continuing discussion on the multidimensional properties of adaptive landscapes; Gavrilets, 2004). This especially concerns intuitive considerations about the distance to the border of the neutral space, since most of the volume of truly multidimensional objects might be located near the borders and not near the centre. Another source of difficulties in the above mentioned considerations is that the different areas of the neutral space can have different mutation rates, which can effectively *freeze* some populations in the regions of low mutability, and, *vice versa*, *diversify* populations in the regions of high mutability. Also, complex functions of an organism rarely depend on one single gene but rather on relatively large sets of interacting genes (networks). Therefore, the notions of neutrality, polymorphism and mutability should be applied not only at the level of genes but also at the level of **signalling pathways*** and networks.

The above mentioned methodological difficulties complement the general problem with the notion of neutrality itself. Wagner (2005) claims that a *puristic* or *essentialist* point of view on neutral mutation as a genetic change, which does not affect an organism's fitness in any environment and genetic background, is useless in any realistic consideration. First of all, most of the biological molecules are multi-functional. Because of this it is rather naive to measure the effect of a mutation with respect to, for example, a well-defined enzyme's activity, for it can affect other less-studied enzymes' functions and have an effect on fitness in some conditions. For example, a classical synonymous codon substitution does not change the structure of a protein but might significantly affect the protein synthesis rate. In practice it is not possible to prove independence of a mutation neutrality to an arbitrary change in the environment. On the contrary, cases when some basic environmental parameters, such as concentration of oxygen, can affect the neutrality of some alleles, with respect to a particular cancer incidence, for example, are well-documented (Astrom et al., 2003).

9.3 Robustness, redundancy and degeneracy

One of the very general principles of evolving biological systems consists in duplicating already working mechanisms (genes or networks or genomes or even organs) and partially specialising them to new functions (Taylor and Raes, 2004). To illustrate the generality of this phenomenon, one can mention that from 25% to 50% of eukaryotic genes have **paralogs***. Studies of the mechanism of evolution through duplication lead to a series of interesting questions about the role of redundancy in explaining systems' robustness and structuring gene regulatory networks. Here we understand *redundancy* as a *repetition* of a part of the system such that the copies perform very similar functions in similar contexts.

Creating several copies of the same gene creates robustness for a system with respect to a deletion of any of these copies. Thus redundancy increases robustness. However, the process of partial specialisation of duplicated gene functions towards new and different functions, conserving partial functional overlap, creates multiple crosstalk between functions, sometimes called *degeneracy* (Edelman and Gally, 2001; Whitacre and Bender, 2010). Two purely redundant components have perfect overlap in the set of functions they support. Two degenerate components have only partial overlap in the set of their functions, but besides this overlap, each of these components is specialised in some qualitatively different functions.

If a set of degenerate components connects a large set of functions (or even all functions of an organism), it serves as a basis for **networked buffering*** (Whitacre and Bender, 2010), which (1) efficiently redistributes resources in a system when some of them are not available, and (2) compensates for the loss of some of the components by recruiting agents (genes, proteins, complexes, *etc.*) from other functions. In such systems where the functions are all connected by a series of overlaps in the sets of agents capable to perform the functions (with various efficiency, of course), one expects to observe a gradual (not catastrophic) decrease in performance when a certain number of agents is removed. This type of robustness is called **distributed robustness*** meaning that the robust system performance cannot be attributed to any particular element of a system but rather to a general pattern of how the system's elements are connected to each other (Wagner, 2005).

It is believed that the role of pure redundancy in supporting robust features of biological systems is relatively modest, unlike engineering solutions (such as control systems of an airplane) where pure redundancy is abundant. By contrast, the role of distributed robustness and degeneracy is more important in biological systems and favoured by selection. This belief is supported by studies of large-scale gene deletion experiments in yeast and studies on functional divergence of paralog genes (Wagner, 2005).

The phenomenon of degeneracy is found at all levels of living matter organisation. At the protein function level, it is exemplified by various combinations of specialised protein domains which couple certain protein functions. At the level of biochemical networks, it is manifested by existence of multiple crosstalks between biochemical cascades. At the organismal level, many organs have overlapping functions, with production of red blood cells or immune response being stereotypical examples.

9.4   Mechanisms of robustness in the structure of biological networks

Cancer is thought to be a *network disease* (see **Chapter 1**), hence, one of the most important issues is to understand a particular aspect of biological robustness related to the functioning of biological networks. Most of this section will be devoted to this question.

During last decades, multiple studies of the structure and dynamics of biological networks allowed identifying several universal emergent patterns; many of them were already associated with appearance of robustness in various senses of this word. The most common patterns are *scale-freeness, bowtieness, functional* and *structural redundancy, degeneracy, modularity, cooperativity, network motifs* and *feedback controls.* We list these patterns of network robustness in the order of the scale of the network structures they shape, starting from the most global such as scale-freeness and bowtie structure, and ending with robust features of more local structures such as network motifs (see **Figure 9.1**).

9.4.1 Scale-freeness

Barabási and Albert (1999) noticed that many biological networks as well as other networks in engineering, computing and social sciences, follow a close to power-law node degree distribution. Such networks were called *scale-free* (see **Box 9.2** and **Figure 9.1A**). These networks are characterised by more frequent presence of highly connected **hubs**[*] than one would expect in randomly wired graphs. It was suggested that a general feature of such networks is robustness towards a random removal of a node, and extreme fragility towards instructed removal of big hubs. This suggestion was supported by an observation that in large-scale yeast gene deletion studies, the most connected genes have the tendency to be essential (knocking them out is lethal for the organism) (Jeong et al., 2001). Also, many genes traditionally associated with cancer susceptibility belong to highly connected hubs in networks of Protein–Protein Interactions (PPI), with TP53 being one of the most connected proteins.

❏ **BOX 9.2: Scale-freeness**
Scale-free graphs are graphs in which node connectivity degree distribution follows a power law. Node connectivity degree is the number of node neighbours. A graph with a power law node degree distribution is a graph in which the number of nodes n having k neighbours is a power function of k, *i.e.* $n = \alpha k^{-\gamma}$, where α and γ are real numbers. For protein-protein interaction networks $\gamma \approx 2$.

The question whether scale-freeness is a naturally selected mechanism insuring robustness, or rather a phenomenon of self-organisation, or even a reflection of a bias connected with our strategy of collecting information about interaction between proteins, was largely discussed in the literature (for example, see Lima-Mendez and van Helden, 2009). Nowadays, the fact of scale-freeness of biological networks and the role of scale-freeness in insuring robust network functioning remain attractive but arguable hypotheses.

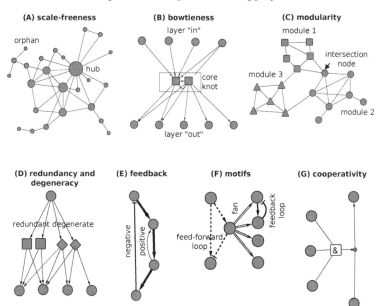

FIGURE 9.1 Network structures responsible for robust network proper-ties. (A) An example of a scale-free network: the size of the node is proportional to the node connectivity degree. (B) An example of a network having bowtie structure: all connections between "in" and "out" nodes are processed through the core net-work; the core network has two states. (C) An example of a modular network with three modules: module 3 is connected to both module 1 and module 2 only through edges, while module 1 is connected to module 2 through a node in the intersection; (D) An example of redundancy and degeneracy: square nodes are fully redundant; any of them can be removed without affecting connections between upper and bot-tom nodes; rhomb nodes are degenerate: they share part of the connections, but have their own specificities; the left rhomb has lost one connection and acquired a new connection (shown by dashed line). (E) An example of feedback: thick arrows show the principal cascade which is regulated by positive and negative feedback in-teractions. (F) An example of a network constructed from the combination of three motifs (elementary subnetworks): feed-forward (dashed lines) and feedback (solid lines) loops and a fan (double lines). (G) An example of cooperativity: a transition from bottom to upper state is possible only if three nodes shown on the left are "active" simultaneously.

9.4.2   Bowtie structure

Bowtie structure is a special type of relatively large-scale organisation of metabolic, regulatory and signalling biological networks, characterised by converging a wide range of inputs to *a synthetic core* (also called *a knot*), consisting of relatively few operational elements (see **Figure 9.1B** and Csete and Doyle, 2004). After convergence, a variety of outputs propagate from the core.

The nature of the core can be different in various contexts, but the general principle of compressing the dimensionality of the input signals to a core with a small number of degrees of freedom always applies. One of the sources for appearance of the bowtie structure lies in the fact that the cell uses a relatively small number of *universal cell currencies* such as energy carriers (ATP, NADH, NADPH, *etc.*) and precursor metabolites (glucose 6-phosphate, fructose 6-phosphate, pyruvate, *etc.*). In metabolism, a huge variety of nutrient sources are catabolised first into a handful of such currencies with further synthesis into a large number of structures used by the cell. Universal or close to universal use of the polymerases, **ribosomes***, codon usage for most of the genes also determine bowtie organisation of intracellular networks on a large-scale. Signalling pathways are also commonly organised in bowtie structures with the core often made from bistable proteins such as G-proteins (Polouliakh et al., 2009; Oda and Kitano, 2006).

The advantages of bowtie organisation are the following. Firstly, it is the facility of control. Central molecules of the core of the bowtie structure act as master regulators of a large number of processes. In other words, there is little democracy in the most critical decisions that the cell must be able to make. Secondly, the bowtie network organisation is robust with respect to the variations of the input sources of nutrients or signals, for example, an ability of supporting lack of certain nutrients that can be compensated from other sources. Thirdly, it is the facility of modifying the peripheral molecular processes by plugging and unplugging them through commonly defined protocols and interfaces. On the other hand, the danger of the bowtie organisation is that it creates fragility against targeted damage of the core mechanism which can be, and is, exploited by cancer cells (Csete and Doyle, 2004).

An alternative for the bowtie structure is *ad hoc* and independent processing of signals and nutrients by parallel pathways without convergence to a common core. In an exaggerated view, this would correspond to the situation when for the transcription of any gene, a separately designed type of polymerase would be implemented. Parallel processing of information and mass flows by independently implemented routes is indeed another pattern of organisation of cellular networks (see discussion about redundancy below). Thus, both parallel and convergent designs serve for solving various trade-offs between robustness, efficiency and adaptability and are preserved by evolution. The most optimal (and favoured by nature) pattern seems to be the following one: cellular networks contain a relatively small number of highly conserved and optimised compact core mechanisms that solve some relatively simple tasks (such as switching from one state to another). These core mechanisms, just as a computer CPU, can be implemented using a parallel architecture for providing better stability, but the number of their internal degrees of freedom should be much less than the number of input signals coming to the core (Csete and Doyle, 2004).

9.4.3   Modularity

Molecular biology established a while ago that cellular functions, such as signal transmission, are carried out by *modules* (see **Figure 9.1C** and Hartwell et al., 1999). Modularity is believed to occur at all levels of biological organisation, from gene and protein sequences to body tissues and organs. Modularity is found also at the level of biological network structures. Exact definitions of a module are numerous and usually depend on the context or a problem posed (for example, see Radulescu et al., 2006; Calzone et al., 2008, and also examples of modular network analysis in the **Section 4.9**).

One of the ideas in this field relies on an analogy with engineering devices: it consists of stating that, in a modular design, a random mutation or a modification of a network will affect and will be limited to one particular module, and would not propagate to the whole network. This simple argument is, however, rather weak, because if an affected module is essential then the mutation could still be lethal. On the other hand, modularity should promote a certain degree of evolvability to a system by allowing specific features (*i.e.* subnetworks) to undergo changes without substantially altering the functionality of the entire system. Each module is free to evolve within itself, as long as the interfaces between modules remain conserved. Modularity with conserved and well-designed protocols also allows an organism to borrow and adapt modules that perform some precise functions from other organisms by some kind of horizontal transfer. One of the most striking examples of this phenomenon is the existence of mitochondria in eukaryotic cells or horizontal transfer of antibiotic resistance modules between bacteria, or even between bacteria and higher organisms. Another advantage of modularity is an ability of duplicating certain critical functions and creating redundant (and, further, degenerate, see **Section 9.3**) backups of critical subnetworks (this requires that the network components should be co-localised on the genome). Therefore, one can think that modularity can enhance robustness of an organism on a long time scale, connected to significant evolutionary changes.

The role of modularity in insuring robustness of biological networks remains an open question. In some developmental and heat shock response studies, it was suggested that robust network topologies have to be modular (Ma et al., 2006; Kurata et al., 2006). However, more theoretical and experimental works are needed to confirm the universality of this relation.

9.4.4   Network redundancy and degeneracy

The above described general biological features of redundancy and degeneracy (see **Section 9.3**) also shape the structure of biological networks to a large degree and significantly contribute to their robustness (see **Figure 9.1C**).

Degenerate nodes of the network compensate for one another under conditions where they are functionally redundant, thus providing robustness against component or pathway failure. Because degenerate components are not identical, they tend to harbour specific sensitivities so that a targeted attack such as a specific inhibitor is less likely to present a risk to all components at once. For instance, gene families can encode for diverse proteins with many distinctive roles, yet sometimes these proteins can compensate for one another during lost or suppressed gene expression.

One of the classical and the best studied examples of pathway degeneracy associated with cancer robustness is a family of HER receptors which comprises four homologous members HER1 (EGFR), HER2 (ERBB2), HER3 (ERBB3) and HER4 (ERBB4) (Citri and Yarden, 2006). This family demonstrates highly degenerate complex many-to-many relations between the receptors and the ligands able to activate them. Nematodes contain only one member of this family. It was suggested that two events of duplication and partial specialisation of the receptors could be explained by natural selection towards more robust signalling network design (Citri and Yarden, 2006). As a consequence, therapies targeting EGFR are thwarted by the co-activation of alternate receptor tyrosine kinases that have partial functional overlap with EGFR , but are not targeted by the same specific EGFR inhibitor (Stommel et al., 2007).

Another example of pathway degeneracy directly related to cancer is the cell cycle regulation mechanism involving RB and E2F (see Calzone et al. (2008) and **Chapter 2**). The RB protein belongs to a family of pocket proteins (RB, p107, p130) with partially overlapping functions. There are eight genes belonging to the E2F family of transcription factors (E2F1 to E2F8). The E2F transcription factors are involved in many cellular functions including regulation of cell cycle, apoptosis and DNA repair with various specialities among the family members. Some members are known to be activators of the cell cycle: E2F1, E2F2 and E2F3a, while other members are referred to as inhibitors of the cell cycle: E2F3b, E2F4, E2F5, E2F6, E2F7, E2F8. Among the cell cycle activating members, only E2F3 was clearly suggested to be associated with invasiveness in bladder cancer and with poor survival in prostate cancer, while the role of E2F1 is shown to be controversial and different from one cancer to another. The E2F family members form a nontrivial network of mutual regulation (see **Section 4.9.2**).

The degenerate nature of biological networks creates a specific type of distributed robustness, which can be a subject of mathematical modelling. Some mathematical aspects of distributed robustness will be discussed in **Chapter 10**.

9.4.5 Feedback control

We have already discussed the role of feedback loops in biological networks in **Section 7.4**. Feedback is a common design principle in constructing most of the engineering devices (see **Figure 9.1E**). The basic functions of the feedbacks are to (1) insure the *control of execution* of a command, (2) maintain optimal levels of some signals in the system, (3) amplify response to a signal,

(4) discretise the system's response, and (5) filter noise. It is a well-established fact that the biological networks ubiquitously use feedback controls in insuring robust performance (Kitano, 2004a).

Multiple layers of feedback loops and associated gene-regulatory events are involved in the robustness characteristics of tumours at the levels of intracellular and tumour-host interactions. At the cellular level, feedback controls can give rise directly to robustness against **chemotherapy***. For example, tumour cells that turn on the expression of the multi-drug-resistance 1 gene (*MDR1*) acquire multidrug resistance by exporting drugs out of the cell through an ATP-dependent efflux pump, P-glycoprotein (P-gp), encoded by *MDR1*. This is a simple, but effective, feedback control mechanism to minimise cytotoxin levels. Another example is tumour overexpression of *MDM2*, which causes degradation of TP53, effectively blocking apoptosis. The MDM2-TP53 interaction functions as a negative feedback loop to maintain optimal levels of TP53, and creates certain dynamics (pulsed or oscillatory) of TP53 expression levels (instead of sustained expression) after serious DNA damage.

9.4.6   Network motifs

Analysis of the structure of several types of biological networks and the transcription regulatory networks in particular suggested that these networks can be built from some elementary blocks called network motifs (see **Figure 9.1F**). Network motifs were discovered as recurrent regulatory patterns from statistical analysis of certain subgraphs frequencies in the graphs representing transcription networks. It was suggested that such regulatory units might provide particular dynamical functions, including noise filtering, reduction of cell-cell variation of protein concentrations, acceleration of cellular response, creation of sign-sensitive delays (Alon, 2007b; Shoval and Alon, 2010).

Experimental and theoretical studies have shown a connection between motif dynamics and robustness in several aspects. Firstly, it was shown that the dynamical function of a motif is often barely sensitive to variation of the parameters of the interactions comprised in the regulatory motif (Prill et al., 2005). This type of robustness is connected to the *structural stability* (briefly discussed in **Section 10.5.3**). The motifs that have fewer loops have the tendency to have more structurally stable dynamics, and this was suggested as an arguable explanation for their overrepresentation in biological networks (Prill et al., 2005).

Secondly, some types of motifs can create a buffering effect allowing the biological system to deal with fluctuations in the environment (Alon, 2007b). Thus, a negative auto-regulation motif permits to stabilise the expression of a gene in variable conditions, while a coherent Feed Forward Loop (FFL) motif permits to neutralise and filter out short and not-persistent fluctuations in the concentrations of signal molecules. There is an interesting hypothesis that a feed forward motif including a transcription factor, a microRNA (miRNA) regulated by this factor, and their common target gene can play a fundamental role in buffering stochastic noise in protein synthesis (Osella et al., 2011). We described functioning of some of the regulatory motifs in more details in **Chapter 7**.

9.4.7 Cooperativity

The regulation of a molecule or a function cooperatively by one or more regulatory molecules makes yet another very common pattern of biological robustness (Wagner, 2005). Cooperation here means that the efficiency of regulators mutually depends on their simultaneous presence, or combination of regulators in some specific configurations (see **Figure 9.1G**). One of the basic examples of cooperativity is given by transcriptional control with the presence of multiple copies of binding sites for the same factor in the promoter region of a gene. Cooperativity is created in the situation when the factor, being bound to a site in the promoter, is able to facilitate binding of other molecules of the same factor to nearby sites. As a result, the expression of a gene depends very little on the concentration of the transcription factor in a wide range but there exists some relatively small window of concentration values in which the expression of a gene changes drastically, in a switch-like manner.

Strength of cooperation between regulators can provide a necessary degree of *digitalisation* of the input signal. Very strong cooperativity provides ON or OFF responses to a change of regulator concentrations, while absence of cooperation provides a rather gradual response, when more proportional system response is needed. Digitalisation is a characteristic of most of developmental programs, whereas gradual response is more frequent in inflammation (connected to NFkB transcription factor) and apoptosis signalling (such as a response of *TP53* transcription levels to UV radiation).

In addition, it can be shown by mathematical modelling that, if a biological function is accompanied by a mechanism of regulation involving cooperative behaviour of many regulators, then the robust properties of such networks can evolve by stabilising selection (Wagner, 2005).

The structure of networks determines their dynamical properties only to some extent. On top of the network structure, the distribution of kinetics-related parameters plays an equal if not dominant role in determining the robust features of the biological networks. For example, the multiscale nature of kinetic parameter distribution guarantees some robust network dynamics properties, and determines some very general design principles of the reaction networks. In **Chapter 10**, we consider these aspects in more detail.

Together with network robustness, other universal molecular machineries contribute enormously to sustaining the functioning of living cells in aggressive and variable environments. These machineries include: DNA repair, RNA editing, heat shock response and stabilisation of partially unfolded proteins, kinetic proofreading, multidrug resistance, immune response, polyploidy of the genome where each gene is backed up with its copy, tissue organisation with rarely dividing stem cells and rapidly dividing transient progenitor cells, and many others. Some of them play a role of buffering and masking mutations, and by this, serving as *evolutionary capacitors* (notion described in the next section).

9.5   Robustness, evolution and evolvability

A large corpus of the literature is devoted to the discussion of the relation between robustness and evolution (Masel and Trotter, 2010). *How can a robust system evolve?* This question has been a subject of debate for at least the last few decades. In fact, this question can be split in two different questions:

- Does robustness prevent biological systems' potential for evolution, does it reduce evolvability?

- Can robustness appear as a result of natural selection?

Both of these questions have a direct relation with the question of cancer robustness because it seems logical to assume that the process of natural selection is continuously taking place during **tumour progression***. The evolutionary-based approach is very important for the general understanding of the principles of tumour appearance as *a genetically heterogeneous cell population with rapid processes of natural selection for the best fitness (survival)* (see also **Chapter 2**).

Is there a contradiction to resolve between two statements: (1) a biological system is robust and (2) a biological system can evolve?

9.5.1   Environmental robustness and mutational robustness

It happens that for the discussion, it is crucially important to distinguish between environmental robustness as the ability of an organism to survive environmental stress, and mutational robustness as the ability of an organism to maintain a relatively invariant phenotype despite significant genetic variation (mutations). As already mentioned, this distinction is relative and is not always possible, especially in the case of cancer cell populations where the environmental stress can increase the genetic variation. However, we will keep these definitions for clarity as a first approximation.

Having this in mind, the notion of environmental robustness represents an advantageous trait that should be selected by natural evolution in stressful and rapidly changing environments. However, robustness can consume so much of the organism's resources that this would make developing robustness disadvantageous (a question of trade-offs, see **Section 10.4**). So, the first conclusion is that, in the simplest scenarios, the environmental robustness does not and cannot preclude evolution.

9.5.2 Evolvability and mutational robustness

Probably, the most exciting and related to cancer questions arise from considering the relation between mutational robustness and evolvability (see **Box 9.3**). The first insights on mutational robustness were proposed by Waddington (1957) in his book *The Strategy of the Genes*.

❏ **BOX 9.3: Evolvability**

Evolvability, just as robustness, is used in too many meanings. Among them:

- Evolvability can be expressed as a synonym of adaptability, *i.e.* the ability of modifying a property (for example, a regulatory network) and adapting it to a changed or new environment.

- *"Evolvability is the ability of a population to generate heritable phenotypic variation that could be adaptive in some contexts"* in the middle evolutionary time scale (there are three such time scales: (1) the short time scale of a single generation, (2) the middle time scale of many generations, and (3) the long geological time scale during which major morphological changes and evolutionary innovations happen).

In the second definition of evolvability (see **Box 9.3**), evolution is distinguished from evolvability as variation (observed differences in a property) is distinguished from variability (propensity of a property to vary, regardless the fact that it varies or not). Most of the modern literature converges towards the opinion that robustness to mutations reduces phenotypical variation but, as a result, increases phenotypical variability by increasing cryptic genetic variations, accumulated by random and almost neutral genetic drift (Masel and Trotter, 2010; Lesne, 2008; Draghi et al., 2010).

In the last years, discussions on the tension between robustness and evolvability led to the notion of an *evolutionary capacitor* (see **Box 9.4**) as a mutation buffer. The concrete details of the evolutionary capacitor implementation are not so important. For example, it can fire the genetic variation as a response to stressful conditions, or it can do it without any particular reason, randomly. An important aspect to consider, though, is the separation of time scales. First of all, the average duration of the capacitor in the robust state should be longer than the time needed for the appearance of a single mutation. Then the duration of the capacitor in the non-robust state should be longer

❑ **BOX 9.4: Evolutionary capacitor**

An evolutionary capacitor is a molecular mechanism that is characterised by two properties:

- It provides mutational robustness by buffering mutations and preventing them from causing phenotypic changes. Thus, if the evolutionary capacitor is switched on, then genetic variation accumulates.

- It is capable of switching off and revealing the accumulated cryptic genetic variation leading to a burst in phenotypic variation.

Thus, an evolutionary capacitor (and hence mutational robustness) promotes evolvability.

than a typical time needed for assimilation of new phenotypes, *i.e.* relaxation time, for the selection to happen.

Repeating stages of accumulating and releasing genetic variation help to solve several common problems in evolutionary theory. In particular, it brings at least one possible solution to the problem of *jumping over* low fitness valleys separating regions of genotype space with high fitness. Such a jump requires several mutations to happen simultaneously and not consequently.

The phenomenon of the evolutionary capacitor can also contribute to the hypothesis on punctuated evolutionary equilibrium which is a model for discontinuous tempos of change in the process of speciation and the deployment of species in geological time (see also a discussion on self-organised criticality in the next section).

Several concrete molecular implementations of evolutionary capacitors are known, the most famous being the heat shock protein HSP90. HSP90 acts as a chaperone, *i.e.* a protein helping folding or unfolding and the assembly or disassembly of other macromolecular structures. HSP90 is essential for activating many signalling proteins in the eukaryotic cell. Experiments in *Drosophila* and *Arabidopsis* have demonstrated three key properties of HSP90:

- It suppresses phenotypic variation under normal conditions and releases this variation when functionally compromised.

- Its function can be saturated by extensive environmental stress, thus, leading to its compromised behaviour.

- It exerts pleiotropic effects on key developmental processes. This function is assumed to be conserved in other organisms, potentially influencing the pace and nature of evolution (Bergman and Siegal, 2003).

The evolutionary capacitor mechanism seems to be very general. Experimental results on systematic knock-out mutations in *Saccharomyces cerevisiae* identified more than 300 gene products with exceptionally high capacities to stabilise morphological variation. Silencing any of these gene products will potentially release the region of cryptic genetic variation it masked and allowed to accumulate.

9.5.3 Evolvability and natural selection

There is a hot debate in the literature on how, and if, evolvability itself can be a subject of natural selection. One of the main obstacles here is that increasing evolvability does not seem to give an immediate benefit for fitness of an organism but rather protects it from the future stresses. Note that the same remark concerns robustness.

To illustrate this question, let us imagine a simple scenario when two haploid subpopulations only differ by one gene. Let us assume that this difference does not affect the function of the gene's product but rather its robustness to a random mutation. Similarly, let us call one subpopulation *robust* and the other one *fragile*. Since the function of a gene's product is not affected, both robust and fragile individuals must have the same fitness during the periods when the gene is not affected by a random mutation. These periods can last for many generations during which the polymorphism of the population will be neutral. When a random mutation in the gene of an individual happens, it is more likely to be deleterious in the fragile individual, hence the frequency of a robust allele has a chance to increase and to be fixed. Notice that the scenario describes a polymorphic population (with respect to a neutral robustness mutation) which can exist for long periods of time in the case $N\mu \gg 1$, where N is the size of the population and μ is a mutation rate (see also **Section 9.2**). In the case when the populations are monomorphic for most of the time, selection for robustness is not possible.

Another topic on the relation between robustness and selection touches rather delicate questions of the possibility of group selection (Okasha, 2001), robustness of populations rather of the individuals (Lesne, 2008), congruent evolution, properties of high-dimensional adaptive landscapes (Gavrilets, 2004), and others (see **Box 9.5**). Most of these studies allow concluding that there are no unresolvable contradictions and paradoxes between robustness and evolvability. On the contrary, mutational robustness often promotes evolvability and *vice versa*, even though these processes can be separated in time.

All above mentioned considerations are directly connected to the question of cancer robustness. There is a common point of view that cancer cells are characterised by higher mutation rates (see **Section 8.1.6**) than normal cells and represent growing cell populations (Weinberg, 2007). This means that tumours must be characterised by clonal heterogeneity at least starting from some sufficient tumour size. This heterogeneity will in turn promote selection for robustness both with respect to mutagenesis and various stresses including therapeutical actions on cancer cells (see **Section 2.3**). This concerns the time scale of tumour development. On the other hand, considering longer time scales of many organism generations, there should exist mechanisms of

---

❏ **BOX 9.5: Robustness evolution**

In the theories describing evolution of robustness, several key terms are frequently mentioned. Among them:

Group selection is a mechanism of evolution when a particular allele becomes fixed because of the benefits it brings to the population rather than to the fitness of individuals (for example, altruism).

Congruent evolution is a scenario when evolvability appears as a byproduct of other evolutionary processes (for example, the Haldane's hypothesis that mutational robustness is a byproduct of selection for environmental robustness).

Adaptive landscape or fitness landscape is a metaphor visualising relations between genotype and reproductive success (fitness). It is assumed that one can introduce a distance between genotypes and that each genotype is characterised by a fitness value which is represented as a height of the landscape.

---

selection preventing normal cells from developing cancer, since high risk of cancer can decrease the organism's fitness. Therefore, some arguments about group selection can also be applicable here. The coexistence of these multiple time scales and their interplay should be taken into account in the future evolutionary models of cancer cell populations.

9.6   Cancer cells are robust and fragile at the same time

9.6.1   Oncogene addiction

A naive question can be formulated: should we consider cancer cells more robust than normal cells? To answer this question, we should specify features and perturbations to compare them in terms of robustness.

One of these features is the *survival of individual cells*. On one hand, normal cells should be sensitive to the deprivation of growth signals and the presence of anti-growth signals in the extracellular space, whereas the cancer cells are free of such constraints (see **Chapter 2**). Cancer cells develop various mechanisms to diminish growth control, in particular, by overexpressing growth receptors and inducing autocrine signalling loops. However, in doing so, they might create specific fragilities such as *oncogene addiction*, a term coined in Weinstein (2002).

It is an experimental observation that cancer cells may be more dependent on the activity of specific **oncogenes*** and more sensitive to the growth inhibitory effects of specific **tumour suppressor genes*** than normal cells (Weinstein, 2002). One of the models explaining this phenomenon consists of assuming that the activation of some oncogenes leads not only to accelerating proliferation in cancer cells but also to induction of anti-growth and cell death programs. The hypothesis is that these mechanisms were already activated in cancer cells when adapting to the oncogenic signalling. When the oncogene is active, it is able to compensate and overrun these programs. When the activity of the oncogene is abruptly stopped, cell death programs start to dominate and eventually kill the cancer cells. One of the earliest convincing examples of oncogene addiction comes from an experiment on genetically modified mice with an implanted mechanism of *MYC* modulation. In a transgenic mouse model, switching on the *MYC* oncogene in the hematopoietic cells led to the development of T-cell and myeloid **leukaemias***. However, when this gene was subsequently switched off, leukaemia cells stopped dividing and displayed differentiation and apoptosis. Dependence on the continued expression of other oncogenes (*EGFR*, *HER2*, *KRAS*) for the maintenance of the **neoplastic*** state has also been seen in other tissues in murine models and human cell lines. Some evidence that the oncogene addiction can be the Achilles Heel in specific cancers comes also from clinical trials (Felsher, 2008).

9.6.2   Robustness of individual cancer cells to various stresses

Another specific fragility induced by cancelling growth control is the *replication and nutrient deprivation stress*. Rapidly growing and dividing cells are more fragile than the quiescent ones with respect to many environmental stresses, because they are more dependent on availability of basic construction units, in particular, nucleotides and aminoacids. Cancelling checkpoint mechanisms potentially leads to accumulation of errors in replicated DNA, saturating the capacity of the DNA repair machinery, which in turn can lead to genomic instability and radical genome reshuffling, for example, as a result of survival in the mitotic catastrophe. For each individual cancer cell, the result of genomic instability is lethal with high probability.

Individual cancer cells can detach from the **primary tumour***. Travelling through the body increases the risk of dying without the support of special local ecosystems in which the cancer cells evolved. Individual cancer cells are extremely fragile with respect to changing their native microenvironment to completely different surroundings. Thus, for each individual cell, the detachment from the primary tumour is lethal with high probability.

In order to cope with fragilities induced by stress, some cancer cells might activate various defence mechanisms, increasing their capability to withstand external perturbations. One example of this is the already-mentioned activation of multidrug resistance by inducing a MDR1 protein, which is a cellular pump able to remove a variety of toxic chemicals from the cell including anti-cancer drugs (Nooter and Herweijer, 1991). Another example of a very generic response of cancer cells to stress is overexpression of chaperones (Whitesell and Lindquist, 2005).

Rapidly growing populations of cancer cells suffer from specific stresses, **hypoxia*** being one of the most universal ones. To overcome this stress, the cancer cell switches to production of ATP through upregulation of glycolysis (referred to as the Warburg effect) and develops its own vascularisation very early in tumorigenesis. Thus, cancer cells become dependent on the availability of glucose and less dependent on the presence of oxygen. On the other hand, hypoxia itself plays a protective role for cancer cells: hypoxic tumour cells display increased resistance to radiation and drugs as well as an increased incidence of both apoptosis-resistant and invasive clones.

9.6.3 Robustness of cancer cell populations and evolvability

Another feature is the survival of *cell populations*. For normal tissue functioning, the control mechanisms activating events such as programmed cell death should be considered as safety mechanisms, providing robustness to normal cell populations with respect to invasion by a mutant population. In this way, functioning of the *TP53* gene clearly decreases the robustness of individual normal cells with respect to mutations, but increases robustness of the tissues and the whole organism. Normal cell populations have other active defence mechanisms against mutant invaders, including triggering immune responses targeting cancer cells (*cancer immunosurveillance* mechanism) (Dunn et al., 2002).

There are multiple evidences that cancer cells try to mask, switch off or bypass these normal cell mechanisms which evolved in order to control mutability. As a result, cancer cells become more robust with respect to individual survival, but inevitably less robust against the invasion of mutations, *i.e.* they are not as stable as genetically homogeneous populations.

Arguably, the most important difference in the way the normal and the cancer cell populations function is in their ability to evolve. As a matter of fact, cancer cells might achieve higher evolvability as a result of their own problems, such as increased number of mistakes during DNA replication caused by uncontrolled cell division and deficit of basic cell building blocks. Let us mention that in principle the mutagenesis can be also enhanced by deliberate downregulation of DNA repair machinery, as it is the case in bacterial populations. However, in many aggressive forms of cancer the DNA repair machinery seems to be upregulated, which arguably means that cancer cells do not explicitly exploit the bacterial strategy for adaptation but rather are not able to satisfy the demand for DNA repair during the replicative stress.

In their struggle for survival, cancer cells can develop cooperativity with normal stroma, but also can rely on cooperation with each other. For example, it has been hypothesised that cancer cell populations can develop subclones specialised in producing growth signals for the rest of the tumour cells (Axelrod et al., 2006).

Cancer cells can acquire the capability to actively withstand normal cells in the competition for living space. In particular, one of the interesting hypotheses is that as a result of their glycolytic phenotype, cancer cells can induce microenvironmental acidosis which is not physiological for normal cells (Gatenby and Gillies, 2004). Such a strategy requires further adaptation of cancer cells through somatic evolution to phenotypes resistant to acid-induced toxicity.

Cancer therapy represents a major stress for rapidly dividing cancer cells. Some of them can eventually develop a therapy resistant phenotype. It would be interesting to understand in detail how a cancer cell develops resistance mechanisms in the process of their evolution in the context of therapy application. There are several examples where such a process is relatively well documented. One of the most studied examples is the development of resistance to cisplatin, which is one of the most used platinum-based compounds that exerts clinical activity against a wide spectrum of solid **neoplasms**[*] (Galluzzi et al., 2011). Cisplatin exerts anticancer effects via multiple mechanisms, with the best understood mode of action involving the generation of DNA lesions followed by activation of DNA damage response and induction of apoptosis. Cisplatin often leads to an initial therapeutic success associated with partial responses or disease stabilisation. Nevertheless, an important fraction of already-sensitive tumours eventually develop chemoresistance. The mechanisms of resistance to cisplatin can be classified as pre-target, on-target, post-target and off-target mechanisms. Thus, an example of a *pre-target resistance* mechanism is activating multidrug resistance genes such as *MRP2*. An example of *on-target resistance* is the increased activation of some of the DNA repair pathways, in particular, NER, which can mitigate the DNA damage effect of cisplatin. *Post-target effects* are usually associated with inactivation of the apoptosis pathway, for example, by overexpressing BCL2-like proteins. Finally, an example of *off-target resistance* mechanism is the activation of non-cisplatin-specific survival mechanisms such as autophagy. All these examples illustrate the variety of strategies a cancer cell can use in order to protect itself from the action of a cytotoxic drug. Hence, it is difficult to imagine a single universal magic bullet by which the cisplatin treatment should be accompanied in order to overcome cellular resistance (Galluzzi et al., 2011).

The above mentioned comparison between cancer and normal cells in terms of robustness with respect to survival, creates a complex picture. Robustness of cancer cells can be a result of selective pressure but also it can be a byproduct of their stressful conditions. It would be correct to say that the cancer cells have their own specific spectrum of fragilities not common to that of the normal cells. Compared to normal cells, individual cancer cells can be extremely fragile, sick and suffering from multiple defects. On the contrary, genetically heterogeneous cancer cell populations characterised by genetic instability can be extremely robust due to significantly increased evolvability and, hence, possibilities to adapt and survive stresses induced both by aggressive environment (for example, as a result of anticancer therapy) and the destructive consequences of increased mutagenesis.

9.7 Cancer resistance, relapse and robustness

A major problem in treating cancer concerns its relapse. Available cytotoxic therapies can efficiently reduce the bulk of tumour cells, many solid tumours can be removed by surgery, targeted therapies can specifically kill the cancer cells overexpressing certain receptors or possessing specific mutations, *etc.* However, the efficiency of these treatments is often mitigated by re-appearance of drug-resistant tumours or **metastasis***. From an evolutionary perspective, two major hypotheses are considered in the current literature explaining the phenomenon of tumour relapse: (1) The *clonal genetic heterogeneity* of tumours accompanied by the Darwinian selection of the resistant traits and (2) *The cancer stem cell* hypothesis.

9.7.1 Clonal genetic heterogeneity

Nowell (1976) summarised multiple observations on the tumorigenic karyotypes with the following hypothesis: *"acquired genetic lability permits stepwise selection of variant sub-lines and underlies tumour progression."* By this, it was hypothesised that natural selection for the most fit (aggressive and treatment-resistant) cell clones is constantly happening inside tumours. Several years later, Goldie and Coldman (1979) developed a mathematical model which predicted that tumour cells mutate to a resistant phenotype at a rate dependent on their intrinsic genetic instability. The probability that a cancer contains drug-resistant clones depends on the mutation rate and the size of the tumour. According to this hypothesis, even the smallest detectable cancers would contain at least one drug-resistant clone.

Recent experimental studies prove the hypothesis of coexistence of multiple clones in leukaemia and many solid tumours (Marusyk and Polyak, 2010). In some studies, it is estimated that the total number of mutations and subclones in clinically detectable tumours can reach billions (Klein, 2006). Such a variability creates a basis for robustness of cancer cell populations against most of the cancer treatments (Gerlinger and Swanton, 2010).

Loeb et al. (1974) proposed that cancer cells should acquire much higher mutation rates than the normal cells in order to achieve such a level of heterogeneity. This is related to the increased, by many orders of magnitude, the number of errors during DNA replication (*genomic instability*). This state of a cell is called a *mutator phenotype*, an important notion in the cancer robustness theory (see also **Section 8.1.6**). The theory claims that there must exist a trade-off in cancer cells: on one hand, genomic instability creates selective disadvantages by creating numerous lethal mutants; on the other hand, genomic instability promotes evolvability of cancer cells which can be advantageous in environments with high DNA damage rates (*e.g.* during chemotherapy). Mathematical modelling provided several important insights in this direction. Firstly, it predicts that a mutator phenotype should be acquired relatively early in tumorigenesis (Beckman and Loeb, 2006). Secondly, the prediction is that the mutator phenotype creates advantages for tumorigenesis only if more than two causal oncogene mutations are required for converting a normal cell to a cancer cell (which means, for example, that the mutator phenotype is not advantageous for **retinoblastoma*** but is advantageous for the prostate cancer; Beckman and Loeb 2006). Interestingly, mathematical modelling predicts that for successful cancer cell proliferation it might be advantageous to change this trade-off in time from maximal genomic instability to maximal stability (Komarova et al., 2008).

The clonal heterogeneity hypothesis inspires clinical strategies for cancer treatment to overcome the problem of tumour relapse. The Goldie-Coldman model suggests that the best chance for a cure would be to use all effective chemotherapy drugs. In practice, this has meant using two different non-cross-resistant chemotherapy regimens in alternating cycles. This idea was further developed by Day (1986) who suggested *"the worst drug first"* rule. The rule consists in applying first an effective, but the least active, drug which will eliminate the bulk of the tumour cells and will significantly decrease tumour heterogeneity by leaving first drug-resistant cells for the action of a second effective more active drug. However, the worst drug rule has not yet shown efficiency in clinical trials. More systematic mathematical models have reconsidered it by taking time and cross-resistance into account and justifying *"best drug first, worst drug longer"* strategy (Katouli and Komarova, 2011).

Evolutionary modelling approaches led to the proposal of a new therapeutic strategy called *adaptive treatment* that aims to maintain a stable tumour population instead of trying to achieve the maximal cell kill (Gatenby et al., 2009). This prevents the elimination of sensitive tumour clones which should, in theory, suppress the growth of the therapy resistant clones in a competitive manner. A basic assumption of this strategy is that the resistant clones have a lower fitness than sensitive clones because they commit more resources to maintain the resistant phenotype. This strategy has been tested in animal models of ovarian cancer treated with carboplatin but proof of the principle in humans is not yet available (Gatenby et al., 2009).

In order to reduce the genetic heterogeneity in cancers by therapy, we still have to better understand evolutionary and molecular mechanisms that govern it. Do mechanisms of canalisation exist in tumours as in normal cells? Can they be enhanced? What is the role of genetic variation buffering in tumorigenesis? For example, it is known that the classical evolutionary capacitor HSP90 or other chaperones are frequently overexpressed in cancer cells (Whitesell and Lindquist, 2005). Inhibitors of *HSP90* are currently investigated for application in combination with cytotoxic drugs (such as cisplatin). However, it remains unclear how application of these inhibitors will affect tumour heterogeneity, and how the role of chaperones such as HSP90 changes in tumorigenesis with the patient's age. All these questions are central to understanding how to control the clonal genetic heterogeneity.

9.7.2 Cancer stem cells

The second hypothesis postulates existence of a minority of relatively slowly proliferating cells able to self-renew among the total cancer cell population. By analogy with the function of normal stem cells, they are called **Cancer Stem Cells (CSC)*** (see **Chapter 2**). Though genetically identical to those cells that actively proliferate and differentiate, it is assumed that the CSCs play a more important role in the long-term sustenance of the tumour and dissemination of cancer cells throughout the body.

Moreover, in some models, it is assumed that the CSCs can be localised in protective hypoxic niches formed by an extracellular matrix with a special microenvironment controlling their division. These niches protect CSCs from the action of anti-proliferative and cytotoxic drugs. Also the CSCs are less affected by most of the cytotoxic drugs (acting most effectively at dividing cells) because they are dormant most of the time. Hypoxia in the niches can protect the CSCs from the effect of radiation therapies because radiation requires oxygen to induce DNA damage. Some studies show that CSCs can use DNA repair mechanisms more efficiently and activate multiple drug resistance and anti-apoptotic genes. In other words, CSCs are assumed to be much more robust than the majority of cancer cells with respect to tumour treatment, capable to survive it and eventually to restore the eliminated tumour mass.

Currently, CSCs are defined as those cancer cells that are capable of inducing xenograft mouse tumours by transplantation. Technically, the cancer cells are divided in subpopulations with differential expression of some surface markers (such as CD24 or CD44), and each subpopulation is tested by transplantation into a mouse for its ability to seed a tumour. Due to this complicated test for detecting stemness, a number of questions remain about the definition, the role and the genesis of CSCs. For example, two alternative hypotheses currently coexist; one assumes that CSCs are stable in time (*hierarchical model*), the second suggests that any cancer cell can fluctuate between stem and non-stem states (*stochastic model*). Most of the properties of the CSCs such that they represent a minor, dormant, resistant to therapies subpopulation, have not yet been unambiguously proven experimentally. For example, some models suggest that, at later stages of tumorigenesis, the majority of cancer cells can possess stem properties (Clevers, 2011).

Nevertheless, several therapeutical strategies are in development to target the CSC population rather than the rapidly dividing tumour cells. They include destruction of the protective niche, inhibition of cell migration ability, activation of cells from dormancy, inhibition of DNA repair, increase of Reactive Oxygen Species (ROS), inhibition of survival mechanisms and induction of differentiation (Trumpp and Wiestler, 2008). One of the obstacles on the way to develop specific drugs against CSCs populations is current standards for estimating the efficiency of anticancer treatment. Most of the treatment success criteria are now based on measuring the total mass of the tumours, while hypothetical specific anti-CSC therapies will not affect them in the first place.

9.7.3 Unifying two theories

From a theoretical point of view, it is possible to unify the two explanations for cancer resistance in one consistent view (Tian et al., 2011). To do this, one has to consider *epigenetic heterogeneity* as a part of clonal heterogeneity. That way, CSCs represent a particular subpopulation of cells (genetically identical but epigenetically different from the rest of the clone with particular robustness properties). Furthermore, with increasing tumour aggressiveness, the epigenetic landscape of the cancer cells can become more shallow, with higher probabilities of stochastic switching between states. This can increase the frequency of transitions between stem-like and non-stem-like epigenetic states. Thus, both hierarchical and stochastic models can be unified, with the first one playing a more important role at the earlier stages of tumorigenesis (when epigenetic cellular states are better separated from each other), and the latter being more important at the later and more aggressive stages (when the majority of cancer cells acquire a high-level of epigenetic plasticity).

9.8 Experimental approaches to study biological robustness

There are several well-known experimental studies dedicated to measuring and explaining the robustness of certain biological mechanisms. Arguably, the best experimentally-studied robust properties in living organisms are segmentation patterns in *Drosophila menalogaster* development and bacterial chemotaxis systems. A lot of effort has also been invested in understanding the robustness of mechanisms of cell cycle and DNA repair in yeast because the knowledge of those systems can be utilised in understanding human cell biology.

9.8.1 Robustness of *Drosophila* development

Canalisation in development is a type of biological robustness conceptually developed by Waddington (1957). Using the famous metaphor of a ball rolling down an epigenetic landscape, he hypothesised that there must only be a finite number of distinct developmental trajectories possible. Along the trajectories, cells make discrete cell fate decisions. Each trajectory, called a *chreod* (meaning, necessary path), must be stable against small perturbations. Waddington underlined that the canalisation property is a result of natural selection and of adaptation of the development to the environment. Note that mutant phenotypes usually show much greater diversity than the wild type ones.

Studies on *Drosophila melanogaster* development, and, in particular, *Drosophila* embryo patterning, confirmed Waddington's hypothesis. The existence of embryo variation-prone positional information in some of the protein gradients, such as the Hunchback protein, proved to be an extremely robust system. It was shown that despite large environmental variations, the position of the first segmentation point has a variation of 1% of the total embryo length (Houchmandzadeh et al., 2002).

Waddington's canalisation was also demonstrated at the level of gene expression variability in *Drosophila*. The variation of the zygotic segmentation gene expression patterns is markedly reduced when compared to the variation in the levels of maternal gene expression by the time gastrulation begins. For instance, this variation is significantly lower than the variation of the maternal protein gradient Bicoid (Manu et al., 2009). The genetic network whose properties form the epigenetic landscape, was identified and analysed through mathematical modelling (von Dassow et al., 2000).

9.8.2   Gene knock-out screenings in yeast

Robustness of normal cells with respect to viability is extensively studied in yeast. One of the most studied questions is: What genes should be significantly perturbed (by knocking them out or by overexpression) in order to make the cell nonviable? Of course, the answer to this question depends on the environment in which the cell is cultivated. Usually the experiments are done in the standard laboratory conditions. For these conditions, it is known that only about 20% of *Saccharomyces cerevisiae* genes are required for viability (Costanzo et al., 2011). Among the 80% remaining, some genes genetically interact forming *synthetic lethal* pairs of genes (see **Box 9.6**).

The genome-wide scale mapping using Synthetic Genetic Array (SGA) methodology in yeast was accomplished in 2010 in which more than 5 millions of gene pairs were checked and about 170,000 genetic interactions (3% of all checked pairs) were identified, approximately balanced for negative and positive ones. With respect to growth, a genetic interaction is classified as positive (respectively negative) when the fitness of the double mutant is higher (respectively lower) than expected. Negative genetic interactions often indicate functional redundancy between two genes, with the most pronounced case being synthetic lethality when simultaneous deletion of two otherwise non-essential genes leads to cell death. Among those genes which have the tendency to have mostly negative interactions, many genes required for normal progression of the cell division cycle are identified, indicating the most fragile part of the yeast cellular physiology with respect to viability and growth.

> ❏ **BOX 9.6: Synthetic lethality / Synthetic dosage lethality**
> Synthetic interactions are identified if mutations in two separate genes produce a different phenotype from either gene alone, and indicate a functional association between the two genes. Two genes have a synthetic lethal relationship if mutants in either gene are viable but the double mutation is lethal. Synthetic dosage lethality is a type of genetic interaction which is detected when overexpression of a gene is lethal only if another, normally nonlethal, mutation is present.

Synthetic lethality and synthetic dosage lethality (see **Box 9.6**) studies in model organisms and human cells give hope to develop cancer drugs that would kill cancer cells very selectively: if a cancer cell has a characteristic deletion or **amplification*** of a gene, then inhibiting or overexpressing another nonessential gene forming a synthetic pair, will lead to specific lethality of cancer cells. For example, one particular type of breast cancer is characterised by loss-of-function mutation in *BRCA1* gene involved in DNA repair. The *PARP1* gene forms a synthetic lethal pair with *BRCA1* in cellular models and, therefore, inhibitors of *PARP1* for treating *BRCA1*-deficient breast cancer were developed and went to clinical trials (Helleday, 2011), but with no confirmation of success yet. There is a belief that not only synthetic lethality pairs can be tried to selectively kill cancer cells but also the *synthetic lethality cocktails*, *i.e.* drugs or drug combinations affecting several targets simultaneously, or taking into account more than one cancer genome modification (see also **Section 11.4**). *In silico* modelling of robustness of metabolic networks in yeast suggests that one can find a significant number of k-tuples of genes such that deletion of any $k-1$ genes in a k-tuple would not give a lethal effect while knocking-out all k genes is lethal (so called k-robustness) (Deutscher et al., 2006). These predictions remain to be experimentally confirmed.

9.8.3 Statistical and epidemiological analyses

Statistical analysis of epidemiological data can also be considered as a measure of cancer robustness. One of the major questions here is: How many oncogene (or tumour suppressor) mutations are potentially needed to convert a normal cell to a cancerous cell? Knudson (1971) performed a statistical analysis on retinoblastoma and suggested his famous two-hit model: two mutational events are necessary to fully inactivate two working copies of the retinoblastoma gene (*RB*). *RB* later became known as the first tumour suppressor gene (see **Chapter 2**).

A single genomic **translocation*** creating a chimeric oncogene

EWS/FLI1 in Ewing's **sarcoma*** seems to be sufficient to cause cancer in adolescents. More generally, many paediatric cancers are believed to be caused by very few genetic modifications. However, such a low number of mutations sufficient to induce tumorigenesis is an exception. Armitage and Doll (1954) suggested a statistical model explaining the experimental observation that, in industrialised nations, the frequency of most common cancers seems to increase proportionally to the sixth power of age. This correlation could be explained by assuming that the outbreak of cancer requires the accumulations of six consecutive mutations. Similar studies showed that about 12 mutational events are necessary to induce prostate cancer (Cook et al., 1969). These works led to the development of the *multistep theory of tumorigenesis* according to which the normal tissues are transformed into cancerous ones by means of a series of discrete stages: **somatic mutations***, broad genomic rearrangements, or changes in tissue interactions and environment.

9.8.4   Gene knock-out screenings in mammalian organisms

In summary, multiple experimental observations have showed that the normal human cells are relatively safely protected from transformation to cancer cells. In a typical scenario, multiple mechanisms should be violated to induce tumorigenesis. However, there exist specific fragilities which are exploited by various cancers to accelerate their development. On other hand, cancer cells develop their specific fragilities.

In order to find these fragility points in cancer cells with respect to the cytotoxic treatments, large-scale gene knock-out screenings have been introduced. In these experiments, many genes are affected one by one by infecting the cell with a specific small interfering RNA (siRNA) such that the expression of the affected gene is significantly (but usually not completely) reduced. With respect to chemotherapy, the objective is to find the genes whose inhibition would sensitise the treatment and make it more efficient and more specific to cancer cells. For example, a large scale siRNA screening was done in order to find genes sensitising the cisplatin treatment. These screenings revealed potential genes that synergistically interact with cisplatin enhancing its action on cancer cells (Bartz et al., 2006).

9.9   Conclusion

To finalise this chapter we can still ask ourselves: Why does it make sense to talk about cancer robustness? To summarise, the current answers to this question are the following:

1. Since cancer cells are living cells derived from normal predecessors, they

can benefit from all biological mechanisms developed by evolution for maintenance of cellular and organismal function integrity and stability. Some of these mechanisms (such as multiple control feedbacks, multidrug resistance) allow cancer cells to actively withstand attempts to eliminate them by using drugs.

2. Cancer cells deal with specific challenges with which their normal counterparts deal to much less extent. These are stresses related to active proliferation (DNA replication stress, nutritional stress, hypoxia), genomic instability and action of anticancer drugs. The main defence mechanism that cancer cell populations utilise to be robust in these stressful conditions is their ability for fast evolution by reprogramming biological networks inherited from the normal predecessors. In this battle, cancer cells become more robust to some perturbations and more fragile to the others.

3. Cancer as a disease has a significant chance to appear during a typical human lifespan. When it appears, it is able to robustly sustain itself, in the absence of treatment often leading to lethal consequences for the host organism. The major manifestation of cancer robustness and complexity is the ability of tumours to relapse after a visibly successful treatment. The phenomenon of tumour relapse is currently explained by clonal tumoral heterogeneity and long-term sustenance of tumoral clones by cancer stem cells (CSCs).

✎ **Exercises**

- List and define major patterns that can potentially make biological networks robust.
- What are the conditions on long-term existence of a polymorphic cell population?
- Explain how robust living systems can evolve.
- List major biological mechanisms of tumour relapse.
- What would be the danger in making an analogy between robustness in engineering and biological systems?

⇨ **Key notes of Chapter 9**

- Living systems are characterised by robust properties because they are able to perform their functions despite perturbations in their environment and genetic background.

- Cancer is robust as a disease because it has a significant chance to appear during a typical human lifespan, and is able to sustain itself and withstand therapeutic interventions.

- Because cancer cells deal with specific challenges not common with their normal predecessors, they are expected to have specific fragilities that can be used for developing new cancer therapies.

- Robust biological systems, including cancer cell populations, can evolve by accumulation and release of genetic diversity with further selection for better fitness.

Chapter 10

Cancer robustness: Mathematical foundations

Robustness is a nontrivial biological phenomenon difficult to study using pure reductionist approach (see **Chapter 1**): it would be very naive to think that robustness is assured by a few selected elements of the system (however, it can happen in some special cases such as mechanisms of protein refolding). By contrast, robustness is rather frequently distributed everywhere in the system, guaranteed by the whole system organisation and cooperation between its various parts.

In this chapter, an attempt to formally describe biological robustness is made by trying to associate some of the known robust properties of mathematical objects with robust properties of biological systems. A mathematical definition of robustness measure will be given and it will be shown that simple and commonly-used definitions of robustness have caveats in their practical use. Several well-known examples of mathematical objects possessing robustness properties will be described. Finally, a tentative unifying view on robustness in mathematical terms will be proposed. This chapter contains some mathematical material for which basic training in calculus and differential equations is needed.

We will only evoke several general ideas concerning robustness through various mathematical theories without going deeply in them. The aim of the chapter is to highlight possible connections of these theories to the problem of robustness and to provide starting references for further study. We will nevertheless consider several notions in more details than others, mainly because of their high citation index and the recent interest to them.

10.1 Mathematical definition of biological robustness

10.1.1 Defining robustness as variance contraction

Intuitively, we can formulate the simplest mathematical definition of robustness as a measure of the average variation of a property of a biological system from its reference (nominal) state when the conditions in which the system exists vary.

Thus, to define robustness, one must specify:

- A system S

- A property M of the system that can be a quantity, a state, a structure or a function

- A set of all possible conditions P in which the system S can be found

- A probability distribution $\phi(p), p \in P$ defined over the set of conditions P

A way to quantify the property M (which can be represented by several numbers) of the system S in any condition p should be defined, *i.e.* M represented as a vector field $\mathbf{f}_M = \mathbf{f}_M(p)$. Together with this, a way to measure the change of M between two different conditions p' and p'' should also be defined. The corresponding measure is denoted $D(\mathbf{f}_M(p'), \mathbf{f}_M(p''))$.

Let us define a reference condition of the system $p^* \in P$, in which the robustness is measured. Then the *relative local robustness* $R_S^M(p^*)$ is defined as an average deviation of the property M from its reference state in the condition p^*:

$$R_S^M(p^*) = \int_{p \in P} \phi(p) D(\mathbf{f}_M(p), \mathbf{f}_M(p^*)) dp. \tag{10.1}$$

After the local robustness is quantified, the *global robustness* can be computed as an average over all possible system conditions:

$$\bar{R}_S^M = \int_{p \in P} \phi(p) R_S^M(p) dp.$$

With this formulation, different values of R_S^M can be compared but it cannot be concluded if a particular R_S^M is big or not.

To measure the *absolute robustness*, a measure of change between two different conditions p' and p'' needs to be introduced. First, consider a distance function $d(p', p'')$. Then, it is possible to define the scale of condition variance and to compare R_S^M to it, provided the units of variation of D and d are comparable. The latter can usually be achieved by various normalisations such as logarithmic scales. In this case, let us define the *absolute local robustness* as:

$$\hat{R}_S^M(p^*) = \frac{R_S^M(p^*)}{\int_{p \in P} \phi(p)(d(p, p^*)) dp}. \tag{10.2}$$

When $\hat{R}_S^M(p^*) \ll 1$, the property M of system S in condition p^* is *robust* with respect to perturbations from P. In other words, the meaning of **Equation 10.2** is that the variance of a robust system property should be much smaller than properly quantified variance of conditions in which the system exists.

Notice that the system is supposed to be robust and stable only in the most typical conditions (with high $\phi(p)$). Meanwhile in some relatively rare conditions, it might be fragile and nevertheless considered to be robust *on average*. The system can even be lethally sensitive to some rare disastrous conditions to which it has never had a chance to adapt. In this case, the probability of such conditions will determine the average system lifetime.

As a typical example, let S be a dynamical system (*e.g.* a model of a biological pathway), and let M be some scalar property $M = M(X_1, X_2, \ldots, X_k)$ which depends on a set of system parameters $\{X\}$. A particular combination of parameters will be our condition $p = \{x_1, x_2, \ldots, x_k\}$. Let us define how the parameters will vary by defining probability distributions $\phi_i(x_i), i = 1 \ldots k$ for each parameter independently. For example, we can say that a parameter i will be distributed log-uniformly in some allowable range around its reference value x_i^*, while other parameters will be fixed at their nominal values $x_j^*, j \neq i$. Let D be simply $D = [\log M(x_1, x_2, \ldots, x_k)/M(x_1^*, x_2^*, \ldots, x_k^*)]^2$. Then:

$$R_S^M(\mathbf{x}^*) = Var(\log M),\tag{10.3}$$

where Var is the variance.

To make its meaning absolute, R_S^M can be normalised on the maximal log-variance of any individual parameter:

$$\hat{R}_S^M(\mathbf{x}^*) = \frac{Var(\log M)}{\max_{i \in \{1 \ldots m\}}\{Var(\log x_i)\}}.\tag{10.4}$$

The definitions of robustness in **Equation 10.3** and **Equation 10.4** were proposed in Gorban and Radulescu (2007), where, in the spirit of Wagner (2005), it was called **distributed robustness*** (because it might be dependent on the whole distribution of parameter values and not necessary on one single parameter) and further refined.

For example, we will call a property M *r-robust*, if, among k parameters X, one needs to change at least r parameters in order to violate the robustness condition $\hat{R}_S^M \ll 1$. If, in addition, these r parameters are chosen by blind random choice then the property is called *weakly r-robust*.

Kitano (2007) introduced a definition of robustness similar to **Equation 10.1**, relating it to the average degradation of a function caused by a perturbation.

The definition of robustness given above is by no means universal. In many specific studies, it is reasonable to introduce other *ad hoc* definitions. For example, Wagner (2005) in his book defines robustness of a genotype (RNA sequence) as the number of single-mutated variants of the genotype having the same phenotype (secondary RNA structure). In modelling metabolic pathways in yeast and bacteria, robustness of metabolism is often defined as a fraction of *nonessential reactions*. A nonessential reaction is a reaction which removal does not preclude the cell to produce necessary metabolites from the nutrients available in a given environment.

10.1.2   Properties of the variance-based robustness measure

There are flaws in the definition of robustness based on variance (or on log-variance) such as in **Equation 10.4**: unfortunately, it is not invariant with respect to nonlinear transformations of the variables x_i, *i.e.* for the same phenomenon, different conclusions will be made depending on what are the parameter definitions. Let us consider a simple example:

$$M(k_1, k_2, k_3) = \frac{k_3 - k_1}{k_2 - k_1}, k_1 = 10^{3s}, k_2 = 10^{2s}, k_3 = 10^{s},$$

where s is uniformly distributed in the interval $[-1; 1]$. According to the definition in **Equation 10.4**, $\hat{R}_S^M = 0.23$ which is an indication of M being robust. At the same time, the function:

$$M(k_1, k_2, k_3) = \frac{e^{-k_3} - e^{-k_1}}{e^{-k_2} - e^{-k_1}}$$

produces $\hat{R}_S^M = 242$ which is an indication of M being (very) nonrobust.

If we think of k_i as parameters of a dynamical model of some chemical reaction network then it means that the results of measuring robustness are drastically dependent on what we use as parameter values: the kinetic rate constants or activation energies. However, kinetic rate constants and activation energies are connected by a simple exponential function. Gorban and Radulescu (2007) underlined that the definition of robustness in **Equation 10.4** is well suited only for some classes of functions M, for example, for positively homogeneous functions of degree one having the property $M(\alpha x_1, \alpha x_2, \ldots, \alpha x_3) = \alpha M(x_1, x_2, \ldots, x_3)$, which is a reasonable limitation for a large class of measured dynamical system characteristics. In practice, it means that the function for which the robustness is studied should be checked for its scaling behaviour when all its parameters are multiplied by the same constant.

Yet another interesting aspect concerns the scaling of robustness $\hat{R}_S^M$ with respect to the number of arguments of the function M. In other words, this is a question of how robustness of a property scales with the dimension of the parameter space where it is defined.

In mathematics an interesting and nontrivial observation was made by Millman and Gromov (1999) which was called *the phenomenon of concentration of measure*. It states that any Lipschitz function (see **Box 10.1**) depending on many arguments is almost constant, *i.e.* the variance (or the log-variance) of the function is much smaller than the variance (or log-variance) of the distribution of input vectors.

The simplest example of such a concentration comes from considering a uniform distribution of points on a n-dimensional sphere with unit radius.

❑ **BOX 10.1: Lipschitz function**

Lipschitz function (called also 1-Lipschitz function) is a function of limited growth. A function of many arguments $M(\mathbf{x})$ is called a Lipschitz function if there exists a constant K such that for all pairs of vectors $\mathbf{x}$ and $\mathbf{y}$:

$$\frac{|M(\mathbf{x}) - M(\mathbf{y})|}{d(\mathbf{x}, \mathbf{y})} \leq K, \tag{10.5}$$

where $d(\mathbf{x}, \mathbf{y})$ is the distance between $\mathbf{x}$ and $\mathbf{y}$.
The smallest K is called the Lipschitz constant. If $0 < K < 1$, then the function is called *a contraction mapping*.

Imagine that this distribution is projected on any 2-dimensional plane embedded in $\mathbb{R}^n$. If n is large, then almost all projections are located in a very tight vicinity of the centre of the projected sphere, with a variance much less than one decreasing exponentially with n.

The phenomenon of robustness (contraction of variance) is thus an expected property if: (1) the system under study exists in a multi-dimensional space (described by many independent variables and parameters) and (2) the property M of the system is measured by a Lipschitz function which depends on many variables or parameters. The latter is usually the case in many practical applications.

In the next section, we will show several examples when concentration of measure phenomenon leads to robustness. The idea of concentration of measure as one of the generic underlying mechanisms of biological robustness was first suggested by Gorban and Radulescu (2007). Talagrand (1996), a mathematician, claimed: *"The idea of concentration of measure (which was discovered by V. Milman) is arguably one of the great ideas of analysis in our times."*

10.2 Simple examples of robust functions

10.2.1 Mean value

Many functions used in mathematics and statistics show robust behaviours. The simplest one is the formula for calculating the mean value:

$$M(x_1, x_2, \ldots, x_n) = \frac{1}{n} \sum_{i=1}^{n} x_i.$$

From the law of big numbers, it is known that the variance of the mean value scales with n as $1/\sqrt{n}$ and becomes less and less sensitive to the variations of individual x_i given that, among x_i, there are no *dominating* variables. An averaging of some kind is an abundant source of robustness in mathematical modelling. Median function is another, even more robust, example. It is less sensitive to the exact values of a limited number of atypical measurements (*i.e.* outliers).

10.2.2   Rational functions

Let us consider another example which is a simple rational function (Gorban et al., 2010):

$$M(k_1, k_2, ..., k_n) = \frac{P(k_1, k_2, ..., k_n)}{Q(k_1, k_2, ..., k_n)}.$$

In some cases, this function can show robust properties similar to those of the mean value function with growing number of parameters. Let us consider a case when all k_is have very different scales, *i.e.* for any pair i and $j \neq i$ one has $k_i \ll k_j$ or $k_j \gg k_i$. In this case we will say that k_is are *well-separated*.

Among the monomials from which P and Q consist, one will find few which dominate and determine their values. In the extreme case, the whole expressions for P can be approximated by a single monomial $P \approx A \prod_{i=1}^{n} x_i^{\alpha_i}$ with some positive integers α_is, with many of them equal to zero for a typical polynomial. The same is true for Q. The ratio of two monomials will be a monomial $M \approx B \prod_{i=1}^{n} x_i^{\beta_i}$, where β_is are (positive or negative) integers. Frequently, many or even all β_is are equal to zero, which means that the value of M effectively depends only on a small number of parameters or does not depend on them completely.

Possible asymptotic values of M, in the case when all k_is are well-separated, can be obtained by introducing new variables $\psi_i = \epsilon \log k_i$ such that $k_i = e^{k_i/\epsilon}$, letting $\epsilon \to 0$ and studying the limit $\lim_{\epsilon \to 0} M(k_1, k_2, \ldots, k_n)$. Typically, this limit will depend on very few k_is or will not depend on them at all.

For the function considered in the previous section, $M(k_1, k_2, k_3) = \frac{k_3 - k_1}{k_2 - k_1}$, if we assume that $k_3 \ll k_2 \ll k_1$ then $M \approx 1$ with high accuracy (the accuracy will be approximately equal to k_2/k_1). In other words, M depends very weakly on parameters, hence, it is robust.

10.2.3   Examples from chemical kinetics

Rational functions appear everywhere in mathematical modelling of biological networks using chemical kinetics approach. One example of a rational function is the saturating function (for example, associated with Michaelis-Menten kinetics, see **Chapter 7**):

$$r(x) = \frac{V_m x}{K + x},$$

For big values of $x \gg K$, the function very weakly depends on x and approximately equals V_m, *i.e.* become robust with respect to variations in x (to those variations when x remains sufficiently big, of course).

To illustrate the difference between several types of robustness, let us consider a simple cyclic chain of irreversible reactions:

$$A_1 \overset{k_1}{\to} A_2 \overset{k_2}{\to} A_3 \overset{k_3}{\to} \ldots \overset{k_{n-1}}{\to} A_n \overset{k_n}{\to} A_1, \qquad (10.6)$$

where k_is are reaction kinetic rate constants. The formula for the steady state flux through such a chain is:

$$F = \frac{A}{\frac{1}{k_1} + \frac{1}{k_2} + \frac{1}{k_3} + \cdots + \frac{1}{k_n}},$$

where $A = [A_1] + [A_2] + \cdots + [A_n]$ is the sum of all component concentrations in the pathway which is conserved.

Let us assume that the kinetic rate constants are sampled from a normal distribution with the mean $\bar{K}$, *i.e.* in a typical sample, they are all similar by the orders of magnitude. Then, one can say that the longer the chain, the less the flux F depends on individual variations of k_i values. With growth of n, F becomes more tightly distributed around its mean value $\frac{A \times \bar{K}}{n}$ with variation scaled as $1/\sqrt{n}$. This type of robustness is called the *cubic robustness* (Gorban and Radulescu, 2007).

Now assume that the kinetic rate constants are sampled from a log-normal or a log-uniform distribution. In this case, the rate constants are all of different orders of magnitude, *i.e.* with high probability, there exists the smallest one k_s which is well-separated from the others: $k_s \ll k_i, i \neq s$. In this case the flux F equals, with good accuracy, to $k_s A \times \left(1 - \sum_{i \neq s} \frac{k_s}{k_i}\right)$. Let us vary a randomly chosen kinetic rate constant. If the variable is not the rate constant of the limiting reaction step k_s, then the effect of this perturbation on the flux will be negligible. If this perturbation affects $k_s n$ then the flux will depend on this variation linearly. However, if the amplitude of the perturbation is such that it violates the initial inequality $k_s \ll k_i, i \neq s$ then the perturbation will not have any effect anymore because the flux will be limited by another rate constant. In other words, when k_s increases by an arbitrary number of orders of magnitude, the flux F can be increased only by a factor $\frac{k_{s_2}}{k_s}$, where k_{s_2} is the second slowest kinetic rate constant in the chain. The longer the chain, the smaller the mathematical expectation for this factor. As a result, the variation of F with respect to sampling k_is from a fixed log-uniform distribution scales as $1/n$. This *simplex* type of robustness is potentially even stronger than the one connected with averaging (Gorban and Radulescu, 2007).

10.3   Forest-fire model: Simple example of evolving robust system

10.3.1   Forest-fire percolation lattice model

Carlson and Doyle (2002) suggested several interesting examples in which simple mathematical systems demonstrate complex behaviours similar, in a sense, to the ones of biological organisms. These similarities come, more particularly, from considering the strategies of dealing with trade-offs between robustness and fragility. Here, we will consider in detail the metaphor of forest-fire in the percolating lattice models. The forest-fire model itself was suggested by Henley (1989) and Drossel and Schwabl (1992).

Let us suggest the following very primitive model of forest-fire. Consider a regular rectangular lattice $N \times N$ of sites (lattice vertices). Position of each site will be designated by a pair of indices i, j. Each site can be occupied by a tree or can be empty. Two trees are called *neighbours* if their positions i_1, j_1 and i_2, j_2 are different only in one vertical or horizontal step, *i.e.* $|i_1 - i_2| + |j_1 - j_2| = 1$. Any tree configuration is characterised by its density $\rho = N_{trees}/N^2$. A *cluster* refers to a connected subset of trees such that there exists a path from any other tree in the cluster to any tree through neighbourhood relations. Based on the percolation theory, it is known that for a randomly uniform distribution of trees, the cluster size distribution demonstrates a phase transition behaviour: before the critical density $\rho_c \approx 0.59$, there is a very little chance to obtain large clusters (subcritical phase) while for $\rho > \rho_c$ (supercritical phase) there will be one large cluster containing most of the trees with high probability (see **Figure 10.1**).

Now we will consider the event of ignition, *i.e.* when the fire starts at a lattice site. If ignition is taking place at a site (i, j), then the whole cluster containing (i, j) will be burnt. Let us define a probability distribution for ignition as $I(i, j)$. Carlson and Doyle (2002) suggested to consider two ignition distributions: uniform $I(i, j) = f = const$ and exponentially skewed P with a peak in the left upper corner (for example, $I(i, j) \sim \exp(-\frac{6(i+j)}{N})$), as in Zhou et al., 2005).

For each configuration of trees, an act of ignition at position (i, j) is characterised by *an yield* Y, which is defined as the density ρ after the fire, *i.e.* it is proportional to the total number of trees before the ignition minus the size of the burned cluster containing (i, j). For the uniformly random tree distribution, the mathematical expectation of yield grows approximately linearly for $\rho < \rho_c$ and gradually drops to zero for $\rho > \rho_c$. It is assumed that the time of the cluster burning is instantaneous (the original forest-fire model considers gradual propagation of fire, but, here, we only consider the simplest scenario of instantaneous elimination of the whole cluster). For analytical considerations, the lattice is considered to be of infinite size.

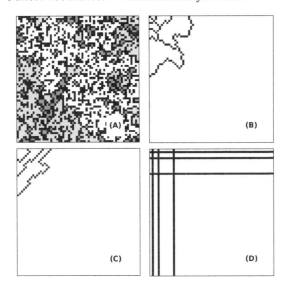

FIGURE 10.1 Forest-fire lattice percolation model and mechanisms of SOC and HOT. (A) The percolation forest-fire model at the critical forest density. Black spots are not occupied by trees. Tints of grey colour clusters of trees (the largest, percolated cluster is white). HOT configurations obtained by (B) simulation of Darwinian evolution, (C) local incremental algorithm (see details in Carlson and Doyle, 2002), and (D) grid design (manually designed lattice). Reproduced from Carlson and Doyle (2002) with the publisher's permission. © 2002 National Academy of Sciences, USA

This simple probabilistic cellular automaton game allows the demonstration of two nontrivial phenomena: Self-Organising Criticality (SOC), a notion coined by Bak et al. (1987) and Highly-Optimised Tolerance (HOT) coined by Carlson and Doyle (1999). Both of them are frequently mentioned in the recent discussions on biological robustness (Kitano, 2004a). Some characteristics of these scenarios can be computed analytically, which makes them a useful toy case study. (Bak et al., 1987; Drossel and Schwabl, 1992).

10.3.2 Self-organising criticality

In the SOC scenario, we allow the forest to recover after every fire. At each empty site, a new tree will appear with probability p. If we consider the uniform ignition distribution $I(i, j) = f$, then the model contains two parameters, with a dimension of time (number of update steps): $1/p$ is the average time in which a tree grows, $1/f$ is the average time between two ignition incidences. Hence, p/f is the number of trees grown between two ignition incidences. The dynamical behaviour of ρ converges to a steady value $\hat{\rho}$, which is remarkable in the following aspects.

Provided a separation of time scales $\frac{1}{p} \ll \frac{1}{f}$, *i.e.* the trees grow much faster than ignitions occur, the density $\hat{\rho}$ does not depend on exact values of p and f. $\hat{\rho}$ is also the minimal density possible, *i.e.* the number of growing trees and of burned trees is maximum which corresponds to the maximum energy dissipation in the system.

The resulting family of tree configurations is characterised by the following *critical properties*:

- The system shows long-range correlations (such as percolation) in fire incidents.

- The distribution of cluster sizes shows power-law distribution.

- The shapes of the clusters are fractal-like, *i.e.* are self-similar at various scales.

The concrete parameters of this criticality are highly nontrivial (Pruessner and Jensen, 2002).

The conclusion is that there is a big region of the system parameters in which the forest-fire dynamics naturally converges to a *critical state* of the system. This is why this phenomenon is called SOC. The phenomenon of criticality is extremely robust: it does not require parameter fine-tuning as it is the case in many critical systems in physics. The separation of time scales $\frac{1}{p} \ll \frac{1}{f}$ is suggested to be a common mechanism in dynamical systems showing SOC. In this forest-fire model, there are three well-separated time scales: time of cluster burning (which is considered instantaneous here), time of tree growth, and time between two ignition events. The dynamics can be described as a balance of two processes: (1) slowly accumulating stress (larger clusters) and (2) fast relaxation of stress (ignitions eliminating clusters) which is relatively infrequent.

10.3.3   Highly-optimised tolerance

The HOT mechanism works in a quite different manner, mimicking the process of natural selection, where the mathematical expectation of yield Y given certain ignition probability distribution $I(i, j)$ is considered to be an explicitly optimised *fitness function*. A simple scenario is the following. One generates a population of percolation lattices, each containing some randomly generated configuration of trees. At every optimisation step, each lattice (an individual) gives rise to an *offspring*, with a certain probability of **mutation*** per site (mutation here is converting a tree into an empty site, or converting an empty site into a tree). Hence, the population doubles. For each individual, its fitness is calculated by generating an ignition from the probability distribution $I(i, j)$ and calculating the yield. The yield depends on the individual tree configuration (genotype). Half of the individuals with the least fit are removed from the population, *i.e.* they die. The process is repeated until convergence in which typically only one most fit tree configuration survives. This configuration is characterised by the appearance of barriers of empty sites separating dense clusters of trees and preventing large losses of trees (see **Figure 10.1**).

The pattern of these barriers is very specific to $I(i,j)$. The distribution of cluster sizes reflects $I(i,j)$ and will follow the power-law provided that $I(i,j)$ follows the power law. The point is that this configuration allows achieving very high expected yields Y given the distribution $I(i,j)$. This is why the configuration is called Highly-Optimised. For the sites with high $I(i,j)$ (typical ignitions), the yield will also be very high, while for the sites with low $I(i,j)$ (unusual locations of ignitions) the yield can be extremely low corresponding to catastrophic burning off huge tree clusters.

Experiments with artificial evolution were continued by Zhou et al. (2005). In particular, they modelled the effect of coexistence of two habitats (populations of lattices) characterised by different types of $I(i,j)$: uniform and skewed. The individuals well fit to the uniform $I(i,j)$ are called *generalists* while those best fit to the skewed $I(i,j)$ are called *specialists*. In the model, there is a possibility of an invasion attempt (when some individuals are transferred) from one habitat to another. Simulations show that most of the time the generalists cannot invade the skewed habitat and cannot fix their genotype in it since they are less fit for typical perturbations from $I(i,j)$. On the other hand, specialists cannot invade the uniform habitat since they are not well adapted to it and suffer from low yields. For each population in its habitat, this situation is called *stability to invasion* and represents one type of robustness characterising evolving systems. However, modelling showed that a rare sequence of atypical ignitions can completely eliminate specialists in their skewed habitat. In this sense, specialists' populations do not have the so-called *internal* stability (*i.e.* ability for sustainable existence) another type of robustness, on a large time scale. Generalists' populations in uniform habitat have internal stability and, hence, their genotype is immortal. Generalists have a chance to invade the skewed habitat and successfully proliferate there during the epochs of specialists extinction. However, in a long run, their genotype inevitably becomes more specialised to the skewed $I(i,j)$.

10.3.4 Interpretation for living organisms

Zhou et al. (2005) underlined that the percolation lattice model can be given a wider interpretation than only forest-fires. The connected clusters can symbolise highly coupled functions of an organism such that a failure in one function would cause cascading failures of all functions in the cluster. In this interpretation, an organism has to deal with a trade-off between integrating all functions and separating them by protective barriers to prevent catastrophic cascades of failures. Need in supporting the existence of barriers is the price for reliable functioning.

Both SOC and HOT prototype mechanisms were suggested to play an important role in the evolution of living organisms, thus, leading to robust system behaviours. For example, Bak and Sneppen (1993) suggested SOC as a mechanism leading to the *punctuated equilibrium* effect in biological evolution. Punctuated equilibrium is a hypothesis suggested by Eldredge and Gould

(1972) stating that biological evolution takes place in terms of intermittent bursts of activity separating relatively long periods of quiescence (stability of phenotype), rather than in a gradual manner (that was the initial spirit of Darwin's work).

SOC corresponds to self-organisation while HOT is more explicitly related to imitating natural selection. Note that robustness in SOC and HOT has rather different meanings. SOC robustly leads to the appearance of certain generic (not sensitive to small structure modifications) system configurations characterised by some optimal properties (such as maximum energy dissipation). HOT leads to the appearance of robust yet fragile system configurations which are highly resistant (tolerant) to typical perturbations, but can be very fragile to unusual ones. These optimal-in-a-given-context configurations are highly structured, can be heterogeneous and sensitive to small changes in their structure. The SOC forest-fire tree configuration is rather (equally) fragile to any ignition probability distribution $I(i,j)$ while HOT can be more resistant to some $I(i,j)$ and even more fragile to atypical $I(i,j)$s.

10.4   Robustness/fragility trade-offs

10.4.1   Mathematical modelling

The concept of trade-offs between the functioning of living systems and their robustness is one of the most important in the field of cancer robustness. The living cell possesses only a limited amount of resources and tends to be evolutionary adapted to use and distribute these resources in the most efficient way (a principle that, nevertheless, can be argued in many cases). Hence, one can try to apply some of the simple ideas from the well-developed mathematical optimisation theory to make predictions on the complicated relations between robustness of a function and the efficiency of the function itself.

One of the simplest mathematical considerations is the following (compare to the formalism proposed in Wagner, 2005). Assume that a quantitative measure of fitness of an organism w is gradually selected by evolution towards its maximum value. Let us consider that the fitness w depends on an efficiency f of some biological function and on the robustness of this function r, *i.e.* $w = w(f, r)$. We assume that the fitness is increased with increase of both f and r, *i.e.* $\frac{\partial w}{\partial f} > 0, \frac{\partial w}{\partial r} > 0$. If f and r are completely independent variables, then the organism will eventually maximise w by evolving to the maximum values of f and r, *i.e.* $w_{max} = w(f_{max}, r_{max})$.

The situation is different when the robustness depends on the function. This dependency can take various forms: for example, robustness can be limited by the function value. However, since we assume selection for the maximum possible values of both f and r, we can say that $r = r(f)$. Consider the simplest case when this function is monotonous, *i.e.* either growing $\frac{dr}{df} > 0$ or descending $\frac{dr}{df} < 0$.

In the case of growing dependence $\frac{dr}{df} > 0$, there is a synergy between the function and its robustness: this is a case of *robustness for free*. To give a simple example, imagine a function dependent on an enzyme concentration e such that a bigger enzyme concentration gives a more efficient function, though, the efficiency gradually saturates: $f(e) = f_{max}e/(K + e)$. Then, the evolution will optimise the fitness by increasing e. At the same time, the dependence of f on variations of e will drop with the growth of e, hence, the robustness of the function will be automatically and freely augmented. The maximum fitness is again $w_{max} = w(f_{max}, r_{max} = r(f_{max}))$.

Let us consider the case of descending dependence $\frac{dr}{df} < 0$. Since $\frac{dw}{df} = \frac{\partial w}{\partial r}\frac{dr}{df} + \frac{\partial w}{\partial f}$, the following condition on the growth of fitness, when the function increases, is:

$$-\frac{\frac{\partial w}{\partial f}}{\frac{\partial w}{\partial r}} < \frac{dr}{df} < 0\,, \tag{10.7}$$

i.e. that robustness should not drop too fast with the increase of the function efficiency, but this threshold can change with f. There are several situations here:

1. Robustness never drops very fast, **Equation 10.7** is always satisfied (see **Figure 10.2A**). Then one has $w_{max} = w(f_{max}, r_{min} = r(f_{max}))$. As a result, the system will be selected for the minimum value of robustness (the most fragile system will be the most efficient).

2. Robustness always drops very fast (see **Figure 10.2B**), **Equation 10.7** is violated for any f. Then $w_{max} = w(f_{min}, r_{max} = r(f_{min}))$. The system is selected for the maximum value of robustness (it is advantageous to be more robust even if less efficient).

3. For some values of f, **Equation 10.7** is satisfied, for some other values, it is not (see **Figure 10.2C**). There will be a maximum of fitness for some intermediate values of function and robustness that, however, will not be the maximum possible values:

$$w_{max} = w(f_{opt}, r_{opt} = r(f_{opt}))\,, f_{min} < f_{opt} < f_{max}.$$

In the last case, we can talk about a trade-off between robustness and function or robustness and fragility. For the values of f close to f_{opt}, one expects that increase in the function gives a fast drop-off of the robustness and the resulting fitness is decreasing. On the contrary, any improvement in robustness will cost too much in terms of function degradation and will be eliminated by selection.

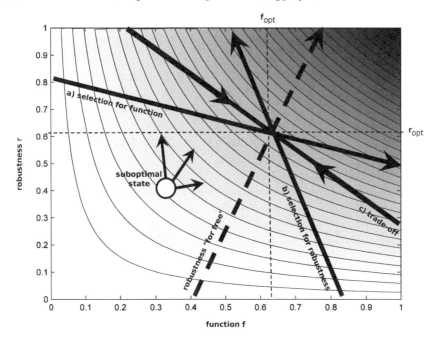

FIGURE 10.2 Simple mathematical view on robustness/fragility trade-off. Shading shows values of the fitness function $w(f, r)$ of function value f and robustness value r. Four straight lines show examples of dependencies of r on f with arrows showing the direction of the increase of w along the $r(f)$ line. A circle below the solid lines represents a suboptimal state in which the fitness can be increased in any direction, increasing both robustness and function values (so-called *violation of trade-off* scenario).

10.4.2   Examples of trade-offs

One of the simplest examples of such a trade-off was provided by Wagner (2005) as the amount of gene overlap in viruses or bacteria. One can say that if there is a possibility to pack genes with overlap, then this can enhance the replication speed, co-regulation of coupled gene products, and maintenance of a shorter genome (which might be crucial for small bacteria and viruses). On the other hand, overlapping genes are more prone to mutations, since a single mutation in the overlap region can affect several genes. In this case, the trade-off solution leads to an optimal partial gene overlap.

The degree of genomic instability in a tumour is also an important example of the robustness/fragility trade-off. Indeed, some mathematical models show that there should be an optimal level of genomic instability in cancer cells. This optimal level provides a balance between clonal heterogeneity and evolutionary advantages connected to it on one hand, and the negative selection associated with lethal consequences of genomic instability on the other hand (Komarova, 2005).

Other mathematical models of trade-offs between robustness and optimal functioning come from the modelling of metabolic networks by Flux Balance Analysis (FBA) and application of constrained linear programming optimisation techniques (Palsson, 2006).

10.4.3   Implications to cancer

In the cancer robustness field, there is an actively promoted idea that in complex systems, robustness is always accompanied by specific fragilities against some well-targeted perturbations (Kitano, 2007). This idea was inspired, first of all, by the above mentioned theory of HOT, which claims that improving systems' robustness by evolutionary selection will produce configurations that can be extremely robust but only for relatively frequent perturbations. Rare types of perturbations to which the system did not have time to adapt can affect the system in extreme and lethal ways. Thus, this theory states that universal systems able to withstand any perturbations (generalists) are in average less efficient and robust than those able to withstand only sufficiently frequent perturbations (specialists).

One can think of this statement either as of a proven theorem (which is only possible to formulate and prove for some model situations such as the forest-fire model) or as of a general law of complex systems, summarising many experimental observations. This idea even motivated some researchers to formulate *the law of robustness conservation*. Other researchers consider control coefficients (see **Equation 10.11** and **Equation 10.12**) of biochemical networks as measures of fragility (Westerhoff et al., 2010) and talk about *conservation of fragilities* manifested by the summation rules (see **Equation 10.13**). Simple ideas borrowed from the application of optimisation theory in economics such as Pareto optimality allowed speculating on the fact that the robustness/fragility trade-off can be violated if the system is not optimised (see **Figure 10.2** and Kitano, 2010). The dependence $r = r(f)$ is valid for the Pareto-optimised system, *i.e.* any increase in robustness leads to a decrease in the function performance. A point on the $f \times r$ plane, which is below this dependence, can move and increase fitness in many possible directions, including the one corresponding to the pure improvement of robustness without decreasing functioning, or improving both robustness and function. As a result, the system will shift towards its Pareto-optimal frontier, represented by $r = r(f)$ curve.

Postulating the necessity of the robustness/fragility trade-off inspires many projects on finding new drug targets against cancer (see **Chapter 11**). However, the general theory is not constructive and does not yet provide concrete strategies on how to identify and take advantage of such fragilities. Some constructive approaches are developed for the robustness/fragility analysis of biological networks, in particular, coming from the control theory and from the theory of model reduction in reaction networks (see more details in **Section 10.7**). However, these methods still need to be adapted to the specifics of the networks involved in cancer, taking into account tissue specificity and large networks for which many parameters are badly determined.

10.5   Robustness and stability of dynamical systems

The theory of dynamical systems provides a solid mathematical basis for studying many natural phenomena including biochemical mechanisms of molecular biology. Much of this theory is devoted to studying stability of dynamical systems, stability of their behaviour or of particular states. This field of science is immense. In this section, we only mention some of the existing directions which can be helpful in understanding sources of robustness of molecular mechanisms. We will describe in more detail some of these approaches without trying to be comprehensive or to give a somehow systematic introduction into the field. We will mention some of the most basic ideas in the field that are required in any constructive discussion on robustness.

Before providing a short review of the ideas in the field of stability of dynamical systems, we should notice that it would not be correct to equal the notion of stability to the general notion of robustness. Robust functioning can be associated with switching from one stable state to another, more adapted, state in changed conditions. In this case, *robust* is more related to the fact of robust switching between two stable states, which is usually connected to some specific instabilities built into the system (Kitano, 2007).

In this section, we will deal with the best-studied class of dynamical systems, *i.e.* systems described by continuous first-order differential equations:

$$\dot{\mathbf{x}} = f(\mathbf{x}, t, K), \qquad (10.8)$$

where t is time, $\mathbf{x}$ is a vector describing the state of the dynamical system, K is a set of parameters.

Mathematical chemical kinetics permits one to model processes of intracellular biochemistry. Networks of biochemical reactions can be modelled by **Equation 10.8** with the right-hand side $f(\mathbf{x}, t, K)$ constructed according to well-defined rules (see **Section 7.3.2**). In this formalism, $\mathbf{x}(t)$ describes time evolution of concentrations or absolute amounts of chemical species. Some of the ideas described in this section are related to the general form of **Equation 10.8**, some are limited to the particular forms of $f(\mathbf{x}, t, K)$, associated with reaction networks, or, even with narrower classes of chemical kinetics equations (such as pure mass action law equations).

There are two types of parameters in the dynamical models. Firstly, there are parameters K, explicitly present in the equations, which are usually kinetic rate constants or concentrations of some chemical species that are considered to be fixed The second ones are parameters representing the values of invariants of the dynamics, *i.e.* the quantities that do not change along any trajectory:

$$b^i(\mathbf{x}(t)) = b_0^i = const, i = 1 \ldots m\,,$$

where m is the number of such invariants. In chemical kinetics, typically, $b^i(\mathbf{x})$ represents the linear conservation laws, *i.e.* the linear combinations of chemical species concentrations that do not vary. Values b_0^i can explicitly be present in f or not. We will call them *balance parameters* or *constraint values*. Let the number of dynamical variables be n. All dynamic system trajectories $\mathbf{x}(t)$ exist in the nonnegative orthant of $\mathbb{R}^n$. The existence of m linearly independent conservation laws reduces the effective dimension of the **phase space**[*] by m: all trajectories are confined to a hyperspace of dimension $n - m$.

Given a vector of initial conditions $\mathbf{x}(t = t_0) = \mathbf{x}_0$, **Equation 10.8** can be solved numerically, sometimes analytically or semi-analytically. The resulting solution can be geometrically thought of as a trajectory in the multi-dimensional phase space. Studying one single trajectory is rarely interesting, it is much more informative to study a geometrical picture coming from sets of trajectories obtained by varying $\mathbf{x}_0$ or the parameters K of the system in **Equation 10.8**. As it follows from our general definition of robustness in **Equation 10.4**, we will study how variation of $\mathbf{x}_0$ and/or K change the set of trajectories and in what sense. This is the theory of stability of dynamical systems.

The trajectories of the system in **Equation 10.8** follow the tangent directions of the function f in the right-hand side of **Equation 10.8** that defines *a vector field* in the phase space. Instead of studying variability in the sets of trajectories, one can study variability of the vector field $f(\mathbf{x}, t, K)$ which depends on the set of parameters K. Notice that this is a difficult but purely geometrical problem and does not require integration of differential equations.

One of the most fundamental characteristics of **Equation 10.8** is the set of all stationary (or steady state) solutions, defined by the system of algebraic equations $f(\mathbf{x}, t, K) = 0$. Let us denote this finite (maybe empty) or infinite set as $\{\mathbf{x}\}_s$. Studying this set of vectors, its dependence on parameters K and the behaviour of the vector field f in its neighbourhood, generates the first set of questions regarding stability (hence, robustness) of the stationary system functioning.

10.5.1 Internal stability of a stationary state

The first category of these questions is related to the so-called *internal stability* of the stationary state. A system is *internally stable* if it is capable to maintain its stationary state despite perturbations that drive the system out of it but do not change the system's parameters. This set of questions goes back to the works by Lyapunov (see **Box 10.2**).

Studying stability of a trajectory and a steady state is mathematically equivalent. To study internal stability of dynamical systems, Lyapunov (1966)

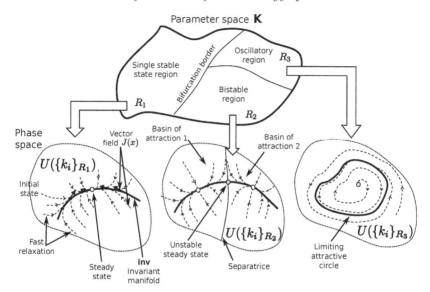

FIGURE 10.3 A schematic illustration of the geometrical representation of a dynamical system. Parameter space (with the possibility of bifurcations, bistability, oscillations, single stable state region), phase (or state) space, attractor, initial conditions, attracting manifolds are represented.

❏ **BOX 10.2: Stability by Lyapunov**

Accordingly to Lyapunov, a stationary state $\mathbf{x}_s$ is *(uniformly) stable* if any trajectory starting in the δ-neighbourhood of $\mathbf{x}_s$ will remain in the ϵ-neighbourhood (where $\epsilon = \epsilon(\delta)$) of $\mathbf{x}_s$ for infinite time. If the trajectory returns back to $\mathbf{x}_s$ then it is called *asymptotically stable* (Lyapunov, 1966).

---

suggested a mathematical apparatus of Lyapunov function, *i.e.* a function that does not decrease along any system trajectory. Lyapunov function can be understood by thinking of a physical system that can be characterised by the energy function. If the system loses energy over time and the energy is never restored, then eventually the system converges to a final state of energy minimum. Abstract dynamical systems, such as describing functioning of molecular pathways, are usually not naturally equipped with such a function, but if it exists then the system's internal stability is guaranteed.

It is also important to mention the notion of *internal stability by the first approximation* or *linear or local stability*, when perturbations are so small that one can consider linear approximation of the dynamics in **Equation 10.8**:

$$\dot{\mathbf{x}} \approx \frac{\partial f}{\partial x}(\mathbf{x}_s)\mathbf{x} + g(\mathbf{x}),$$

(10.9)

where $\frac{\partial f}{\partial x}$ is a matrix of partial derivatives called *Jacobian*. Very roughly, one can say that for perturbations that are small enough, stability of the linearised system leads to the stability of the initial system in **Equation 10.8**. The stability of the linearised system in **Equation 10.9** is studied by computing the eigenvalues of the Jacobian in x_s.

When a dynamical system is capable to return to the same stable state despite perturbations changing its state (position of vector x in the phase space), it is called internally stable. In terms of biological interpretation, it means that an internally stable dynamical system can demonstrate a robust and predictable behaviour with respect to varying environment, modelled as changes of the initial conditions of the dynamical system.

Provided the internal stability one can study how the dynamical system reacts to modification of its structure. Mathematically speaking, it is modelled by changing the right-hand side $f(x, t, K)$ of **Equation 10.8** which represents the vector field in the phase space driving the system dynamics. There are several possible problem statements here:

1. If we change a set of parameters K for some other set $K' = K + \delta K$, will the system lose its internal stability?

2. How does the vector field (*i.e.* $f(x, t, K)$ function) depend on variation of parameters K? Mathematically, it can be stated, for example, as: How does $\frac{\partial \log f(x,t,K)}{\partial \log K}$ behave in different regions of the phase space?

   There is a related question to the previous one: how does some function $p_f(K)$ of parameters K, and dependent on f, change with respect to parameter variations? A typical example is: a steady state $x_s(K)$ is a solution of equation $f(x_s, t, K) = 0$ which of course depends on K. How sensitive $x_s(K)$ is to variation of K?

3. How do perturbations of the vector field itself change certain system characteristics, *i.e.* what will change if we substitute the function f by $f + \delta f$, where δf is small? Possible examples of studies can be: would the stationary states of the system remain close to the unperturbed ones? Would the number of states or their stabilities change after such a perturbation? If the qualitative features of the dynamics remain untouched by δf, then this type of robustness is called *structural stability*.

10.5.2 Sensitivity analysis

Related to the second question just mentioned, the problem of how sensitive the steady states are to parameter variations is traditionally studied in the field of control theory by the series of mathematical methods called *parameter sensitivity analysis*. One possible definition of sensitivity is:

$$S(x) = \frac{\partial x}{\partial k} \times \frac{k}{x} \equiv \frac{\partial \ln x}{\partial \ln k}, \qquad (10.10)$$

where $k \in K$ is a system parameter and x is a system property that can be quantified (such as a steady state value, a flux through particular reaction, relaxation time). This definition is close to our definition of robustness in **Equation 10.4** and all remarks that we have made, for example, concerning noninvariance of **Equation 10.10** with respect to nonlinear transformation of ks remain valid.

The sensitivity coefficient is known as *the control coefficient* in the metabolic control theory dealing with modelling of metabolite transformations through connected pathways of reaction steps where the speed of each reaction step is controlled by a particular enzyme. A control coefficient measures the relative steady state change in a system variable, *e.g.* pathway flux J or metabolite concentration S, in response to a relative change in a parameter, *e.g.* enzyme activity or the steady state rate v_i of step i. For example, the flux control coefficient is:

$$C_{v_i}^J = \frac{\partial \ln J}{\partial \ln v_i}, \tag{10.11}$$

and the concentration control coefficient is:

$$C_{v_i}^S = \frac{\partial \ln S}{\partial \ln v_i}. \tag{10.12}$$

Two important summation rules are known for the control coefficients:

$$\sum_i C_{v_i}^S = 0, \sum_i C_{v_i}^J = 1. \tag{10.13}$$

These summation rules in the context of metabolic control theory were derived independently by Henrick Kacser with Jim Burns and Reinhart Heinrich with Tom Rapoport in 1970s (see references in the textbook Burns et al., 1985) but in other fields of science they have much deeper roots, coming back to the Kirchhoff's rules in the theory of electrical circuits.

10.5.3 Structural stability

The third question mentioned in the end of the **Section 10.5.1** concerns the notion of structural stability. It has a long and exciting history coming back to the *catastrophe theory* by Thom (1975). Recently, the notion of structural stability was revived in the context of regulatory motifs (see **Section 9.4**) found overrepresented in various biological networks. These motifs are hypothesised to be simple functional elements of the biological networks. In Prill et al. (2005), it was suggested to characterise a simple network, modelled by chemical kinetics equations, by its Structural Stability Score (SSS). SSS is a probability to have a linearly stable steady state independently of the kinetic parameter values (in other words, it is a probability of the Jacobian eigenvalues to have negative real parts and close to zero imaginary parts). It was shown that the statistical overrepresentation of motifs and their SSS are correlated, suggesting some natural selection for structural functional blocks which have a tendency to be stable for a maximally wide range of parameters.

All these questions are relevant to robustness studies and require various and sometimes quite complicated mathematical methods to tackle them. It is out of the scope of this chapter to provide their detailed review.

10.5.4 Qualitative properties of dynamical systems

Studying stability properties for a general dynamical system in **Equation 10.8** is a daunting task: the general form of function f can give too many possible dynamical behaviours. Only in rare cases, and for very general results, it is possible to take all of them into account. Any practical method has to deal with a limited functional class of the right-hand part in **Equation 10.8**. For example, Thom (1975) assumed that f should be a gradient of some other scalar function g, *i.e.* $f(x) = -\nabla g(x)$, which led to appearance of the catastrophe theory.

Starting from 1970s, in mathematical chemistry, there was a hope that it will be possible to develop a rich mathematical theory for a special class of functions f that naturally appear in modelling reaction networks. Each reaction speed is characterised by its kinetic law and associated kinetic parameters. In most of the applications, f becomes a rational function: this includes, for example, dynamical models of Michaelis-Menten and Hill kinetic reaction networks. In the simplest case of the mass action law equations, f is a sparse polynomial of not a very high degree (see **Equation 7.1**).

Particular attention was paid to the results on the properties of dynamical systems of reaction networks (such as multiplicity and stability of stationary states) that can be obtained without knowledge of the exact values of kinetic parameters. In other words, mathematicians tried to connect the structural properties of the reaction graph with some qualitative properties of the dynamics. For example, this problem is formulated as the main challenge of the Chemical Reaction Network Theory developed by Feinberg (1987).

If we were to make constructive conclusions about qualitative features of dynamics, looking only at the structure of the reaction graph then, evidently, this would mean that these features are robust, *i.e.* do not depend on parameters. This question is interesting for the case of large reaction graphs (containing at least 10 reactions) with a nontrivial structure, containing feedbacks, branching points, *etc.*

Unfortunately, not many practical results were obtained in this direction. Horn, Jackson and later Feinberg (for references, see Feinberg, 1987) studied properties of dynamical systems describing reaction networks characterised by mass action law. One of the principal results of their theory is known as *deficiency zero* and *deficiency one* theorems.

In simple terms, deficiency represents a number of degrees of freedom which distinguish a given reaction network model from a reaction network with nice relaxation properties and a possibility of detailed balance. Deficiency zero networks are equipped with a law analogous to the second law of thermodynamics: the total entropy of the concentrations of chemical species does not decrease along any trajectory. Hence, any deficiency zero reaction network has a natural Lyapunov function and converges to a unique nontrivial (with positive concentrations of chemical species) equilibrium, which is asymptotically stable. It is also possible to prove some results on the uniqueness of the equilibrium state for deficiency one networks (Feinberg, 1987). The problem is that the networks of deficiency zero and one are not among the most interesting networks for describing biological pathways, where more interesting and nonequilibrium phenomena are thought to be essential. Thus, these results can serve as a part of the toolbox for qualitative dynamical analysis of reaction networks, but for a relatively modest class of systems.

Nevertheless, in some particular cases, the mathematical results connected with deficiency give insights into how robust properties of some molecular mechanisms are achieved. The absolute concentration robustness notion was introduced by Shinar and Feinberg (2010). Absolute concentration robustness is the complete insensitivity of some concentrations of chemical species to the values of balance parameters.

10.6   Dynamical robustness and low-dimensional dynamics

The notion of attractor (see **Box 10.3**) plays one of the central roles in the analysis of long-term and stable behaviours of dynamical systems. Analysis of attractors and their basins of attraction together with their dependence on system parameters is one of the most used methods in analysing the stability of dynamical systems. In the context of robustness, several questions are important here: What is the dimension of the attractor itself? What is the size (in some measure) and the configuration of its basin of the attraction? How stable are these properties with respect to parameter variations? *etc.* Low-dimensional attractors (points, limit cycles) with large regions of attractivity can be associated with stability of certain homeostatic biological system properties. Moreover, possibility of switching between several attractors as an adaptive response to a perturbation can be a good mathematical metaphor of certain robust biological functions such as glycolitic shift in tumour metabolism (Kitano, 2007).

10.6.1   Dynamical robustness

In many biological applications not only the final attractive state of a dynamical system matters but also the way and the speed with which the system

---

❏ **BOX 10.3: Attractors**

An attractor is a subset of the phase space characterised by the properties of:

- *positive invariance* under the system dynamics: if a trajectory starts from a point on the attractor, then it never leaves it,

- existence of *a basin of attraction* such that all points from this basin will eventually enter the attractor and *minimality, i.e.* there is no any subset inside the attractor possessing the properties of invariance and attractivity.

Attractors can be points, curves and manifolds, or more complicated objects such as *strange attractors* (complicated sets with a fractal structure and characterised by a noninteger dimensionality).

---

trajectories approach it. Many biological networks determine cell behaviours through time-varying signals operating during a transient activated state that ultimately returns to a basal steady state. This process can be called *relaxation* towards an attractor. For example, the final state of the dynamics of a DNA repair pathway can be repaired DNA. Nevertheless, if this process takes too long, it can be lethal for the cell. Some models of apoptosis interpret apoptosis as a passage of system trajectories through nonphysiological and lethal regions of the phase space (Aldridge et al., 2006). Weak dependence on parameters or initial conditions of not only the final stable system properties but also of its relaxation processes is called *dynamic robustness*, a term coined in Gorban and Radulescu (2007).

Relaxation processes are highly affected by the *hierarchical organisation of time scales*, which is a common observation in most biochemical pathways. For example, changes in gene expression programs can take hours and even days, post-translational protein modifications can take minutes and protein complex formation can take seconds. Protein half-lifetimes can vary from minutes to days. This important observation applies not only to time scales but also to concentration values of various species in these networks. mRNA copy numbers can change from some units to tens of thousands, and the dynamic concentration range of biological proteins can reach up to five orders of magnitude.

10.6.2 Low-dimensional invariant manifolds

Separation of time scales in dynamical systems leads to the phenomenon of existence of *lower-dimensional invariant manifolds* in the phase space of a dynamical system. An invariant manifold is embedded in the phase space such that if a trajectory starts on it, it never leaves it. The manifold is called slow and attractive if a typical trajectory of the system in the phase space can be divided in a fast motion phase towards the manifold and relatively slow motion close to the manifold and along it. Invariant manifolds of higher dimensions can contain the invariant manifolds of lower dimensions forming an *invariant flag*: a hierarchy of invariant sets embedded into each other. All attractors of the system are evidently located on the invariant manifold. If the invariant manifold is slow then the system trajectories spend most of the time in the close vicinity of invariant manifolds and only transiently appear at large distance (provided some distance measure) from it.

The high-dimensional system dynamics can be projected, *i.e.* reduced, onto the manifold and studied on it. This is the essence of many methods of *model reduction* dealing with separation of time scales. Quasi-steady state and quasi-equilibrium approaches construct an approximation to the slow manifold under certain assumptions, whereas other methods allow finding explicitly the slow invariant manifold for some classes of dynamical systems and study the system dynamics restricted to the manifold. This is the most consistent approach due to invariance (Gorban, 2005).

It is important to understand that in the case of a dynamical system characterised by a hierarchy of time scales, not all events that occur in any time scale can be observed in experiment. Very fast time scales correspond to *immeasurably fast processes* (repeating Nobel laureate Manfred Eigen). Very slow time scales correspond to the changes that can be considered almost constant during the time of the experiment. Only processes taking place on the time scales comparable to the time scale of making measurements are observable. Moreover, only the system parameters that contribute to these processes are important for explaining the system's behaviour, and can be estimated from experimental data. The invariant manifold, if it exists, indicates that some fast time scales of the dynamical system can be neglected after which the system becomes simpler and more tractable.

Existence of a low-dimensional invariant manifold assures the dynamical robustness in two aspects:

- The system weakly depends on changing the initial conditions, since big regions of the phase space will be projected onto a relatively small area on the invariant manifolds. That way, an invariant manifold can be understood as an implementation of Waddington's notion of *homeorhesis* for dynamical systems. Homeorhesis can be defined as the phenomenon of trajectory stability as opposed to **homeostasis***, *i.e.* steady state stability.

- The system weakly depends on the parameters determining the processes of fast relaxation towards the invariant manifold and not responsible for the manifold shape.

Both of these types of dynamical robustness will be more pronounced if the intrinsic dimension of the invariant set is getting smaller.

In this sense, one can suggest that the higher the variability of the initial conditions and of the system parameters and the smaller the intrinsic dimension of the system dynamics, the stronger the robustness properties one would expect to observe. The intrinsic dimensionality of the dynamics is not determined by the structure of biochemical reaction networks in a simple way: depending on the distribution of kinetic parameters, even complex systems with many components and connections can possess surprisingly low-dimensional intrinsic dynamics confined to invariant manifolds. Moreover, our observation is that this is true for most of the mathematical models of biochemical pathways (for example, see Radulescu et al., 2007).

10.7 Dynamical robustness and limitation in complex networks

10.7.1 Static and dynamic limitation

Another important and general phenomenon that can contribute to a particularly robust response of biochemical reaction networks is the *static and dynamic limitation*. The idea of limitation is that, even in very complex dynamics of a reaction network, at every particular moment of time, there might be a limiting place in the network which completely determines the dynamics locally.

Limitation is inspired by the notion of a rate limiting reaction step in a chain of linear reactions (as opposed to the cyclic chain of reactions introduced in **Equation 10.6**). If among all the reactions, the slowest one has a rate constant well-separated from all the others then the flux through the cyclic chain of reactions depends only on this rate constant and will be completely insensitive with respect to (sufficiently small) variations in other constants. Hence, this system of reactions is extremely robust yet controllable. The simple fact that the relaxation time (see **Box 10.4**) of this cycle is determined by the second slowest kinetic constant can be easily demonstrated (Gorban et al., 2010).

10.7.2 Asymptotology of reaction networks

During the 20th century, the concept of the limiting step was revised several times. The simple idea of a narrow place, *i.e.* the least conductive step, could be applied without modification only to a simple cycle or to a chain of first-order irreversible steps. In more complex situations, various difficulties arise for its proper definition and use. The notion of the limiting step was outmoded; however, several modifications of this notion were suggested (for a review, see Gorban et al., 2010).

In Gorban et al. (2010), it was suggested that the correct generalisation of

---

❑ **BOX 10.4: Relaxation time**

The relaxation time is the characteristic time needed for a dynamic variable to change from the initial condition to some close vicinity of the stationary state. Most naturally the relaxation time is introduced in the case of linear relaxation dynamics. For example, if a variable follows simple dynamics in the form $x(t) = A\left(1 - e^{-\lambda t}\right)$, where A is the steady state value of x, then the relaxation time is $\tau = \frac{1}{\lambda}$ and it is the time needed for x to increase from the zero initial value to approximately $1/e \approx 63\%$ of the A value. Measuring the approximate relaxation time in practical applications consists of fitting the linear dynamics to the experimental time curves and estimating λ. The relaxation time is a relatively easily measurable quantity, sometimes essential, as in the field of *relaxometry*.

---

the limiting step is the notion of *dominant system*. A dominant system is an auxiliary minimal reaction network which defines the main asymptotic terms of the stationary state and the relaxation times in the limit of well-separated time scales. For complex networks, this dominant system cannot be reduced to a single reaction step. Moreover, in nonlinear (with nonmonomolecular reactions) networks, this limiting place can change with time. In monomolecular reaction networks with separation of time scales, the dominant system (limiting place) does not change in time and an algorithm for its construction has been developed (Gorban et al., 2010).

Finding dominant asymptotic solutions is useful in at least two aspects: (1) these solutions are often simple and even trivial and tractable analytically, and (2) these solutions often depend on much smaller number of parameters than the initial system. Typically, the asymptotic solutions depend on the ordering of kinetic constants rather than on their exact values.

Asymptotic analysis explores the dynamical properties of complex networks by listing all possible dominant systems under given qualitative constraints on parameter values, by analysing these solutions and by associating the observed biological system behaviours with these solutions. Automatically, all these solutions are robust because they depend on a small number of parameters. Those parameters or parameter combinations that are conserved in the asymptotic solutions are *system control parameters* allowing manipulating the system. Those parameters that are not included in the asymptotic solutions can be considered *insensitive*. From this analysis, one can also predict which parameter relations (inequalities in the form $k_i \gg k_j$) should be changed to switch from one type of asymptotic behaviour to another, thus providing predictions of what should be the most critical interventions with the most desired robust outcome.

10.7.3   Limitation and design of robust networks

Analysis of robustness of the systems with limiting *places* predicts that robustness of the system can grow with the length of the pathways or cycles in the network. In other words, complex networks are predicted to be more robust given that they follow some general principles of evolutionary design. These robustness principles are described in Gorban and Radulescu (2007) in the following way: *"...We can obtain design principles for robust networks. Suppose we have to construct a linear chemical reaction network. How to increase robustness of the largest relaxation times for this network? To be more realistic let us take into account two types of network perturbations: (1) random noise in constants; (2) elimination of a link or of a node in reaction network. Long routes are more robust for the perturbations of the first kind. So, the first recipe is simple: let us create long cycles! But long cycles are destroyed by link or node elimination. So, the second receipt is also simple: let us create a system with many alternative routes! Finally the resources are expensive, and we should create a network of minimal size. Hence, we come to a new combinatorial problem. How to create a minimal network that satisfies the following restrictions (1) the length of each route is bigger than L; (2) after destruction of D_l links and D_n nodes there remains at least one route in the network. In order to obtain the minimal network that fulfils the above constraints, we should include bridges between cycles, but the density of these bridges should be sufficiently low in order not to affect significantly the length of the cycles. Additional restrictions could be involved. For example, we can discuss not all the routes, but productive routes only (that produce something useful). For acyclic networks, we obtain similar receipts: long chains should be combined with bridges. A compromise between the chain length and number of bridges is needed. We can also mention the role of degradation reactions (reaction with no products). Concentration phenomena are more accentuated when the number of degradation processes with different relaxation times is larger. Thus, one can increase robustness by increasing the spread of the lifetimes of various species."*

10.7.4   Model reduction

Extracting both the low-dimensional dynamics of a complex dynamical system and its limiting place is included in the toolbox of analytical and computational methods referred to as *model reduction*. Model reduction aims at simplifying the description of large and complex models and is tightly linked to the study of robustness. The main principle remains the same as previously described: the parameter combinations (typically some complex ratios of them) that one finds in the reduced model are the control parameters to which the system is the most sensitive. Other parameters can be largely (frequently, by several orders of magnitude) varied without having a significant effect on the most essential observable system properties such as its attractors and the relaxation processes towards attractors. By systematically applying model reduction, one can construct a hierarchy of models of decreasing complexity (see an example of this approach with *NFkB* pathway modelling in Radulescu et al., 2008). It allows the comparison of models of various complexity levels and to derive the model of *the minimal possible complexity, i.e.* a model that still approximates well the available experimental data but cannot be simplified further without making the approximation worse. That way, only the essential model parameters entering the minimal complexity model will be fitted from experiment, thus avoiding the problem of *model nonidentifiability* and *parameter overfitting*.

10.8 A possible generalised view on robustness

It would be extremely difficult to develop a universal view on robustness of biological systems. In this section, we try to generalise our view on robustness and summarise the conclusions listed previously in this chapter.

One general observation on biological systems is that more robust organisms are more complex in the sense that they usually contain more elements, more reactions, more levels of regulation and control: they exists in spaces of higher number of dimensions. It can be shown that there is no consistent trend to complexification (progress) in the history of biological organism evolution. Periods of increasing complexity bursts are rather a consequence of a weak purifying selection (Koonin, 2009). So, we do not claim that the biological systems tend to complexity, or that complexity evolves, we only observe that when complexification takes place, it often leads to creation of robustness generators such as redundancy and degeneracy, distributed robustness, bowtieness, multiscaleness, *etc.*

Theoretically, we can think of three types of complexity: reducible, self-averaging and wild (Gorban, 2009). Wild complexity is intractable with our current scientific tools while reducible and self-averaging represent two approaches trying to understand the complexity from two very different perspectives.

To understand this classification, we can think of a complex phenomenon as an object existing in a multi-dimensional space. Our perception of this object is inevitably low-dimensional (because for instance, our mind is organised by the presentation of our motion in three-dimensional space and the convenient static visualisation is two-dimensional). We can represent our perception as a projection of the object from high-dimensional to low-dimensional space. A biological function can be also considered as a projection of its high-dimensional microscopic detailed description onto a low-dimensional space where it is manifested at macroscopic level. Let us try to imagine what one can observe through such a projection.

10.8.1   Reducible complexity

The reducible complexity model states that despite the fact that the complex object is embedded in high-dimensional space, intrinsically, it remains low-dimensional with a relatively small number of degrees of freedom. Changing and choosing a right angle of view or a screen, would reduce this type of complexity to a much simpler and low-dimensional view. A right metaphor for this is a cloud of points in a multi-dimensional Euclidean space, distributed in the vicinity of some nonlinear low-dimensional surface. The cloud exists in many dimensions, but effectively it has only few degrees of freedom, few true coordinates. Projecting the cloud on a randomly chosen low-dimensional screen will show an image, difficult to interpret and potentially very different from one projection to another. However, knowing the intrinsic coordinates of the system would reveal the internal low-dimensional structure of the cloud, and would allow the observation of all its features without distortions. Reducible complexity of dynamical systems is manifested in low-dimensional intrinsic structure of their attractors or existence of low-dimensional invariant manifolds described in the previous section.

Another frequent type of reducible complexity is a system's structure following some relatively simple organisational principle. One of the most common principles is the *hierarchical organisation*. A good mathematical metaphor can be a cloud of points in a high-dimensional space organised in a system of embedded clusters: there are high-level clusters, inside which there are clusters of points of the second level, inside which there are clusters of the third level, *etc.* Importantly, at each level, the number of clusters is relatively small compared to the total number of points. In biological networks, this type of hierarchical reducible complexity is revealed in the existence of modules, compartmentalisation, multiple concentration and time scales. In physiology, it can be seen as the construction of an organism from organs, tissues and cells. The term *self-simplification* with respect to evolving biological systems was coined by Pattee (1972) to describe the complexity with some internal hierarchical structure. Indeed, existence of relatively simple hierarchies is frequently associated with self-organisation.

10.8.2   Self-averaging complexity

Generally speaking, we say that the reducible complexity is associated with a possibility of a compact or a low-dimensional description of a complex object. Self-averaging complexity, by contrast, is usually associated with truly high-dimensional objects that do not possess any intrinsic low-dimensional simple structure. However, after projection on most of the low-dimensional screens, they will look very similar. A good mathematical metaphor for this type of complexity is a multi-dimensional hypersphere uniformly sampled by points. After projection on any two-dimensional plane, most of the points will be located very close to the centre of the sphere, and, very rarely, they will be projected at the distance of the sphere radius from the centre. Moreover, if we look at this cloud of projected points with a magnifying glass, we discover that their distribution is very close to a normal one (*i.e.* Gaussian).

In statistical physics, this corresponds exactly to the well known Maxwellian distribution and in general, this is an example of Gromov's measure of concentration phenomenon: truly high-dimensional objects look very small (concentrated) after projection onto a low-dimensional space and most of the distributions become almost normal after projections on the low-dimensional subspaces (see **Section 10.1**, where it is briefly discussed). In statistical physics, this creates a possibility for relatively simple low-dimensional macroscopic description of high-dimensional complex microscopic processes.

10.8.3   Wild complexity

Wild complexity cannot be simplified neither by reduction nor by averaging. There is no good low-dimensional screen to observe this object and any projection will give a different view of the object, with new features. One cannot dissect this complexity with levels because there is no clear separation between them, they all coexist and penetrate into each other. There is no time or space scale separation; most of the processes are happening at the same time and everywhere with strong between-scale coupling. In wild complex systems, many local perturbations produce global effects which might be very different from one perturbation to another similar perturbation. As a matter of fact, the objects possessing wild complexity would be unimaginable and indescribable, *i.e.* not allowing a description in a compact and abstract form.

Due to this, we do not have many examples of wild type complexity, because its definition comes from a negation: it is a complexity that cannot be reduced or averaged whereas any simple illustrative example will be already reduced or averaged. Probably, one of the few examples can be found in the collective neuron excitation dynamics of our brain. Izhikevich and Edelman (2008) developed a computer brain model with one million multi-compartmental spiking neurons and a billion synapses. The model is calibrated to reproduce known types of responses recorded *in vitro* in rats. Computer simulations of this model show overwhelmingly complex dynamics characterised by global excitation-like responses, spontaneous activity, sensitivity to changes in individual neurons, functional connectivity on different scales. The complexity of this model can be tentatively characterised as wild.

A close by wording (but not by its meaning) term *irreducible complexity* in evolutionary theories is historically associated with intelligent design ideas (in particular, by Michael Behe). The irreducibly complex systems were defined as *"composed of several well-matched, interacting parts that contribute to the basic function, wherein the removal of any one of the parts causes the system to effectively cease functioning,"* *i.e.* as extremely fragile systems. The notions of wild and irreducible complexities are quite different in our understanding.

10.8.4 Complexity and robustness

We, cautiously, claim that reducible and self-averaging complexity models can lead to robustness and to the possibility of control, while wild complexity models are hypersensitive to small perturbations. They are neither robust nor easily controllable.

In other words, we claim that all types of robustness can be associated with two principal causes: (1) with the fact of intrinsic simplicity of a visibly complex and multi-dimensional system: an object has much less degrees of freedom than parameters, or (2) with the built-in and automatic robustness of low-dimensional projections of high-dimensional objects: an object will look very similar being seen from many possible projections. Robustness generators such as redundancy, bowtieness, or buffering are consequences of the reducible complexity. Sources of robustness such as distributed robustness, cooperativity, or robustness of many dynamical features such as oscillation periods, steady state fluxes, *etc.* can be consequences of self-averaging complexity. When neither reduction nor averaging is possible, the wild complexity leads to hypersensitivity and fragility as a principle of organisation.

This theoretical suggestion needs to be more elaborated in order to provide concrete recipes to measure or to manage robustness in concrete examples. However, some simple ideas can be already implemented. For example, in Radulescu et al. (2007), it was suggested to compare the robustness of dynamical models of *NFkB* pathway by estimating the intrinsic dimension of the invariant manifold. It was shown that this dimension is much smaller than the dimension of the phase space and is almost invariant with respect to application of model reduction techniques. Moreover, it was shown that adding or removing some specific regulations can significantly affect this dimension leading to appearance or disappearance of new intrinsic degrees of freedom in the system. We can predict that this is a common feature of most of the existing mathematical models of biochemical pathways, including those involved in cancer.

10.9 Conclusion

The mathematical interpretation of biological robustness is still in its infancy. There are some promising mathematical objects with the robust properties resembling those of biological mechanisms. For example, simple evolutionary models establish principles of self-organisation and optimised tolerance in evolving cell populations. Robustness of cellular biochemistry can be associated with robustness of the corresponding dynamical models of cellular networks, where dynamical systems stability theory and control theory are the most developed approaches. Mathematical models and ideas borrowed from optimisation theory can serve as a basis for understanding trade-offs between efficiency and robustness that cancer cells have to resolve. Model reduction techniques, including asymptotic approaches from chemical kinetics, give a possibility to identify the most sensitive combinations of parameters of biochemical reaction networks. For the future mathematical theory of cancer robustness, all these methods should be made much more specific to the problems met in mathematical modelling of the mechanisms involved in **tumorigenesis**[*].

✎ **Exercises**

- Give an example of a robust mathematical function and compute its robustness.

- Why can increasing network size automatically lead to more robust behaviours?

- What is the principal difference between the mechanisms of SOC and HOT in terms of robustness to invasion of a mutant individual in a population?

⇨ **Key notes of Chapter 10**

- The theory of dealing with biological robustness, how to measure and modify it, can contribute to cancer treatment strategies. By systematic identification of fragile points and by hitting them with right strengths, in the right periods of time and in the most efficient combinations, one can try to develop new cancer therapies. One of the most important conclusions of the theory of cancer robustness is that such fragilities should inevitably exist due to necessity for resolving trade-offs between robustness and other functions of the cancer cells.

- Complex systems optimised for performing certain functions usually have to resolve trade-offs between robustness and efficiency.

- From the mathematical point of view, robustness is often an expected property of multi-dimensional systems connected with self-averaging and dimension reduction.

Chapter 11

Finding new cancer targets

Anti-cancer drugs should ideally kill cancer cells, and leave normal cells in peace. This paramount goal is extremely challenging because cancer cells have few demonstrable biochemical differences with normal cells, and because of the large diversity among cancer cells (see **Chapter 2**). Historically, cancer drugs have targeted proteins active in rapidly proliferating cells, such as metabolic enzymes active in cell division or DNA polymerase and topoisomerase important for DNA replication. Unfortunately, these processes are not specific to cancer cells and the corresponding drugs also hit all normal cells having a rapid turnover such as skin, hair, gastrointestinal and bone marrow, leading to the many common side effects associated with cancer chemotherapy.

More recently, insight into hormone signalling have led to the targeting of nuclear hormone receptors, and the elucidation of **signalling pathways*** has highlighted the relevance of signalling proteins, including growth factor receptors and kinases, as promising cancer drug targets. The rapid accumulation of cancer genomic data, and parallel progress in our understanding of the molecular basis of cancer, have increased the hope that more effective and less toxic therapies can be discovered less serendipitously than in the past. This requires the development of specific computational models to identify new candidate cancer targets from the analysis of large omics datasets and mathematical models of **tumorigenesis***, which we discuss in this chapter.

In particular, we review how systematic comparison of cancer and normal cells at the molecular level may help pinpoint new specific targets by analysing lists of *interesting* candidate genes (see **Section 11.1**). Alternatively, since cancer can be considered a network disease (see **Chapter 2**), many recent methods to identify cancer drug targets rely on the analysis and modelling of the network responsible for sustaining tumorigenesis (see **Section 11.2**) and on efficient ways to disrupt its functioning (see **Section 11.3**). This may ultimately lead to modelling frameworks able to identify *combinations* of targets with high efficacy and specificities on cancer cells (see **Section 11.4**).

11.1   Finding targets from a gene list

A typical systems biology project aiming at identifying drug targets starts with collection of data, such as genomic or transcriptomic data obtained with high-throughput technologies (see **Chapter 3**). After proper data normalisation and statistical analysis (see **Chapter 4–Chapter 6**), one frequently gets a list of genes and proteins which may contain good candidates to identify new drug targets. The list typically contains differentially expressed genes between cancer and normal cells, or genes located in a genomic region found to be often amplified or deleted in cancer samples. One can then try to find the most promising targets within this list by *prioritising* the genes on the list (see **Section 11.1.1**), or identify good candidate targets outside of the list by identifying master regulators responsible for the deregulation of the genes on the list (see **Section 11.1.2**).

11.1.1   Gene prioritisation

Traditional linkage analysis or study of chromosomal aberrations in DNA samples can lead to the identification of genomic regions containing one or several cancer genes, whose disruption causes or allows tumorigenesis. The proteins coded by these cancer genes have an obvious potential as new therapeutic targets for cancer. The genomic regions identified, however, typically contain tens to hundreds of candidate genes. Similarly, the analysis of cancer or normal samples with gene or protein expression techniques often allows one to identify many *interesting* proteins, among which only a few are causal and may become cancer targets. In both cases, since experimental validation of candidate cancer genes is a long and expensive process, it is important to be able to identify, among a list of candidate genes, which ones are the most promising cancer genes, or at least to prioritise the genes from the most likely to be a cancer gene to the less likely.

Gene prioritisation among a list of candidate genes is typically based on what we already know about the genes: the most promising candidate genes are typically those which are known to play a role in some biological process important for cancer cells, or which share similarities such as coexpression with other known cancer genes. Since the knowledge and information we have about genes and proteins is nowadays fragmented in different forms in a multitude of databases (see **Section 4.5.1** and **Section 4.7**), computational approaches to integrate heterogeneous data and knowledge have emerged recently as promising tools for gene prioritisation (Giallourakis et al., 2005). For example, some methods try to automatically compare the known functional annotations of each candidate gene to the description of the disease in order to automatise the process of cancer gene hunting (Perez-Iratxeta et al., 2002; Turner et al., 2003; Tiffin et al., 2005). Many other approaches, varying in the algorithm they implement and the data sources they use, try to identify candidate genes sharing similarities with known disease genes, as reviewed by Tranchevent et al. (2010). For example, Endeavour (Aerts et al., 2006; De Bie et al., 2007) use state-of-the-art machine learning techniques to integrate heterogeneous information and rank the candidate genes by decreasing similarity to known disease genes, while PRINCE (Vanunu et al., 2010) uses label propagation over a Protein–Protein Interaction (PPI) network and borrows information from known disease genes of related diseases. ProDiGe is a recently proposed, state-of-the-art method to combine heterogeneous data, including genomic, transcriptomic and protein network data, in a single machine learning model for gene prioritisation (Mordelet and Vert, 2011). An obvious limitation of this *guilt-by-association* strategy is that only genes sharing similarity to those we already know can be discovered, limiting their potential for cancers with no or few known causal genes. Interestingly, Mordelet and Vert (2011) showed in a retrospective study that sharing of information across different diseases, in particular different cancers, is possible and can lead to the discovery of cancer genes even for cancers with no or few known causal genes.

11.1.2 Drug targets as master regulators of genes and proteins with altered expression

When a list of candidate genes results from the analysis of gene expression data, as typically a list of differentially expressed genes between cancer and normal tissues, it may be the case that looking for potential targets within the list using techniques mentioned in **Section 11.1.1** is useless. Indeed, this list of genes alone is not sufficient to claim that affecting differentially expressed genes will revert or disrupt the tumorigenic phenotype. As a matter of fact, most of these molecules will probably be deregulated as a consequence rather than a cause of tumorigenesis. Moreover, the molecules that are the most important *drivers* (see **Chapter 2**) of the changes in cancer cells can be absent in the differential expression list, simply because their signal could be amplified by cell signalling cascades, producing the most visible effects downstream of them. Metaphorically speaking, the cause of a snow avalanche can be a slight movement of a small stone, which will be completely lost in the catastrophic falling down of big stones as a consequence of the avalanche.

One idea to control deregulated cellular signalling in the most efficient way is to target molecules that are rather *upstream* of cellular signalling. These upstream causal genes can be also called *master regulators* of the genes and proteins with altered expression. They can be potentially identified by applying graph theoretical approaches to regulatory networks.

Using software and databases such as JASPAR (Portales-Casamar et al., 2010), Allegro (Halperin et al., 2009), Weeder (Pavesi et al., 2004), Pscan (Zambelli et al., 2009) and the commercial network analysis pipelines such as $ExPlain^{TM}$ from BIOBASE (Kel et al., 2008) and $geneXplain^{TM}$, the master regulators are identified in the following fashion. From a set of differentially

(A)

(B)

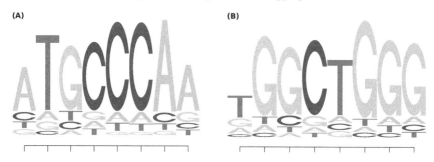

FIGURE 11.1 Identification of overrepresented regulatory motifs. From a set of differentially expressed genes, overrepresented regulatory motifs have been identified using JASPAR (Portales-Casamar et al., 2010) and Weeder (Pavesi et al., 2004). (A) Gene expression has been compared between breast cancer cell lines over-expressing *MYC* and the same cell lines depleted in *MYC* using a small interfering RNA-mediated knock-out (Cappellen et al., 2007). (B) Gene expression has been compared between tumours mutated and not mutated for *TP53* on breast cancer. (Bertheau et al., 2007). Image adapted from Meng et al. (2010).

expressed genes identified with statistical and prioritisation methods (see **Section 6.3** and **Section 11.1.1**) the promoters of these genes are analysed for the presence of overrepresented regulatory motifs (*i.e.* DNA transcription factor binding sites) to identify a potential set of transcription factors able to regulate this particular gene set.

The positions of these transcription factors in the global gene regulatory network are then determined and serve as *anchors* for the master regulator set. The regulatory network is further analysed by tracing the paths going upstream of the anchor points. The master regulator is defined as a node towards which these paths tend to converge after a certain number of upstream steps, *i.e.* those nodes that maximise the number of reachable anchor nodes at certain number of step downstream of a master node. Of course, a careful statistical analysis is performed to ensure that the potential master regulators represent unexpectedly significant path convergence points and the corresponding *p*-values are estimated. More subtleties come from trying to establish preferable routes when going upstream of anchor nodes. This is done by using data on gene expression or introducing the notion of a molecular cascade, *i.e.* a set of signalling pathway steps frequently observed to act together in the same cell and experimental conditions.

The master regulators are usually better candidates to be targeted by drugs than simply the deregulated nodes because they are more probable driver players causally involved in the changes observed in cancer cells.

11.2 Prediction of drug targets from simple network analysis

In this section, we consider the situation where we have a network model of a process importance for cancer, such as the RB pathway described in **Section 4.9.2** or the subnetwork identified by differential analysis of gene expression data discussed in **Section 6.5.2**, and we wish to identify important nodes in this model as potential drug targets. Design of interventions in the network functioning with the desired outcome is a nontrivial problem, especially considering that the network contains complex combinations of feedback regulatory loops. With a little amount of prerequisite information, the simplest methods aim at disrupting the network connectivity by attacking nodes with properties of **hubs*** and **routers***.

Indeed, it can be shown that the proteins with the most essential function in a living cell have tendency to be hubs, *i.e.* to be connected to many other proteins via PPIs and various regulatory mechanisms (Jeong et al., 2001). Hubs are intuitively good targets to damage a network since removing a hub from a network can drastically change its properties such as its connectedness (or degree of connectivity).

In addition, from graph theory we know that the connectedness of a graph (the property of being connected in one component) can be affected in the most pronounced way not only by removing nodes with the highest connectivity (hubs), but also by removing nodes with the highest **centrality*** (router nodes). The notion of network router is less well-known that of the hub (see **Figure 11.2**). Intuitively, centrality is a measure of how far a node is situated from the *centre* of a graph. The graph centre can be roughly defined as a node minimising the sum of distances (for example, path lengths) to all other nodes of the graph. In practice, many measures of node centrality exist, *betweenness* being one of the most used. To compute betweenness, one has to find all shortest paths connecting each node in the graph to all other nodes. According to its definition, betweenness of a node is roughly the number of the shortest paths passing through this node. It is easy to understand that nodes with high betweenness are not necessarily highly connected. One can imagine a graph constructed by connection of two densely connected clusters of nodes with a relatively thin *bridge* between them. The nodes in the bridge will have high betweenness but not the highest connectivity. Removing these nodes from the graph can disrupt it into two disconnected parts preventing communication between two graph clusters.

Following this idea, the simplest approach in predicting drug targets from a network of PPIs potentially involved in **tumour progression*** consists of listing the most significant hubs and routers of this network. There are several subtleties, however. For instance, predicting that hub nodes are drug targets, and at the same time significant hubs in the *global network of molecular inter-*

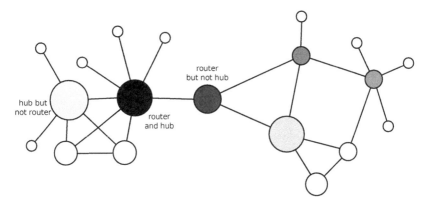

FIGURE 11.2 Network hubs and routers. In this network node sizes are proportional to the connectivity of the node (number of neighbours) and the gray colour reflects betweenness values (white nodes have smallest betweenness and the black nodes have highest betweenness). Hub nodes are not necessarily routers, and routers are not necessarily hubs.

actions is not very promising for cancer therapies: affecting these nodes will most probably disrupt not only the functioning of a network driving tumorigenesis but also the essential functions of normal cells. Hence, this therapy will not discriminate cancer and normal cells and will be most probably too toxic. To take this into account, one can introduce the notion of hubs and routers specific to the cancer network analysed. For example, a *specific hub* would be a node with an unexpectedly large connectivity in the cancer-specific network, given the structure of the genome-scale network of molecular interactions. The easiest way to achieve this is to use relative connectivity and relative centrality as ratios between connectivity in the network to be analysed and the genome-scale global network.

11.3   Drug targets as fragile points in molecular mechanisms

Cancer cells, as well as normal cells, possess certain properties of robustness (see **Chapter 9**). Another way to identify drug targets is to find *fragile points* in cancerous molecular mechanisms Ideally, these fragile points should be specific to cancer cells and not affect normal cells. By definition, a fragile point is a parameter of a molecular network whose change will have the largest effect on a desired network property (for example, production of ATP, activation of apoptosis or arresting cell growth).

Several mathematical approaches for defining the most sensitive parameters of the dynamics of a biological network are reviewed in **Chapter 10** and include: computing parameter sensitivities in metabolic control analysis (see **Section 10.5**), identifying rate limiting *places* in complex reaction networks (see **Section 10.7**), and using Flux Balance Analysis (FBA) to determine enzymes whose activity change can affect the maximum number of vital fluxes in a reaction network (see **Section 10.1**). Let us mention a few examples when network modelling lead to suggesting a novel drug target.

In Faratian et al. (2009), a continuous kinetic model was developed to predict resistance to Receptor Tyrosine Kinase (RTK) inhibitor therapies. The mathematical model included RTK inhibitor antibody binding, HER2/HER3 dimerisation and inhibition, AKT/Mitogen-Activated Protein Kinase (MAPK) crosstalk, and the regulatory properties of *PTEN*. The model was parameterised using quantitative phosphoprotein expression data from cancer cell lines using Reverse-Phase Protein Arrays (RPPA) (see **Chapter 3**). The simulations of the model showed that *PTEN* is a promising drug target, acting as the key determinant of resistance to anti-HER2 therapy. This prediction was further validated in a cohort of 122 breast cancers.

In Sahin et al. (2009), a Boolean model of ERBB signalling coupled with G1/S transition of the cell cycle was constructed. *In silico* analysis of loss-of-function using this model defined potential therapeutic strategies for *de novo* trastuzumab resistant breast cancer. It was shown that combinatorial targeting of ERBB receptors or of key signalling intermediates does not have a potential for treatment of *de novo* trastuzumab resistant cells. At the same time, *MYC* was identified as a novel potential target protein in resistant breast cancer cells.

11.4 Predicting drug target combinations

Both experimental and theoretical studies suggest that the most efficient and, maybe almost always, the only possible way to affect the behaviour of cancer networks is through affecting several fragile points at the same time. In clinics, this approach is known under the name of *combinatorial therapy*. Note that a drug can affect several targets simultaneously, hence, there is a possibility of having one-drug multiple-targets therapy.

Recently, the idea of exploiting the notion of *synthetic lethality* (see **Box 9.6**) in treating cancer gathered a lot of interest. Synthetic lethality is an observation on the experiments of gene knock-outs in model organisms. Knocking-out gene A and gene B separately might not have any effect on vital cellular functions while simultaneous knock-out of genes A and B can be lethal. It is known that cancer genome is modified with respect to a normal genome, hence, some of the genes in a cancer cell can be either lost or mutated with loss-of-function. If, for such a gene with lost function, one can identify another nonessential gene that synthetically lethally interacts with it then this can be a basis for specific cancer therapy (suppressing function of the partner will kill only cancer cells but not normal cells). PARP1 inhibitors to treat BRCA1-mutated breast cancers were suggested following this idea (Helleday, 2011). However, the reality of tumoral cells and their genetically heterogeneous populations (see **Section 2.3**) might be more complex and not limited to synthetically lethal *pairs*. There might be synthetic lethal cocktails containing more than two elements. However, their identification is not amenable to experimental discovery through large-scale screenings due to a large number of possible gene combinations. Therefore, mathematical modelling can help to predict the most promising drug target combinations from the analysis and modelling of gene regulatory networks.

❑ **BOX 11.1: Minimal intervention set**

A minimal intervention set is a combination of knock-outs (deletions of genes or proteins) and knock-ins (overexpression of genes and proteins) inducing a desired signalling network behaviour (Klamt, 2006; Samaga et al., 2010). The desired network behaviour can be fixing some network nodes in particular states (in Boolean modelling), disrupting paths in the network from a set of source to a set of targets nodes (in structural analysis of networks or modelling metabolic cascades), or impeding any feasible steady state flux distributions involving certain reactions (in stationary flux analysis).

The set of interventions is called minimal if there exists no other subset of this set capable of achieving the same goal. Therefore, there might be minimal intervention sets of several sizes (see **Figure 11.3**).

The problem of finding a minimal intervention set is related to finding a minimal cut set and a minimal hitting set.

One of the most utilised concepts in this field is the notion of *minimal intervention set* (see **Box 11.1**) which permits one to identify combinations of drug targets. An example of finding a minimal cut set (a type of minimal intervention set in which only knock-outs are used) for a toy network is provided in **Figure 11.3**. Given a complex network of large size, the task of listing all minimal intervention sets easily becomes computationally intractable. This creates a need for implementation of approximate algorithms aiming at listing only the most promising combinations of drug targets and prioritising them (Vera-Licona et al., 2012).

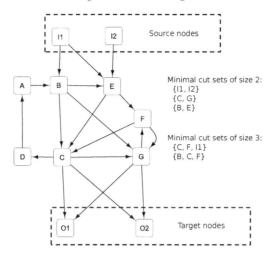

FIGURE 11.3 Finding a minimal cut set to disrupt signalling in a toy network. A toy network contains two source nodes, *i.e.* inputs, and two target nodes, *i.e.* outputs. Here, a minimal cut set aims at disrupting all possible paths from source to target nodes. There exist three minimal cut sets of size 2 and two minimal cut sets of size 3.

✎ **Exercises**

- In **Figure 11.3**, add a connection between node E and node G. How will it affect the minimal cut sets?

- Explain why *PARP1* inhibitors should specifically kill some types of cancer cells but not normal cells.

- Think of in what case a router is a hub at the same time.

⇨ **Key notes of Chapter 11**

- Finding targets using systems biology approaches in cancer biology aims at disrupting functioning of either malignant signalling (leading to stopping proliferation) or survival signalling (leading to death) in cancer cells. To be used in therapy, targets should (1) affect cancer cells to a much larger extent than the normal tissue cells and (2) not induce side effects such as toxicity.

- Methods of gene prioritisation allow one to find the most evident candidates for targets.

- If the structure of the network involved in tumorigenesis is known, one can test *hub* and *router* nodes specific for the network as candidates for targeting the cancer cells.

- Combining gene expression, motif enrichment analysis and network structure allows identifying *master regulator* nodes as potential targets for therapy.

- Dynamical modelling of networks allows finding *fragile points*, using parameter sensitivity analysis. Fragile points can suggest potential therapeutical targets.

Chapter 12

Conclusion

In this book, we have presented the ways that large-scale high-thoughput technologies and computational systems biology have started to revolutionise cancer research and clinical management. We have reviewed in particular how systems biology can help understand the principles of **tumorigenesis**[*] and **tumour progression**[*], and we suggest improvements in cancer treatment through better diagnosis, **prognosis**[*], prediction of response to drugs, identification of new drug targets, and optimisation of therapeutic strategies for killing cancer with drugs. In this last chapter, we briefly discuss alternative approaches for curing cancers, and present a few forthcoming challenges of computational systems biology of cancer.

12.1 Cancer systems biology and medicine: Other paths

The strategy to fight cancer sketched in this book, involving predictive models and rational decision-making to selectively kill cancer cells, is an obvious way to follow for combating the disease. It is however not the only one, and at least two other strategies should be considered. The first one is prevention, that is, taking appropriate measures to avoid the disease; avoiding carcinogens, adopting diet and other life habits that are known to lower cancer risks ... *an ounce of prevention is worth a pound of cure*. The second one is controlling the tumour: preventing metastasis and any major consequences for the patient's health, but accepting not to defeat the cancer. Both strategies are also amenable to systems biology approaches.

12.1.1 Preventing the occurrence of cancer

Prevention programs have shown to be very efficient in decreasing the occurrence of the pathology. This supposes, of course, identifying factors that have strong impact on the risk of cancer. A striking example is the prevention of lung cancer by informing about the risk of tobacco. Adequate surveillance can also decrease drastically the incidence of cancer by early identification of tumours, or pre-cancerous cells. Pap smears, a simple biological test which identifies early lesions in cervix, are used at the population level in many developed countries, and have saved many lives. Surveillance is also adapted for patients who present a genetic risk, like the carriers of *BRCA1* or *BRCA2* mutations who have a relative risk of 10 or more (compared to the general population). Many programs have thus been launched to identify new genes of predisposition, which could explain familial forms, or at least increased risk, of cancer.

The so-called Genome-Wide Association Studies (GWAS) are large-scale studies which proceed by genotyping thousands of affected individuals, and non-affected references, typically using SNP arrays (see **Chapter 3**). Association studies (within populations) identify those mutations that are associated to increased occurrence of disease. The power of the approach resides in the fact that no prior knowledge, and in particular no knowledge of mechanism is required to identify association. It is purely statistical, and blind to the underlying biology which would explain mechanistically the increased risk. Analysis is therefore applied gene-per-gene, or even SNP-per-SNP, which means tens of thousands of tests. Controlling the overall false positive rate supposes to be more stringent for each test, at the cost of reduced statistical power (see **Box 6.2**)

It should be noted that such a gene-per-gene approach does not use any prior knowledge we have about the relationships between genes, and their interplay in biological networks. This traditional SNP-based approach does not use information like gene expression that can also be collected to characterise individuals. Using information like pathway structure or gene expression would open new approaches, that we could regroup and name *systems epidemiology*. In fact, many of the techniques we presented in **Chapter 6** to incorporate prior knowledge including gene sets, networks and heterogeneous data in prognostic and predictive models, can in principle be adapted to systems epidemiology. For example a statistical test could be built to assess association to disease not at the gene level, but at the pathway or network level. The approach would assess if a pathway or a network area (a notion to be precisely defined) is consistently impaired (possibly at different levels) in individuals who developed a tumour. The advantage over the classical gene-per-gene approach would be a gain in power: genes from the same pathway could not reach significance when tested individually, whereas the pathway-level test would be based on more information and could be conclusive. More generally, systems epidemiology can be defined as an approach of molecular epidemiology where assessment of association to cancer occurrence is carried out by integrating in the statistical setting our prior knowledge about the molecular biology of the tumour occurrence. This prior knowledge can be the pathway structure and functioning, some known mechanism of action of a carcinogen (which supposes to know which carcinogens are involved and how the individual has been exposed to it), or a gene expression signature of carcinogenic exposure. The ideas presented here are not new (Lund and Dumeaux, 2008; Thomas et al., 2009), but have not become mainstream yet either, and are still matters of research. Which exact form the modelling will take is still unclear, but we can anticipate that systems biology will soon bring important insights in molecular epidemiological studies of cancer too.

12.1.2 Controlling the tumour

The canonical view about how to treat a pathology is of course considering that the disease should be cured and eradicated from the patient's body. This is all the more true when the disease is embodied in a physical entity, like a microbe, or in the case of cancer, tumour cells. These entities can be physically targeted, and one would expect to do so and kill all of them. Another strategy for treating a disease is to keep it under control, that is, to avoid fatal consequences, or to prevent deterioration of patient health and comfort, while renouncing exterminating all tumoural cells. Accepting to host potential killers might look like a risky strategy, but there is growing evidence that it could in fact be more efficient than trying to eradicate the tumour. This is, of course, in strong opposition to the dominant principle of cancer treatment which has been to give the highest possible dose of the most toxic drug in order to kill as many tumour cells as possible.

In fact, a strategy which controls tumour growth without eradicating the tumour, is already in successful practice for Chronic Myeloid Leukaemia (CML). CML is treated by tyrosine kinase inhibitors (imatinib, and for resistant cases, dasatinib or nilotinib) with excellent patient response, but in most cases some tumour cells persist and the patient will undergo relapse. The same treatment can be given again, several times if needed, and patients have normal life expectancy (Gambacorti-Passerini et al., 2011). This cancer is therefore well controlled: though we do not know how to eradicate all the tumour cells, we prevent the fatal consequences with almost perfect efficiency.

This idea of controlling the disease has also been subject of mathematical modelling, and the rationale and results have been described by Gatenby (2009) (see also references therein) and is exposed below (see also **Section 9.7.1**). Recognising that in most cases we do not know how to eradicate a cancer once it has disseminated, Gatenby proposed to use the principles of evolutionary dynamics of applied ecology and to base therapeutic strategy on tumour evolutionary dynamics: *" . . . efforts to eliminate cancers may actually hasten the emergence of resistance and tumour recurrence, thus reducing a patient's chances of survival."* Some of his examples support the hypothesis that an intensive treatment which kills most cells, opens the road to resistant cells:

- Without treatment the best fit cells, which in general are not the treatment-resistant ones, dominate the tumour composition. Treatment-resistant cells compete with them, but are disadvantaged in the proliferation race, precisely because there is a cost to resistance. They use part of their resources to enable resistance: upregulating a pathway, DNA repair, pumping molecules out, *etc.* They can therefore, at best, maintain a small number of them but, of course, not win the competition with treatment-sensitive cells.

- When the tumour is treated intensively, the treatment-sensitive cells are killed and the treatment-resistant cells can proliferate without competitors for resources.

Gatenby proposed therefore to use adequate treatment level so as not to kill all treatment-sensitive cells and to maintain the tumour volume to a stable and tolerable value. This way the presence of enough treatment-sensitive cells ensures that treatment-resistant cells can not thrive and are maintained in low numbers. Gatenby's mathematical model shows that this would improve patient survival. He also presents an example of such a strategy by treating successfully mice with human ovarian tumour with an adjusted dose of **chemotherapy**[*].

Though issues like long-term toxicity are not solved, this question of modelling tumour evolutionary dynamics for controlling the disease will undoubtedly be an important axis of the systems biology of cancer in the close future.

12.2 Forthcoming challenges

Most subjects we presented in this book are still active fields of research. And they will continue to be major areas of research in cancer systems biology in the future. However, we can anticipate that computational systems biology will also develop in new directions. These directions will be dictated by biological research questions of high clinical relevance, and by progress in investigation technologies.

12.2.1 Tumor heterogeneity

A first direction will concern the tumour heterogeneity. As already exposed, the clonality hypothesis is only a first order approximation. We now know that tumours present internal heterogeneity and a hierarchical organisation (see **Chapter 2** and **Section 9.7.1**). At the centre of this hierarchy are **Cancer Stem Cells (CSC)**[*]. CSC are defined by their ability to self-renew, to father non-CSC progeny and to seed new tumours. CSC also have acquired motility, invasiveness capabilities, and resistance to apoptosis. Therefore, they are thought to be the category of cells which has the highest propensity to metastasise (Chaffer and Weinberg, 2011). They nevertheless only represent a small fraction of the tumour cells. The modelling of this cell hierarchy and its diversity should shed light on its role in the response to a therapeutic treatment, on multidrug resistance, and on the metastatic process. The heterogeneity within the tumour has a direct consequence on personalised medicine. Using **NGS**[*]

and **microarrays***, Gerlinger et al. (2012) observed in renal carcinoma that a single area of the tumour cannot be representative of the landscape of chromosome alterations (*i.e.* mutations, Loss of Heterozygosity (LOH) and ploidy) in the tumour. For example, among all the different **somatic mutations*** found across different areas within the tumour, about two thirds are heterogeneous and are not detectable in every single area. Moreover, a gene expression signature predicting the molecular subtypes can classify the same tumour in *good prognosis* or *poor prognosis* depending on the area of the tumor used for the prediction. Therefore, the statistical methodologies presented in **Chapter 6** will have to consider this heterogeneity issue in order to improve the prediction accuracy. For example, inferring the hierarchical organisation of clonal subpopulations and reconstructing the lineage relation between cancer cells using phylogenetic tree models offer new insight to decipher biomarkers (Navin and Hicks, 2010; Gerlinger et al., 2012). A combination of computational systems biology and high-throughput technology will definitively help to unravel the tumour heterogeneity and to use this information in clinical practice.

12.2.2 Metastasis

The metastasis process itself should also become a major field of study in cancer systems biology soon. How can a cancer cell leave its tumour of origin, proceed to local invasion, intravasate into the blood or lymph stream, migrate safely to a distant tissue, extravasate and settle in a new environment to prosper and form a second tumour? The biology of this phenomenon is still poorly understood, even though it is responsible for most cancer deaths. Basic research and computational models are needed to better understand this process, and propose plans to counter the formation of metastases. These models should include both the description of signalling pathways involved at these different steps, biophysical properties of tissues, cells, and **Extracellular Matrix (ECM)*** which are crucial during invasion and migration, and action of all the tumour microenvironment. Indeed, it is now well established that the tumour microenvironment plays a crucial role both in the development of a tumour, in tumour cell proliferation and in metastasis formation. In particular, the microenvironment is involved in the supply of growth factors, signalling molecules and substrates, and in establishing an inflammatory state which is then involved in **Epithelial-to-Mesenchymal Transition (EMT)***. EMT plays a major role by enabling epithelial cells to generate progeny with metastatic potential (Thiery et al., 2009). Modelling interactions between the tumour and all components of its microenvironment is therefore needed to understand tumour progression (Gatenby and Gillies, 2008; Anderson et al., 2009).

12.2.3   Cancer epigenetics

Another important direction for systems biology in the near future should be the modelling of the role of epigenetics in tumour progression. Recently, our knowledge of epigenetic mechanisms like DNA methylation, various histone post-translational modifications and expression of noncoding RNA (see **Chapter 2**) has made substantial progress (Esteller, 2007, 2008; Rodríguez-Paredes and Esteller, 2011). These mechanisms are involved in many cancers through various modes: DNA hypermethylation silencing tumour suppressor genes; DNA hypomethylation of repetitive sequences, retrotransposons or introns leading to genome instability and aberrant expression of oncogenes; abnormal patterns of histone methylation or acetylation leading to aberrant gene expression. Several initiatives have been launched to establish the epigenomic landscape of normal and tumour cells, for example by the International Cancer Genome Consortium (ICGC), The Cancer Genome Atlas (TCGA), ENCyclopedia Of DNA Elements (ENCODE) or AHEAD consortia (International Cancer Genome Consortium et al., 2010; Consortium, 2004; American Association for Cancer Research Human Epigenome Task Force and European Union, Network of ExcellenceScientific Advisory Board, 2008). Epigenetic markers are promising for cancer detection, diagnosis and prognosis. For example, hypermethylation of *DNMT* is a successful biomarker for the treatment of **glioma*** with temozolomide and radiotherapy (Esteller et al., 2000; Baylin and Jones, 2011). Also, several epigenetic drugs have reached the market, like DNA methyltransferase inhibitors (vidaza and decitabine) for treatment of myelodysplastic syndrome, and histone deacetylase inhibitors (vorinostat and romidepsin) for cutaneous T cell **lymphoma***. The epigenetic mechanisms are key elements of the regulation of gene expression and genome stability. The modelling of the dynamic interplay of the epigenetic actors with genome, transcriptome and proteome is needed to understand their impact on tumour progression and propose therapeutical intervention points and strategies.

12.3   Will cancer systems biology translate into cancer systems medicine?

12.3.1   Is optimism realistic?

After three decades of molecular biology research, our knowledge of the fundamental aspects of tumour progression is becoming accurate and many molecular mechanisms have been elucidated in detail. Also, we possess many sophisticated high-throughput molecular investigation techniques. Almost exhaustive molecular descriptions of tumours at the genome, transcriptome and epigenome levels, are coming with initiatives like TCGA or ICGC. We know how to engineer good genetic mouse models. After ten years, systems biology starts becoming a mature science. Its paradigm has been accepted and computational systems biology approaches are more and more refined and efficient.

This picture of cancer research seems idyllic, but is probably over-optimistic. We have to be careful that enthusiasm for new fields of science is often so strong that claims of future successes might be exaggerated. This could lead rapidly to disappointment, and from disappointment to disinterest, and in a mirror effect, to exaggerated distrust of a discipline. Particular attention has to be given to this question in cancer research, where expectations from the patients and their families are enormous. The past has shown the relative failure of many wars against cancer (for example see Epstein, 1990; Epstein et al., 2002, who analyse the situation in the United States two and three decades after the 1971 National Cancer Act which declared war against cancer). We should therefore acknowledge that there are still many obstacles in front of us in the battle against cancer. Molecular signatures have not delivered all their promises (see **Chapter 6**) and traditional classifications are still used in many cases. Targeted therapies have an unexpected high rate of failures (see Gonzalez-Angulo et al., 2010, and references therein). Robustness of the cancer cells seems much more problematic than anticipated (see **Chapter 9** and **Chapter 10**). A recent example has shown that colon tumours with *BRAF(V600E)* oncogene mutation show no response to *BRAF(V600E)* inhibitors (Prahallad et al., 2012). In contrast, melanomas, with the same mutation do respond. The reason is a negative feedback loop from *BRAF(V600E)* to *EGFR* which is cut by *BRAF(V600E)* inhibition. *EGFR* is then fully activated in colon cancer, and triggers a high level of proliferation. In melanoma where *EGFR* is weakly expressed, the suppression of the negative feedback loop has no consequences and the tumour is sensitive to *BRAF(V600E)* inhibitors. The paradox is therefore that the logical treatment from the knowledge of mutations might worsen the situation. Of course Prahallad et al now propose to combine usage of a *BRAF(V600E)* inhibitor with a *EGFR* inhibitor (a strategy exposed in **Section 11.4**). But it is legitimate to ask whether this second inhibition will cut other feedback loops, thus abrogating the benefit of the inhibitions of *BRAF(V600E)* and *EGFR*. In general, what is the number of intervention points needed for killing the tumours? Which combination will be efficient at an acceptable toxicity level for normal cells? We do not have the answer yet to these questions. Another example of failure of targeted therapy concerns the principle of synthetic lethality (see **Chapter 9** and **Chapter 10**). An example has already been given with genes *PARP* on one hand, and *BRCA1* or *BRCA2* on the other hand. These two combinations show synthetic lethality, but in clinical practice, usage of *PARP* inhibitors to cure *BRCA*-mutated breast tumours has met with failure, because of secondary mutations in *BRCA1* or *BRCA2* which restored the gene open reading frame (Sakai et al., 2008; Edwards et al., 2008; Lord and Ashworth, 2012). Will we be able in such situations to understand rapidly the cause of failure, and to propose another strategy to the patient? Can systems biology follow the pace of mutations in tumour genomes? Can we imagine treatments that will not drive the tumour into untractable patterns of resistance? It is premature to know the answer.

These failures mean that important progresses in computational systems biology of cancer are still needed. We need to be able to model not one pathway, not two, but probably many more in a global approach (see the discussion of **Chapter 8**). We need to identify and describe in detail compensatory mechanisms, feedback loops and crosstalk between pathways. Only then will we understand why a treatment may cure one patient while failing for another.

These progresses will build upon our increased understanding of the fundamental molecular mechanisms that govern the normal and tumour cell behaviour. In other words, omics and systems biology approaches should also let space to classical molecular oncology research if we do not want to accumulate data that we cannot interpret. Progress will also build on the availability of new technologies, like improved exploration of proteome, metabolome and interactome. Investigation at single-cell level (for example single-cell sequencing) and at population level (measuring distribution of variables like protein expression, instead of average expression) will also offer new possibilities for systems biology modelling. Thus the road ahead is open, but nobody knows where it will lead us.

12.3.2   Cost of personalised cancer medicine

Systems biology of cancer aims at designing tailored treatment for each patient. One can also ask if the cost of this personalised medicine is affordable. Progresses based on new technologies are often expensive. Personalised medicine should follow the rule. To start with it will be based on the availability of targeted therapeutic molecules for all intervention points. This means developing a few hundreds of new drugs.

On one hand pharmaceutical companies today are focusing on blockbuster drugs, that are applicable to most people, which means the largest possible market. The model is the average patient, which is at the opposite of personalised medicine. In personalised medicine patients will be enrolled in small clinical trials, tailored to their genetic context. This means higher costs per patient. On the other hand the cost of developing inhibitors for all kinases is only a few days of the planet growth product (around 500 billion dollars). This seems quite reasonable for a crucial health issue such as cancer.

One has also to keep in mind that with systems biology and personalised medicine comes preventative medicine (see **Chapter 1**). This prevention should bring important gains in terms of saving costly treatments, and of course, patient suffering. Making personalised medicine a reality is therefore primarily a matter of political will and organisation.

12.4 Holy Grail of systems biology

Computational systems biology is reviving a long-standing dream of modelling a whole organism *in silico*, of formalising life. For a systemic disease like cancer, the idea is to achieve the construction of a model of the human physiology at the level of the whole body. The novelty lies in the fact that the model could now be fed with an almost exhaustive description of molecular levels, and could be personalised for a given individual, based on his genetics, the specificities of his molecular networks, his history (in particular his immune system), his environment (lifestyle, diet habits, exposure to carcinogens, *etc.*). This model would be used to simulate the action of potential drugs, thus assessing efficiency and anticipating possible negative effects for a given patient. This approach could also save a lot of time and effort in the development of drugs, in clinical trials (*e.g.* by selecting *in silico* the patient to include in a trial or to exclude) and in the design of innovative, multidrug therapeutical strategies. This book has presented the first steps toward this goal. Today, systems biology of cancer is still mainly a research activity, but the first applications in clinical practice show that within a few years it may well become mainstream medicine.

Appendices

A.1 Basic principles of molecular biology of the cell

To understand the mechanism of tumour progression, it is necessary to understand how normal cells work and how they are integrated at the level of the whole organism. A cell is the smallest unit of an organism which is classified as living, and it is sometimes called the building block of life (Alberts et al., 2002, see **Figure A.1**). A living organism can be seen as an ecosystem whose members are cells, reproducing by cell division, and organised into collaborative assemblies or tissues. This ecosystem is very particular since there is no competition between the different cell populations in a healthy organism: each cellular type completes its specialised function which ensures that the organism can live and reproduce. To coordinate their behaviour, the cells send, receive and interpret an elaborate set of signals which serve as social controls, telling each of them how to act. As a result, each cell behaves in a socially responsible manner, resting, dividing, differentiating[1] or dying as needed for the good of the organism and the maintenance of its integrity. In cancer cells, this harmony is broken: the collaboration between cells disappears and a competition and selection between cancer cells appear which can lead to the death of the organism (see **Chapter 2**). To complete its specialised function, the cell follows a specific program which is described in the next section.

A.1.1 Central dogma of molecular biology

The information flow which allows the biological program to be completed inside the cell has been formalised by Crick (1970): *"The central dogma of molecular biology deals with the detailed residue-by-residue transfer of sequential information. It states that such information cannot be transferred from protein to either protein or nucleic acid."* The flow of biological information is presented in **Figure A.2** and the general principles can be summarised as follows:

---

[1]Cellular differentiation is the process by which a less specialised cell becomes a more specialised cell type.

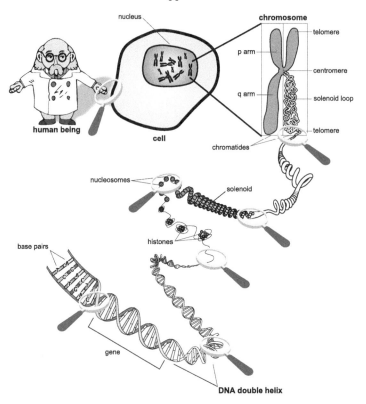

FIGURE A.1 Hierarchical representation of a multi-cellular living organism. A living organism consists of building blocks of life called cells. In a cell, there are chromosomes packing the DNA into a solenoid conformation called chromatin (p and q define the short and long chromosome arms respectively). In the chromatin, DNA is wrapped around nucleosomes. The DNA is the molecule which carries the genetic information. A gene is a DNA segment which encodes for a specific cellular function. A gene is a sequence of bases A, T, C and G.

- **Molecular partners**: *DNA* stores the information in a linear way[2] and can be split into segments or genes which encode for a specific function of the cell. *RNA* can be viewed as the template which allows the synthesis of the *protein*. The protein is the effector in the cell, performing the function encoded by the gene. In cells, the DNA is packed into entities called chromosomes (see **Figure A.1**).

- **Flows**: the step which converts DNA into RNA is called *transcription* and the step which converts RNA into protein is called *translation*. DNA can also be duplicated during the *replication*. This process occurs dur-

---

[2]The sequence of bases A, T C and G.

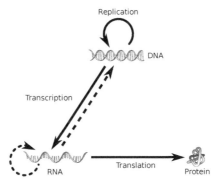

FIGURE A.2 Central dogma of molecular biology. The central dogma of molecular biology holds that information flows from DNA to RNA to protein. *Solid arrows* indicate information flows which occur in all cells, through DNA replication, transcription of DNA into RNA, and translation of RNA into protein. *Dotted arrows* indicate flows which are seen occasionally, through reverse transcription and replication of RNA. Crucially, information cannot flow from protein back into nucleic acid sequence. Image adapted from Crick (1970).

ing the cell cycle in which a parent cell reproduces into two daughter cells. This process allows the conservation of the program information in daughter cells.

Towards a new paradigm: Expanding the *central dogma*. The central dogma of biology states that genetic information normally flows from DNA to RNA to protein. As a consequence, it has been accepted that genes generally encode for proteins, and that proteins fulfil the functions, in all cells, from microbes to mammals. However, the fact that genes encode for proteins may not be the case in complex organisms. Indeed, recent evidence suggests that the majority of the genome of mammals and of other complex organisms is in fact transcribed into RNA which does not encode a protein: such RNA is termed noncoding RNA (ncRNA) but this does not mean that it has no function (Mattick, 2003; Mattick and Makunin, 2006). To distinguish RNA which does not encode protein from RNA which encodes protein, the latter is termed messenger RNA (mRNA). ncRNAs can be divided into two classes: the infrastructural and the small regulatory ncRNAs. Among the infrastructural ncRNAs, there are transfer RNAs, ribosomal RNAs and small nuclear RNAs. They can be involved in regulatory processes. Small regulatory ncRNAs interact with mRNA via RNA interference mechanisms[3] (Mello and Conte Jr, 2004) and inhibit gene expression at the stage of translation. Among the different types of small regulatory ncRNAs, microRNAs (miRNA) are naturally produced in human cells, and small interfering RNAs (siRNA) have also been

---

[3]In 2006, Andrew Z. Fire and Craig C. Mello received the Nobel Prize in medicine for the discovery of the RNA interference process.

identified to be produced endogenously in mouse oocytes (Watanabe et al., 2008). ncRNAs represent a new and additional level of gene regulation. In **Section A.1.2**, we will give more details about the role of miRNAs in gene expression regulation. ncRNAs can be viewed as noncoding genes.

The genetic information carried by a gene is called the *genotype* and when the function encoded by the gene is effective within the cell, it is called the *phenotype*. The genotype corresponds to the genetic description of the cell while the phenotype corresponds to the expression of the function encoded by the gene. It is possible that this function is never expressed if the cell does not need it. Then, how does the cell decide to express or not the function encoded by the gene? This is determined by the interaction between the conditions in which the cell lives and the genetic properties of the cell: the cell has many sensors sensitive to *environmental stimuli*, which are either external (temperature, pH, nutrients, light, pathogen molecules, signals sent from other cells, *etc.*) or internal (DNA damage, length of the telomere[4], osmotic pressure, *etc.*). Thus, the expression of the phenotype is determined by the simple equation *genotype + environmental stimuli → phenotype*. Environmental stimuli depend not only on the stimuli at a given time but also on the stimuli the cell has been receiving throughout its life. As a consequence, the cell has specific characteristics due to its life history[5]. Therefore, the environmental stimuli received by the cell will trigger or not the expression of the phenotype: these stimuli play a key role in the regulation of the information flow. Indeed, although the dogma appears to be a simple sequential information flow, the reality is much more complex as many interactions between the cell and its environment impact the control of replication, transcription and translation. In a cancer cell, the regulation mechanisms are altered and the cell can no longer complete its original program (see **Chapter 2**). We will describe in the next section the mechanisms involved in the regulation of the information flow in a normal cell.

A.1.2   Gene regulation and signal transduction mechanisms

In normal cells, the gene regulation and signal transduction held in the central dogma of molecular biology involve different mechanisms mediated by a large set of molecular entities.

Transcription factors. Groups of genes must be coordinately expressed while other genes must be repressed so that the cells display complex and tissue-specific phenotypes. Such coordination of expression is insured by proteins called *transcription factors*. They regulate the transcription of genes by binding to specific sequences of DNA using DNA binding domains (see **Figure A.3**). They perform this function alone, or in a complex with other proteins. Transcription factors act by increasing (as an activator), or preventing

---

[4]The telomere is the extremity of the chromosome. Its length is an indicator of the number of divisions a cell has undergone.

[5]For example, the cell differentiation signals have led to a specific function.

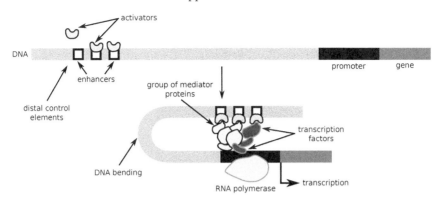

FIGURE A.3 Role of transcription factor in gene expression regulation.
The transcription factor binds to specific DNA sequences of the promoter located in the upstream region of the gene. This allows the formation of a transcription initiation complex including the RNA polymerase which starts the transcription of the gene into RNA.

(as a repressor) the presence of RNA polymerase, a protein which transcribes genetic information into RNA. One transcription factor might have several target genes. Active research is currently going on based on the sequence analysis of promoters in order to discover new target genes for each transcription factor (Tompa et al., 2005). Among transcription factors, TP53, also known as *the guardian of the genome*, plays a key role in preserving the integrity of the genome during the cell cycle. TP53 ensures that genetic material of the cell is correctly transmitted into daughter cells.

Epigenetic regulations. Classical genetics alone cannot explain the diversity of phenotypes within a population. Nor does classical genetics explain how, despite their identical DNA sequences, monozygotic twins or cloned animals can have different phenotypes and different susceptibilities to a disease. The concept of epigenetics offers a partial explanation of these phenomena. It was first introduced by Conrad Hal Waddington in the 1940s to name *"the branch of biology which studies the causal interactions between genes and their products, which bring the phenotype into being"* (Jablonka and Lamb, 2002; Speybroeck, 2002). Epigenetics was later defined as heritable changes in gene expression which are not due to any alteration in the DNA sequence (Esteller, 2008). Epigenetics refers to features such as chromatin and DNA modifications which are stable over series of cell cycles but do not involve changes in the underlying DNA sequence of the organism. These modifications play an important role in gene silencing at the level of transcription. The main epigenetic modifications are the following (see **Figure 2.7**):

- **DNA methylation** is a common epigenetic mechanism of gene silencing. Methylation is a chemical modification of the DNA which can be either inherited, created or modified in response to environmental stimuli without changing the DNA sequence. DNA methylation occurs in cytosines which precede guanines in dinucleotide called CpGs. CpG sites are not randomly distributed in the genome but are located in CpG-rich regions known as CpG islands which span the 5′ end[6] of the regulatory region of many genes. These islands are usually not methylated in normal cells (see **Figure 2.7C1**). DNA hypermethylation is required in particular cases such as **genomic imprinting*** and the X-chromosome inactivation in females[7]. DNA hypermethylation inside repeated sequences could also have a role in the protection of chromosomal integrity, by preventing chromosomal instability and **translocations*** (see **Figure 2.7A1**).

- **Histone modifications** is another common epigenetic mechanism. Transcription of DNA is dictated by the structure of the chromatin. In general, the density of its packing is indicative of the frequency of transcription (see **Figure 2.7B1** and **Figure 2.7D1**). Octameric protein complexes called histones are responsible for chromatin packing, and these complexes can be temporarily or more permanently modified by processes such as methylation and acetylation. These modifications lead to a high degree of packing which prevents genes from being accessible by the transcriptional machinery. Therefore, the modifications act as a gene silencing process. The emerging model is that specific combinations of histone modifications determine the overall expression status of a region of chromatin, a theory known as the *histone code* hypothesis (Turner, 2002).

Post-transcriptional regulations. These regulations occur at the RNA level once the DNA has been transcribed. They are the following:

- **Alternative splicing** is the mechanism in which the exons of the primary gene transcript, the pre-RNA, are separated and reconnected to produce alternative RNA rearrangements. These linear combinations then undergo the process of translation, resulting in isoform proteins. Alternative splicing increases mRNA and protein diversity by allowing generation of multiple RNA products from a single gene (see **Figure A.4**). For a given gene, only some splicing variants exist and not all exon combinations are possible. This is another plausible mechanism for the paradoxical inconsistency between the number of genes transcribed and the diversity of phenotypes. The different ways a gene can be spliced are controlled by the *splicing code* (Barash et al., 2010).

- **Regulation at the mRNA level** involves miRNAs which are about 21-nucleotide-long single-stranded RNA molecules. They are encoded by

---

[6]A DNA sequence is oriented from 5′ end to 3′ end.
[7]In mammalian females, one X chromosome is inactivated.

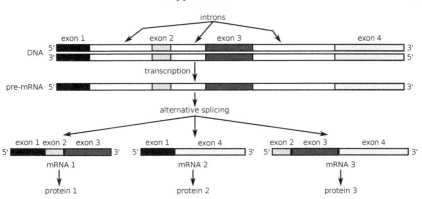

FIGURE A.4 Alternative splicing. A gene consists of several exons and introns. The DNA is transcribed in a precursor mRNA (pre-mRNA) which can be spliced in different possible mRNAs. Only the exons are kept and further translated in different variant proteins. Some of the splicing patterns are specific for certain types of cells.

genes which are transcribed from DNA but not translated into protein. miRNAs are processed from precursor molecules which fold into hairpin structures containing imperfectly base-paired stems. The precursor is processed by enzymes into a mature miRNA which is a single-strand RNA molecule (see **Figure A.5**). Functional studies indicate that miRNAs participate in the regulation of almost every cellular process: in a human a thousand miRNAs are predicted which would regulate about 50% of all protein-coding genes (Filipowicz et al., 2008; Krol et al., 2010). miRNAs control gene expression post-transcriptionally by regulating RNA translation or stability in the cytoplasm: they act similarly to siRNAs operating in RNA interference binding imperfectly to its RNA sequence target. The most stringent requirement is a contiguous and perfect base pairing of the miRNA nucleotides 2–8, representing the *seed* region, which nucleates the interaction with the RNA. miRNAs are generally viewed as negative regulators of gene expression but they could have a much more complex role in gene expression regulation such as gene activation in some cases (Breving and Esquela-Kerscher, 2010).

Signal transduction. As discussed in **Section A.1**, once the DNA is transcribed into RNA, and the RNA is translated into protein, then the protein can play its biological function in the cell. However, some proteins in the cell are present in an inactive state. Post-Translational Modifications (PTM) are needed to allow the protein to perform its function. Why is this mechanism necessary? In the cell, some proteins are present just in case they are needed to ensure a quick and efficient response to environmental stimuli[8]. Indeed, the

---

[8]We remind the reader that environmental stimuli are either external (temperature, pH,

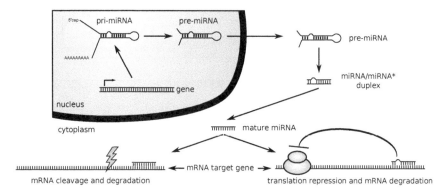

FIGURE A.5 Role of miRNA in a normal cell. In normal tissues, the miRNA is first transcribed into pri-miRNA, then processed into pre-miRNA which is in turn matured in a single strand miRNA. When it binds to its complementary sequences on the target RNA it causes the repression of target-gene expression through protein translation repression (when pairing is imperfect) or altered RNA stability (when pairing is perfect). Image and legend adapted from Esquela-Kerscher and Slack (2006); Breving and Esquela-Kerscher (2010).

transcription and translation machineries take time to complete (on the order of hours or much longer). To ensure fast responses, proteins are present in the cell in an inactive form and can be recruited immediately. Therefore, PTMs (mainly phosphorylations[9]) allow the protein to go from an inactive state to an active state in a process lasting seconds to minutes. Signal transduction is a process that performs a cascade of protein phosphorylations in response to an environmental stimulus which implies the protein is active within the cell. The cascade also allows an amplification of the signal so that a relatively small stimulus elicits a large response: once activated, a protein can activate many other proteins involved in the next step so that the signal grows exponentially (see **Figure A.6**). In this process, *kinase proteins* play a major role because they catalyse the phosphorylation reaction. In contrast, *phosphatases* reverse the reaction and dephosphorylate proteins.

To illustrate the signal transduction mechanism, let us take for example TP53 which is a transcription factor. TP53 only plays a role when the cell has been exposed to a stress such as DNA damage. In unstressed cells, once TP53 has been produced, it binds to another protein called MDM2 which inactivates TP53: when associated with MDM2, TP53 cannot bind to DNA and therefore cannot bind to its target genes. The protein complexes TP53-MDM2

nutrients, light, pathogen molecules, signals send from other cells, *etc.*) or internal (DNA damage, length of the telomere, osmotic pressure, *etc.*). These stimuli are defined by the conditions in which the cell lives.

[9]Phosphorylation is the addition of a phosphate group to a protein molecule or a small molecule.

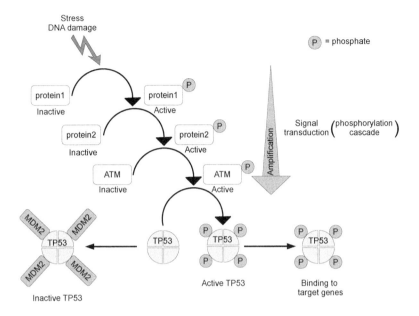

FIGURE A.6 Signal transduction cascade. In response to stress (such as DNA damage) a first phosphorylation reaction activates the function of a first protein which can catalyse the phosphorylation of a second protein and so on. Finally, the phosphorylated ATM protein activates TP53 by a phosphorylation reaction too. Once phosphorylated, TP53 binds to its target genes and initiates the transcription. If not phosphorylated, TP53 binds to MDM2 which prevents its transcription factor activity. Image adapted from Weinberg, 2007, Chap. 9 and Nakamura, 1998.

are exported into the cytoplasm where they are degraded by proteasomes. In some circumstances (especially when cells are suffering certain types of stress or damage), TP53 protein molecules must be protected from MDM2 so that they can accumulate to functionally significant levels in the cell. This protection is achieved by phosphorylation of TP53 by the protein kinases ATM or ATR: they give the signal to TP53 to play its transcription factor activity. The ATM protein is also activated by a phosphorylation reaction as a result of a phosphorylation cascade initiated by stress or DNA damage (see **Figure A.6**).

Different signal transduction modules are involved in response to specific stimuli and are related to specific functions of the cell. These different modules are named *signalling pathways* and complex interactions between them exist as illustrated in **Figure 2.8** with a simplified view of the cell signalling mechanism.

A.1.3 Life of a normal cell

To conclude, there are different stages of the life of a normal cell:

1. The cell performs its specialised function.

2. If needed, it reproduces during the cell cycle.

3. It dies after a limited number of cell cycles, a phenomenon called *senescence*.

These different steps are defined inside a program in which the sequential information flow has been formalised in the central dogma of molecular biology. As we have seen, the execution of the specific program of the cell involves complex regulation mechanisms in response to environmental stimuli: the control of gene expression by the transcription factors and alternative splicing, the epigenetic mechanisms, the regulatory function of ncRNAs and the signal transduction are key processes in the normal behaviour of the cell. In a cell, its specific program also includes permanent monitoring systems to check its ability to always behave in a responsible manner. If this is not the case, the cell must disappear and die in a process called *apoptosis*. Among the monitoring systems we can mention the *cell cycle checkpoints*. Indeed, it is important that after cell division, the daughter cells will be the exact copy of the parent cell in order to complete the same function, otherwise the cell must enter the apoptosis process. The cancer cells which derive from normal cells are not able to perform the original specific program due to a sequential accumulation of events which have disturbed the monitoring system and regulatory mechanisms (see **Chapter 2**). Note that the present section and part of **Chapter 2** have been adapted from Hupé (2008).

A.2 Tools, software and databases

All the resources mentioned in the chapters are listed here.

TABLE A.1 Tools and software.

Tools and software
Adjuvant! Online Estimate the risk of cancer related mortality and relapse, and the benefits of adjuvant treatments http://www.adjuvantonline.com
Allegro Discovery of gene regulatory motifs from high-throughput data http://acgt.cs.tau.ac.il/allegro

Continued on next page

TABLE A.1 – continued from previous page

Tools and software
BiNoM
Biological network manipulation and analysis
http://bioinfo.curie.fr/projects/binom
Bioconductor
Tools for the analysis and comprehension of high-throughput genomic data with R
http://www.bioconductor.org
BioBase
Systems biology databases and software
http://www.biobase-international.com
BioMart
Data integration system
http://www.biomart.org
BioUML
An open integrative platform for systems biology
http://www.biouml.org
CellDesigner
Editor of pathway diagrams
http://www.celldesigner.org
CellSys
Software for growth and organisation processes modelling in multi-cellular systems
http://ms.izbi.uni-leipzig.de/software/cellsys
Cytoscape
Network analysis and visualisation
http://www.cytoscape.org
Galaxy
Workflow manager
http://usegalaxy.org
GeneXplain
Integrative systems biology platform
http://www.genexplain.com
GINsim
Software for creating discrete dynamical models
http://gin.univ-mrs.fr
KNIME
Workflow manager
http://www.knime.org
R
Statistical programming language
http://www.r-project.org
Taverna
Workflow manager
http://www.taverna.org.uk
VANTED
Network analysis and visualisation
http://http://vanted.ipk-gatersleben.de
XPPAUT
Tool for simulating, animating, and analysing dynamical systems
http://www.math.pitt.edu/ bard/xpp/xpp.html

TABLE A.2 Databases.

Database
ArrayExpress Database of public high-throughput experiments http://www.ebi.ac.uk/arrayexpress
BIOBASE Systems biology databases http://www.biobase-international.com
BioCarta Pathway database http://www.biocarta.com
BioModels Lists of published models http://biomodels.net
BioPAX Pathway exchange language for Biological pathway data http://www.biopax.org
BioPortal Ontology database http://bioportal.bioontology.org
BRENDA Enzyme information database http://www.brenda-enzymes.org
Cancer Cell Map Database of signalling pathway related to cancer http://cancer.cellmap.org
CleanEx Database of public gene expression experiments http://cleanex.vital-it.ch/
DGV Database of Genomic Variants http://projects.tcag.ca/variation
DIP Database of protein interactions http://dip.doe-mbi.ucla.edu
DDBK Database of nucleotide sequences http://www.ddbj.nig.ac.jp
EMBL Database of nucleotide sequences http://www.ebi.ac.uk/embl
Ensembl Genome Browser Reference genome for different species http://www.ensembl.org
FGED Standard for high-throughput experiments http://www.mged.org
Gene Ontology Database of gene annotations http://www.geneontology.org
GEO Database of public high-throughput experiments http://www.ncbi.nlm.nih.gov/geo

Continued on next page

TABLE A.2 – continued from previous page

Database
Ingenuity®
Database of molecular interactions
http://www.ingenuity.com
HPRD
Database of protein annotations and interactions
http://www.hprd.org
ICGC
Portal for cancer data
http://www.icgc.org
IntAct
Database of protein interactions
http://www.ebi.ac.uk/intact
JASPAR
Database of transcription factor binding sites
http://jaspar.genereg.net/
JWS
Database of curated models
http://jjj.biochem.sun.ac.za
KEGG
Pathway database
http://www.genome.jp/kegg
MIPS
Database of protein interactions
http://mips.helmholtz-muenchen.de/proj/ppi
miRBase
Database of miRNA
http://www.mirbase.org
MSigDB
Database of molecular signatures
http://www.broadinstitute.org/gsea/msigdb
NCBI Gene
Database of gene annotations
http://www.ncbi.nlm.nih.gov/gene
NCBI Nucleotide
Database of nucleotide sequences
http://www.ncbi.nlm.nih.gov/nucleotide
NCBI OMIM
Database of gene annotations
http://www.ncbi.nlm.nih.gov/omim
OBO
Ontology database
http://obofoundry.org
Pathguide
Database of pathway related resources
http://wwwwpathguide.org
Reactome
Public pathway database
http://www.reactome.com
ResNet®
Database of protein interaction data for Human, Rat, and Mouse
http://www.ariadnegenomics.com

Continued on next page

TABLE A.2 – continued from previous page

Database
SABIO-RK
Database of biochemical reactions and kinetic equations
http://sabio.villa-bosch.de
TCGA
Portal for cancer data
http://cancergenome.nih.gov
TRANSFAC®
Database of transcription factors
http://www.biobase-international.com
TRANSPATH®
Database of molecular interactions
http://www.biobase-international.com
UCSC Genome Browser
Reference genome for different species
http://genome.ucsc.edu
UniprotKB
Catalog of information on proteins
http://www.uniprot.org
WikiPathways
Database of pathways designed for community design
http://wikipathways.org

TABLE A.3 Genes.

Symbol	Synonym	Name	Location
ABCB1	MDR1	ATP-binding cassette, sub-family B (MDR/TAP), member 1	7q21.12
ABCC2	MRP2	ATP-binding cassette, sub-family C (CFTR/MRP), member 2	10q24
ABL1	ABL	c-abl oncogene 1, non-receptor tyrosine kinase	9q34.1
APC		adenomatous polyposis coli	5q21-q22
ATM		ataxia telangiectasia mutated	11q22-q23
BAX		BCL2-associated X protein	19q13.3-q13.4
BCL2		B-cell CLL/lymphoma 2	18q21.3
BCR		breakpoint cluster region	22q11
BRAF		v-raf murine sarcoma viral oncogene homolog B1	7q34
BRCA1		breast cancer 1, early onset	17q21-q24
BRCA2		breast cancer 2, early onset	13q12-q13
CASP3		caspase 3, apoptosis-related cysteine peptidase	4q34
CASP8		caspase 8, apoptosis-related cysteine peptidase	2q33-q34
CCNA2	CCNA	cyclin A2	4q25-q31
CCNB1	CCNB	cyclin B1	5q12
CCNB2		cyclin B2	15q21.3
CCND1		cyclin D1	11q13
CCNE1	CCNE	cyclin E1	19q12
CCNE2		cyclin E2	8q22.1
CDC25B		cell division cycle 25 homolog B (S. pombe)	20p13
CDC25C	CDC25	cell division cycle 25 homolog C (S. pombe)	5q31
CDK2		cyclin-dependent kinase 2	12q13
CDK4		cyclin-dependent kinase 4	12q13
CDK6		cyclin-dependent kinase 6	7q21-q22

Continued on next page

TABLE A.3 – continued from previous page

Symbol	Synonym	Name	Location
CDKN1A	p21CIP1	cyclin-dependent kinase inhibitor 1A (p21, Cip1)	6p21.1
CDKN1B	p27KIP1	cyclin-dependent kinase inhibitor 1B (p27, Kip1)	12p13.1-p12
CDKN2A	p16INK4a	cyclin-dependent kinase inhibitor 2A (melanoma, p16, inhibits CDK4)	9p21
CDKN2B	p15INK4b	cyclin-dependent kinase inhibitor 2B (p15, inhibits CDK4)	9p21
E2F1 [10]		E2F transcription factor 1	20q11
E2F3		E2F transcription factor 3	6p22
E2F4		E2F transcription factor 4, p107/p130-binding	16q21-q22
E2F5		E2F transcription factor 5, p130-binding	8q21.2
E2F8		E2F transcription factor 8	11p15
EGFR	HER1	epidermal growth factor receptor	7p12
ERBB2	HER2	v-erb-b2 erythroblastic leukaemia viral oncogene homolog 2, neuro/glioblastoma derived oncogene homolog (avian)	17q11.2-q12
ERBB3	HER3	v-erb-b2 erythroblastic leukaemia viral oncogene homolog 3 (avian)	12q13
ERBB4	HER4	v-erb-a erythroblastic leukaemia viral oncogene homolog 4 (avian)	2q33.3-q34
EWSR1	EWS	Ewing sarcoma breakpoint region 1	22q12.2
FGFR3		fibroblast growth factor receptor 3	4p16.3
FLI1		Friend leukaemia virus integration 1	11q24.1-q24.3
FLIP	cFLAR	CASP8 and FADD-like apoptosis regulator	2q33-q34
GRB2		growth factor receptor-bound protein 2	17q24-q25
HRAS [11]		v-Ha-ras Harvey rat sarcoma viral oncogene homolog	11p15.5
IGF2		insulin-like growth factor 2 (somatomedin A)	11p15.5
IL1B		interleukin 1, beta	2q14
IL6		interleukin 6 (interferon, beta 2)	7p21-p15
KRAS		v-Ki-ras2 Kirsten rat sarcoma viral oncogene homolog	12p12.1
MAPK1	ERK2	mitogen-activated protein kinase 1	22q11.2
MAPK3	ERK1	mitogen-activated protein kinase 3	16p11.2
MAPK4	ERK4	mitogen-activated protein kinase 4	18q21.2
MAPK6	ERK3	mitogen-activated protein kinase 6	15q21
MAPK7	ERK5	mitogen-activated protein kinase 7	17p11.2
MAPK8	JNK1	mitogen-activated protein kinase 8	10q11
MAPK14	p38	mitogen-activated protein kinase 14	6p21.3-p21.2
MAPK15	ERK7, ERK8	mitogen-activated protein kinase 15	8q24.3
MDM2		Mdm2 p53 binding protein homolog (mouse)	12q13-q14
MLH1		mutL homolog 1, colon cancer, nonpolyposis type 2 (E. coli)	3p22.3
MLL		myeloid/lymphoid or mixed-lineage leukaemia (trithorax homolog, Drosophila)	11q23
MYC		v-myc myelocytomatosis viral oncogene homolog (avian)	8q24
MYCN		v-myc myelocytomatosis viral related oncogene, neuroblastoma derived (avian)	2p24.3
NFKB1	NFkB	nuclear factor of kappa light polypeptide gene enhancer in B-cells 1	4q24
NFKB2		nuclear factor of kappa light polypeptide gene enhancer in B-cells 2 (p49/p100)	10q24

Continued on next page

---

[10]E2F1 and E2F3 are also referred to as E2F in this book

[11]In the book, RAS may refer to HRAS, KRAS, or NRAS

TABLE A.3 – continued from previous page

Symbol	Synonym	Name	Location
PARP1		poly (ADP-ribose) polymerase 1	1q41-q42
PTEN		phosphatase and tensin homolog	10q23
RAD50		RAD50 homolog (S. cerevisiae)	5q23-q31
RAD51		RAD51 homolog (S. cerevisiae)	15q15.1
RASA1		RAS p21 protein activator (GTPase activating protein) 1	5q13
RB1	RB	retinoblastoma 1	13q14.2
REL		v-rel reticuloendotheliosis viral oncogene homolog (avian)	2p13-p12
RELA		v-rel reticuloendotheliosis viral oncogene homolog A (avian)	11q13
RELB		v-rel reticuloendotheliosis viral oncogene homolog B	19q13.32
RPS4X		ribosomal protein S4, X-linked	Xq13.1
RPS6		ribosomal protein S6	9p21
SRC		v-src sarcoma (Schmidt-Ruppin A-2) viral oncogene homolog (avian)	20q12-q13
STAT3		signal transducer and activator of transcription 3 (acute-phase response factor)	17q21
TBC1D9	MDR1	TBC1 domain family, member 9 (with GRAM domain)	4q31.1
TERF2	TRF2	telomeric repeat binding factor 2	16q22.1
TGFB1	TGFbeta	transforming growth factor, beta 1	19q13.1
TNF		tumor necrosis factor	6p21.3
TP53		tumor protein p53	17p13.1
WEE1		WEE1 homolog (S. pombe)	11p15.3-p15.1

Glossary

^{13}C-based metabolic flux data

> is an experimental quantification of the integrated responses of metabolic networks, based on labelling molecules with stable ^{13}C isotopes subsequent gas chromatography mass spectrometric detection of patterns in protein-bound amino acids.

adjuvant therapy

> is a therapy applied after an initial treatment in order to suppress secondary tumour formation and to reduce the risk of future **relapse***.

amplification

> of a gene is a cellular process resulting in multiple copies of a particular gene to amplify the phenotype that the gene confers to the cell.

antagonist drug

> is a drug that blocks binding to a receptor, and therefore receptor activation.

Bacterial Artificial Chromosome

> also referred as BAC, is a plasmid vector in which a DNA sequence has been inserted (from 100 up to 300 kilobases).

biomarker

> is a biological property which indicates the biological state of a cell or a living being (*e.g.* normal state versus pathologic state). A biomarker can be any physical parameter (body temperature, blood pressure, osmotic pressure, *etc.*) or any biochemical substance (expression of a gene, presence of a protein, production of an hormone, *etc.*).

blast

> is any type of immature blood cells.

Cancer Stem Cell

> are defined by their ability to self-renew, to father non-CSC tumoral progeny and to seed new tumours. CSCs also have acquired motility, invasiveness capabilities, and resistance to apoptosis; they are thought to be the category of cells which has the highest propensity to metastasise.

carcinoma

> is a cancer which arises from epithelial cells.

centrality

> is a measure for a node in a graph of how close the node is to the *centre* of the graph. The most popular measure of centrality is the number of shortest paths (from all nodes to all nodes) passing through a node.

chemotherapy

refers to the treatment of a disease by chemical substances, in particular cytotoxic agents for cancers.

chromatin

is the material of which the chromosomes of eukaryotes are composed. It consists of proteins, DNA and RNA.

clinicopathological

refers to both the clinical signs and symptoms of the patient directly observable by the physician and the results of laboratory examination.

comorbidity

is the simultaneous presence of two or more diseases or conditions in a patient.

distributed robustness

is a type of robustness when stable system performance cannot be attributed to any particular element of a system but rather to a general pattern of how the system elements are connected to each other.

emergent properties

are properties of a system containing relatively simple entities that show more complex behaviours as a collective. The notion of emergence first appeared in the theory of self-organisation.

epimutation

is a heritable alteration which does not affect the base pair sequence of DNA.

Epithelial-to-Mesenchymal Transition

is a transdifferentiation program where cells with epithelial phenotype are generating cells with mesenchymal phenotype. It plays an important role in development and morphogenesis, and is reversible (MET). In tumours EMT confers epithelial cells with properties of motility, invasiveness, self-renewal and the ability to seed a new tumour, in one word, to metastasise.

ExtraCellular Matrix

also referred as ECM, is the extracellular part of animal tissue serving as a scaffolding to hold tissues together. It includes the interstitial matrix and the basement membrane, and contains substances produced by cells and excreted to the extracellular space. It forms a complex network of macromolecules with distinctive physical, biochemical, and biomechanical properties.

gel electrophoresis

is a method which separates proteins according to their charge and size. The gel refers to the matrix in which the molecule migrates.

genomic imprinting

is a mechanism by which imprinted genes are expressed in a parent-of-origin-specific manner meaning that only one allele is expressed in the allele but never both in normal phenotype.

glioblastoma

is a brain tumour which arises from glial cells.

glioma

see **glioblastoma***.

grade

of a tumour is a system to classify cancer cells in terms of how abnormal they look and how quickly the tumour is likely to grow.

haploinsufficiency

is a loss-of-function of a gene. It occurs in a diploid organism when a **mutation*** has inactivated one copy of a gene and the remaining functional wild type copy of the gene is not sufficient to produce enough protein to ensure a normal cellular function.

HeLa cells

is the oldest and most commonly used immortal human cell line derived from cervical cancer cells taken in 1951 from the patient Henrietta Lacks.

histological section

is a thin slice of tissue, typically extracted from a tumour, applied to a microscopic slide to be viewed under a microscope.

homeostasis

is the process by which a cell or an organism maintains internal equilibrium by adjusting its physiological processes.

hub

is a node in a network connected to many other nodes.

hyperplasia

is an accumulation of an excessive number of cells having a normal appearance.

hypoxia

is a condition characterised by the lack of adequate oxygen supply.

intravasation

is the invasion of a cancer through the basal membrane and into blood vessels.

leukaemia

is a malignancy of any variety of hematopoetic cell types, including the lineages leading to lymphocytes and granulocytes, in which the tumour cells are non-pigmented (leukaemia means *white blood* in Greek) and dispersed throughout the circulation. Leukaemias are liquid cancers.

lymphoma

is a cancer which originates in lymphocytes (a type of white blood cells in the vertebrate immune system). There exist many types of lymphomas. Lymphomas are solid cancers.

medulloblastoma

is a brain tumour which originates from neurons in the cerebellum.

melanoma

is a tumour arising from melanocytes, the pigmented cells of the skin and iris.

meta-analysis

    refers to statistical techniques to combine the results of several studies that address a set of related research hypotheses.

metastasis

    is a cancerous growth formed by colonisation of cancerous cells from a primary growth located elsewhere in the body.

metastatic

    (adj.) refers to a cancer with one **metastasis*** (or several).

microarray

    is miniaturised sensor tools which consist of a slide glass smaller than two square-centimetres. It can quantify genome-wide molecular parameters such as gene expression, DNA copy number *etc.* A microarray is also called *biochip* or *chip*.

mimetic drug

    is a drug that mimics the structure of a protein or a protein domain (and therefore targets its partners).

mutation

    is a heritable alteration which affects the base pair sequence of DNA.

negative dominance

    is a mutation in one copy of a gene whose mutant protein product interferes with the normal wild type protein produced from the remaining wild type functional copy of the gene.

neoplasia

    is a tumour which consists of cells having an abnormal appearance and an abnormal proliferation pattern.

neoplasm

    see **neoplasia***.

neoplastic

    (adj.) see **neoplasia***.

networked buffering

    refers to a mechanism of emergence of robust properties from a network of interacting agents such that each agent is able to perform more than one function and parts of the agent functions overlap.

neuroblastoma

    is a paediatric extra-cranial solid tumour arising from a sympathetic nervous system tissue.

neutral space

    is a collection of equivalent implementations of the same biological function.

NGS

    stands for next-generation sequencing, a high-throughput technique which enables parallel sequencing of hundreds of millions of short sequences. Originally NGS referred to the second-generation sequencing technique. NGS is now widely used to refer to any high-throughput sequencing techniques from second-generation until the most recent.

off-target effect

refers to the fact that a drug designed to target a specific molecule might also have an effect (*e.g.* bind) to other molecules; this is called off-target effect.

oligonucleotide

is a DNA sequence (typically about 25–60 nucleotides).

oncogene

is a gene whose activation increases the cancer risk.

oncologist

is a physician who treats cancer.

oncology

is the science which studies tumours including their development, diagnosis, treatment, and prevention.

paralog

is another copy of a gene resulted from gene duplication.

pathologist

is a physician who studies and diagnoses diseases through examination of organs, tissues and cells.

phase space

is the space of all system's states, in which all the system's trajectories are localised.

Polymerase Chain Reaction

also referred to as PCR, is the process which allows the copying of DNA or RNA molecules.

primary tumour

is the original site (organ or tissue) where the tumour progression started.

primer

is a DNA or RNA molecule whose $3'$ end serves as the initiation point of DNA synthesis by a DNA polymerase.

prognosis

is a forecast of the likely course of a disease, particularly of the chance of recovery from the disease or the chance of the disease recurring.

prostatitis

is the inflammation of the prostate gland.

proteasome

is a very large protein complex, located in the nucleus and cytoplasm of eukaryotic cells, responsible for the degradation of unneeded or damaged proteins.

recurrence

see **relapse**[*]

relapse

can be defined as:

(1) (n.) a reoccurrence of cancer cells after **remission**[*] of the tumour. A relapse occurs either at the same site of the primary tumour or in another location. (2) (v.) the action of sustaining such a reoccurrence.

remission

is the period during which the symptoms of a disease disappear with the possibility of its eventual reappearance or worsening.

resection

is the removal by surgery of all or part of an organ, tissue.

restriction enzyme

is an enzyme which cuts either double-strand or single-strand DNA at specific recognition nucleotide sequences known as restriction sites.

retinoblastoma

is a tumour which arises from cells of the retina. It is the most common type of eye cancer in children.

reverse transcriptase

are enzymes capable of making a DNA complementary copy of an RNA molecule using the RNA molecule as a template.

ribosome

is the complex made of RNAs and proteins which synthesises protein chains by assembling the amino acid molecules, according to the nucleotide sequence of an RNA molecule.

router

is a node in a network having a high (much bigger than average) **centrality*** measure.

sarcoma

is a cancer of the connective or supportive tissue (bone, cartilage, fat, muscle, blood vessels) and soft tissue. Sarcomas generally occur in young adults.

scaffold proteins

are regulators of signalling pathways. They bind with members of a signalling pathway, tethering them into complexes and favouring their interaction and their localisation.

schwannoma

is a benign nerve sheath tumour arising from Schwann cells.

second-line treatment

is a treatment for a disease after an initial treatment has failed.

signalling pathway

it is set of molecules which control a cellular function (*e.g.* apoptosis). Once the first molecule in the pathway is activated in response to a *stimulus*, it activates another molecule. This process is repeated in an activation cascade until the last molecule is activated and the cell function involved is carried out. For a given cellular function, many different cascades can be connected with crosstalks. The first molecule is in general a membrane receptor (i.e. a protein), and the signal is transduced from the exterior of the cell to its interior.

somatic mutation

is a mutation which appears in the genome of a cell outside of the germ line. By definition such a mutation cannot be transmitted to the next organismic generation.

sporadic

refers to a disease which randomly occurs in a population without any predisposition in contrast to a disease caused by a heritable genetic susceptibility.

stage

of a tumour is a system to classify cancer in terms of how it has spread, based on factors such as the size of the tumour or the presence of metastasis.

translocation

is a chromosome abnormality caused by rearrangement of parts between non-homologous chromosomes.

tumorigenesis

see **tumour progression**[*].

tumour progression

is the process of multistep evolution of a normal cell into a tumour cell. It is also termed tumorigenesis, oncogenesis or carcinogenesis.

tumour suppressor gene

is a gene whose partial or complete inactivation increases the cancer risk.

Bibliography

1000 Genomes Project Consortium. A map of human genome variation from population-scale sequencing. *Nature*, **467**:1061–1073, 2010.

Abbott, L. H. and Michor, F. Mathematical models of targeted cancer therapy. *Br. J. Cancer*, **95**:1136–1141, 2006.

Abbott, R. G. et al. Simulating the hallmarks of cancer. *Artif. Life*, **12**:617–634, 2006.

Achard, F. et al. XML, bioinformatics and data integration. *Bioinformatics*, **17**:115–125, 2001.

Aebersold, R. and Mann, M. Mass spectrometry-based proteomics. *Nature*, **422**:198–207, 2003.

Aerts, S. et al. Gene prioritization through genomic data fusion. *Nat. Biotechnol.*, **24**:537–544, 2006.

Aguda, B. D. and Algar, C. K. A structural analysis of the qualitative networks regulating the cell cycle and apoptosis. *Cell Cycle*, **2**:538–44, 2003.

Aguda, B. D. and Tang, Y. The kinetic origins of the restriction point in the mammalian cell cycle. *Cell Prolif.*, **32**:321–35, 1999.

Aguilera, A. and Gómez-González, B. Genome instability: a mechanistic view of its causes and consequences. *Nat. Rev. Genet.*, **9**:204–217, 2008.

Ahr, A. et al. Molecular classification of breast cancer patients by gene expression profiling. *J. Pathol.*, **195**:312–320, 2001.

Ahr, A. et al. Identification of high risk breast-cancer patients by gene expression profiling. *Lancet*, **359**:131–132, 2002.

Alarcon, T. et al. A cellular automaton model for tumour growth in inhomogeneous environment. *J. Theor. Biol.*, **225**:257–274, 2003.

Albeck, J. G. et al. Modeling a Snap-Action, Variable-Delay Switch Controlling Extrinsic Cell Death. *PLoS Biol.*, **6**:e299, 2008.

Alberts, B. et al. *Molecular Biology of the Cell.* Garland Science, Taylor & Francis Group, LLC, 2002. Fourth Edition.

Albertson, D. G. et al. Chromosome aberrations in solid tumors. *Nat. Genet.*, **34**:369–376, 2003.

Aldridge, B. B. et al. Direct Lyapunov exponent analysis enables parametric study of transient signalling governing cell behaviour. *Syst. Biol. (Stevenage)*, **153**:425–432, 2006.

Alizadeh, A. A. et al. Distinct types of diffuse large B-cell lymphoma identified by gene expression profiling. *Nature*, **403**:503–511, 2000.

Allison, D. B. et al. Microarray data analysis: from disarray to consolidation and consensus. *Nat. Rev. Genet.*, **7**:55–65, 2006.

Alm, E. and Arkin, A. P. Biological networks. *Curr. Opin. Struct. Biol.*, **13**:193–202, 2003.

Aloise, D. et al. NP-hardness of Euclidean sum-of-squares clustering. *Mach. Learn.*, **75**:245–248, 2009.

Alon, U. *An Introduction to Systems Biology: Design Principles of Biological Circuits.* Chapman & Hall/CRC Mathematical & Computational Biology, 2007a.

Alon, U. Network motifs: theory and experimental approaches. *Nat. Rev. Genet.*, **8**:450–461, 2007b.

Alter, O. et al. Singular value decomposition for genome-wide expression data processing and modeling. *PNAS*, **97**:10101–10106, 2000.

Altinok, A. et al. A cell cycle automaton model for probing circadian patterns of anticancer drug delivery. *Adv. Drug Deliv. Rev.*, **59**:1036–1053, 2007.

Altinok, A. et al. Identifying mechanisms of chronotolerance and chronoefficacy for the anticancer drugs 5-fluorouracil and oxaliplatin by computational modeling. *Eur. J. Pharm. Sci.*, **36**:20–38, 2009.

Ambroise, C. and McLachlan, G. Selection bias in gene extraction on the basis of microarray gene-expression data. *PNAS*, **99**:6562–6566, 2002.

American Association for Cancer Research Human Epigenome Task Force and European Union, Network of ExcellenceScientific Advisory Board. Moving AHEAD with an international human epigenome project. *Nature*, **454**:711–715, 2008.

Anderson, A. R. and Chaplain, M. A. Continuous and discrete mathematical models of tumor-induced angiogenesis. *Bull. Math. Biol.*, **60**:857–899, 1998.

Anderson, A. R. A. et al. Microenvironment driven invasion: a multiscale multimodel investigation. *J. Math. Biol.*, **58**:579–624, 2009.

Anderson, R. A. et al. Tumor morphology and phenotypic evolution driven by selective pressure from the microenvironment. *Cell*, **127**:905–915, 2006.

Arakelyan, L. et al. A computer algorithm describing the process of vessel formation and maturation, and its use for predicting the effects of anti-angiogenic and anti-maturation therapy on vascular tumor growth. *Angiogenesis*, **5**:203–214, 2002.

Arino, O. et al. Mathematical modeling of the loss of telomere sequences. *J. Theor. Biol.*, **177**:45–57, 1995.

Arkus, N. A mathematical model of cellular apoptosis and senescence through the dynamics of telomere loss. *J. Theor. Biol.*, **235**:13–32, 2005.

Armitage, P. and Doll, R. The age distribution of cancer and a multi-stage theory of carcinogenesis. *Br. J. Cancer*, **8**:1–12, 1954.

Astanin, S. and Preziosi, L. Mathematical modelling of the Warburg effect in tumour cords. *J. Theor. Biol.*, **258**:578–590, 2009.

Astrom, K. et al. Altitude is a phenotypic modifier in hereditary paraganglioma type 1: evidence for an oxygen-sensing defect. *Hum. Genet.*, **113**:228–237, 2003.

Auer, P. L. and Doerge, R. W. Statistical design and analysis of RNA sequencing data. *Genetics*, **185**:405–416, 2010.

Auffray, C. and Noble, D. Origins of systems biology in William Harvey's masterpiece on the movement of the heart and the blood in animals. *Int. J. Mol. Sci.*, **10**:1658–1669, 2009.

Auffray, C. et al. The Hallmarks of Cancer Revisited Through Systems Biology and Network Modelling. In Cesario, A. and Marcus, F., editors, *Cancer Systems Biology, Bioinformatics, and Medecine.* Springer, 2011.

Axelrod, R. et al. Evolution of cooperation among tumor cells. *PNAS*, **103**:13474–13479, 2006.

Ayala, F. J. Biology as an autonomous science. *Am. Sci.*, **56**:207–221, 1968.

Bach, F. R. et al. Multiple Kernel Learning, Conic Duality, and the SMO Algorithm. In *ICML '04: Proceedings of the twenty-first international conference on Machine learning*, page 6. ACM, New York, NY, USA, 2004.

Bader, G. D. et al. Pathguide: a pathway resource list. *Nucleic Acids Res.*, **34**:D504–D506, 2006.

Bak and Sneppen. Punctuated equilibrium and criticality in a simple model of evolution. *Phys. Rev. Lett.*, **71**:4083–4086, 1993.

Bak et al. Self-organized criticality: An explanation of the 1/f noise. *Phys. Rev. Lett.*, **59**:381–384, 1987.

Baker, M. Next-generation sequencing: adjusting to data overload. *Nat. Methods*, **7**:495–499, 2010.

Balding, D. and McElwain, D. L. A mathematical model of tumour-induced capillary growth. *J. Theor. Biol.*, **114**:53–73, 1985.

Balkwill, F. and Mantovani, A. Inflammation and cancer: back to Virchow? *Lancet*, **357**:539–545, 2001.

Bantscheff, M. et al. Quantitative mass spectrometry in proteomics: a critical review. *Anal. Bioanal. Chem.*, **389**:1017–1031, 2007.

Barabási, A.-L. and Albert, R. Emergence of scaling in random networks. *Science*, **286**:509–512, 1999.

Barash, Y. et al. Deciphering the splicing code. *Nature*, **465**:53–59, 2010.

Barrett, T. et al. NCBI GEO: archive for functional genomics data sets–10 years on. *Nucleic Acids Res.*, **39**:D1005–D1010, 2011.

Bartz, S. R. et al. Small interfering RNA screens reveal enhanced cisplatin cytotoxicity in tumor cells having both BRCA network and TP53 disruptions. *Mol. Cell. Biol.*, **26**:9377–9386, 2006.

Battiti, R. Using mutual information for selecting features in supervised neural net learning. *IEEE T. Neural Networ.*, **5**:537–550, 1994.

Bauer, A. L. et al. Receptor cross-talk in angiogenesis: mapping environmental cues to cell phenotype using a stochastic, Boolean signaling network model. *J. Theor. Biol.*, **264**:838–846, 2010.

Bauer-Mehren, A. et al. Pathway databases and tools for their exploitation: benefits, current limitations and challenges. *Mol. Syst. Biol.*, **5**:290, 2009.

Baumuratova, T. et al. Localizing potentially active post-transcriptional regulations in the Ewing's sarcoma gene regulatory network. *BMC Syst. Biol.*, **4**:146, 2010.

Bayá, A. E. et al. Gene set enrichment analysis using non-parametric scores. In Sagot, M.-F. and Walter, M. E. M. T., editors, *Advances in Bioinformatics and Computational Biology*, number 4643 in LNBI, pages 12–21. Springer-Verlag, Berlin Heidelberg, 2007.

Baylin, S. B. and Jones, P. A. A decade of exploring the cancer epigenome - biological and translational implications. *Nat. Rev. Cancer*, **11**:726–734, 2011.

Beckman, R. A. and Loeb, L. A. Efficiency of carcinogenesis with and without a mutator mutation. *PNAS*, **103**:14140–14145, 2006.

Bengtsson, H. et al. A single-array preprocessing method for estimating full-resolution raw copy numbers from all Affymetrix genotyping arrays including GenomeWideSNP 5 & 6. *Bioinformatics*, **25**:2149–2156, 2009.

Benito, M. et al. Adjustment of systematic microarray data biases. *Bioinformatics*, **20**:105–14, 2004.

Benjamini, Y. and Hochberg, Y. Controlling the False Discovery Rate: a Practical and Powerful Approach to Multiple Testing. *J. R. Stat. Soc. Ser. B*, **57**:289–300, 1995.

Bentele, M. et al. Mathematical modeling reveals threshold mechanism in CD95-induced apoptosis. *J. Cell Biol.*, **166**:839–51, 2004.

Berasain, C. et al. Epidermal growth factor receptor ligands in murine models for erythropoietic protoporphyria: potential novel players in the progression of liver injury. *Cell. Mol. Biol. (Noisy-le-grand)*, **55**:29–37, 2009.

Berger, A. H. and Pandolfi, P. P. Haplo-insufficiency: a driving force in cancer. *J. Pathol.*, **223**:137–146, 2011.

Berger, A. H. et al. A continuum model for tumour suppression. *Nature*, **476**:163–169, 2011.

Bergman, A. and Siegal, M. L. Evolutionary capacitance as a general feature of complex gene networks. *Nature*, **424**:549–552, 2003.

van Berkum, N. L. et al. Hi-C: a method to study the three-dimensional architecture of genomes. *J. Vis. Exp.*, 2010.

Bertheau, P. et al. Exquisite sensitivity of TP53 mutant and basal breast cancers to a dose-dense epirubicin-cyclophosphamide regimen. *PLoS Med.*, **4**:e90, 2007.

Bertoni, A. and Valentini, G. Discovering multi-level structures in bio-molecular data through the Bernstein inequality. *BMC Bioinformatics*, **9 Suppl 2**:S4, 2008.

Bhalla, U. S. and Iyengar, R. Emergent Properties of Networks of Biological Signaling Pathways. *Science*, **283**:381–387, 1999.

Bhattacharjee, A. et al. Classification of human lung carcinomas by mRNA expression profiling reveals distinct adenocarcinoma subclasses. *PNAS*, **98**:13790–13795, 2001.

Bild, A. H. et al. Oncogenic pathway signatures in human cancers as a guide to targeted therapies. *Nature*, **439**:353–7, 2006.

Billerey, C. and Boccon-Gibod, L. Etude des variations inter-pathologistes dans l'évaluation du grade et du stade des tumeurs vésicales. *Progrès en Urologie*, **6**:49–57, 1996.

Billerey, C. et al. Frequent FGFR3 mutations in papillary non-invasive bladder (pTa) tumors. *Am. J. Pathol.*, **158**:1955–1959, 2001.

Bishop, C. *Pattern recognition and machine learning*. Springer, 2006.

Bittner, M. et al. Molecular classification of cutaneous malignant melanoma by gene expression profiling. *Nature*, **406**:536–540, 2000.

Blow, N. DNA sequencing: generation next-next. *Nat. Methods*, **5**:267–274, 2008.

Blows, F. M. et al. Subtyping of breast cancer by immunohistochemistry to investigate a relationship between subtype and short and long term survival: a collaborative analysis of data for 10,159 cases from 12 studies. *PLoS Med.*, **7**:e1000279, 2010.

Boeva, V. et al. Control-free calling of copy number alterations in deep-sequencing data using GC-content normalization. *Bioinformatics*, **27**:268–269, 2011a.

Boeva, V. et al. Control-FREEC: a tool for assessing copy number and allelic content using next generation sequencing data. *Bioinformatics*, 2011b.

Bonabeau, E. Agent-based modeling: methods and techniques for simulating human systems. *PNAS*, **99 Suppl 3**:7280–7287, 2002.

Bonnet, S. et al. A mitochondria-K+ channel axis is suppressed in cancer and its normalization promotes apoptosis and inhibits cancer growth. *Cancer Cell*, **11**:37–51, 2007.

Brazma, A. et al. Minimum information about a microarray experiment (MIAME)-toward standards for microarray data. *Nat. Genet.*, **29**:365–371, 2001.

Brazma, A. et al. ArrayExpress–a public repository for microarray gene expression data at the EBI. *Nucleic Acids Res.*, **31**:68–71, 2003.

Breiman, L. Random forests. *Mach. Learn.*, **45**:5–32, 2001.

Breslin, T. et al. Signal transduction pathway profiling of individual tumor samples. *BMC Bioinformatics*, **6**:163, 2005.

Breving, K. and Esquela-Kerscher, A. The complexities of microRNA regulation: mirandering around the rules. *Int. J. Biochem. Cell Biol.*, **42**:1316–1329, 2010.

Brunet, J. P. et al. Metagenes and molecular pattern discovery using matrix factorization. *PNAS*, **101**:4164–9, 2004.

Buck, M. J. and Lieb, J. D. ChIP-chip: considerations for the design, analysis, and application of genome-wide chromatin immunoprecipitation experiments. *Genomics*, **83**:349–360, 2004.

Bulyk, M. L. DNA microarray technologies for measuring protein-DNA interactions. *Curr. Opin. Biotechnol.*, **17**:422–430, 2006.

Bungaro, S. et al. Integration of genomic and gene expression data of childhood ALL without known aberrations identifies subgroups with specific genetic hallmarks. *Genes Chromosomes Cancer*, **48**:22–38, 2009.

Bunz, F. *Principles of cancer genetics.* Springer, 2008.

Burges, C. J. C. A Tutorial on Support Vector Machines for Pattern Recognition. *Data Min. Knowl. Discov.*, **2**:121–167, 1998.

Burns, J. et al. Control analysis of metabolic systems. *Trends Biochem. Sci.*, **10(16)**, 1985.

Byerly, S. et al. Effects of ozone exposure during microarray posthybridization washes and scanning. *J. Mol. Diagn.*, **11**:590–597, 2009.

Byrne, H. M. Dissecting cancer through mathematics: from the cell to the animal model. *Nat. Rev. Cancer*, **10**:221–230, 2010.

Bürckstümmer, T. et al. An efficient tandem affinity purification procedure for interaction proteomics in mammalian cells. *Nat. Methods*, **3**:1013–1019, 2006.

Cabusora, L. et al. Differential network expression during drug and stress response. *Bioinformatics*, **21**:2898–2905, 2005.

Calin, G. and Croce, C. M. MicroRNA-cancer connection: the beginning of a new tale. *Cancer Res.*, **66**:7390–7394, 2006.

Calinski, R. B. and Harabasz, J. A dendrite method for cluster analysis. *Communs Statist.*, **3**:1–27, 1974.

Calzone, L. et al. A comprehensive modular map of molecular interactions in RB/E2F pathway. *Mol. Syst. Biol.*, **4**:173, 2008.

Calzone, L. et al. Mathematical modelling of cell-fate decision in response to death receptor engagement. *PLoS Comput. Biol.*, **6**:e1000702, 2010.

Campbell, P. J. et al. Identification of somatically acquired rearrangements in cancer using genome-wide massively parallel paired-end sequencing. *Nat. Genet.*, **40**:722–729, 2008.

Cappellen, D. et al. Novel c-MYC target genes mediate differential effects on cell proliferation and migration. *EMBO Rep.*, **8**:70–76, 2007.

Carlotti, F. et al. Dynamic shuttling of nuclear factor kappa B between the nucleus and cytoplasm as a consequence of inhibitor dissociation. *J. Biol. Chem.*, **275**:41028–41034, 2000.

Carlson, J. M. and Doyle, J. Highly optimized tolerance: a mechanism for power laws in designed systems. *Phys. Rev. E. Stat. Phys. Plasmas Fluids Relat. Interdiscip. Topics*, **60**:1412–1427, 1999.

Carlson, J. M. and Doyle, J. Complexity and robustness. *PNAS*, **99 Suppl 1**:2538–2545, 2002.

Causier, B. Studying the interactome with the yeast two-hybrid system and mass spectrometry. *Mass Spectrom. Rev.*, **23**:350–367, 2004.

Chaffer, C. L. and Weinberg, R. A. A perspective on cancer cell metastasis. *Science*, **331**:1559–1564, 2011.

Chen, C. et al. Removing batch effects in analysis of expression microarray data: an evaluation of six batch adjustment methods. *PLoS One*, **6**:e17238, 2011.

Chen, K. C. et al. Kinetic analysis of a molecular model of the budding yeast cell cycle. *Mol. Biol. Cell*, **11**:369–391, 2000.

Chen, S. S. et al. Atomic decomposition by basis pursuit. *SIAM J. Sci. Comput.*, **20**:33–61, 1998.

Chen, W. et al. Mapping translocation breakpoints by next-generation sequencing. *Genome Res.*, **18**:1143–1149, 2008.

Chen, W. W. et al. Input-output behavior of ErbB signaling pathways as revealed by a mass action model trained against dynamic data. *Mol. Syst. Biol.*, **5**:239, 2009.

Cheng, Y. and Church, G. M. Biclustering of expression data. *Proc Int Conf Intell Syst Mol Biol*, **8**:93–103, 2000.

Cheong, R. et al. Understanding NF-kappaB signaling via mathematical modeling. *Mol. Syst. Biol.*, **4**.192, 2008.

Chi, K. R. The year of sequencing. *Nat. Methods*, **5**:11–14, 2008.

Chipman, H. et al. Clustering Microarray Data. In Speed, T., editor, *Statistical Analysis of Gene Expression Microarray Data*, pages 159–200. Chapman and Hall, CRC press., 2003.

Choi, H.-S. et al. Coupled positive feedbacks provoke slow induction plus fast switching in apoptosis. *FEBS Lett.*, **581**:2684 – 2690, 2007.

Choudhary, C. and Mann, M. Decoding signalling networks by mass spectrometry-based proteomics. *Nat. Rev. Mol. Cell Biol.*, **11**:427–439, 2010.

Chowdhury, S. A. and Koyutürk, M. Identification of coordinately dysregulated subnetworks in complex phenotypes. *Pac. Symp. Biocomput.*, pages 133–144, 2010.

Chuang, H.-Y. et al. Network-based classification of breast cancer metastasis. *Mol. Syst. Biol.*, **3**:140, 2007.

Churchill, G. A. Fundamentals of experimental design for cDNA microarrays. *Nat. Genet.*, **32 Suppl**:490–495, 2002.

Cianfrocca, M. and Goldstein, L. J. Prognostic and predictive factors in early-stage breast cancer. *Oncologist*, **9**:606–616, 2004.

Ciliberto, A. et al. Steady states and oscillations in the p53/Mdm2 network. *Cell Cycle*, **4**:488–93, 2005.

Citri, A. and Yarden, Y. EGF-ERBB signalling: towards the systems level. *Nat. Rev. Mol. Cell Biol.*, **7**:505–516, 2006.

Clark, G. M. Do we really need prognostic factors for breast cancer? *Breast Cancer Res. Treat.*, **30**:117–126, 1994.

Clarke, D. C. et al. Systems theory of Smad signalling. *Syst. Biol. (Stevenage)*,

153:412–424, 2006.

Clevers, H. The cancer stem cell: premises, promises and challenges. *Nat. Med.*, **17**:313–319, 2011.

Cloutier, M. Warburg revisited: modeling energy metabolism for cancer systems biology. In Wang, E., editor, *Cancer Systems Biology*. CRC Press, 2010.

Cobleigh, M. A. et al. Tumor gene expression and prognosis in breast cancer patients with 10 or more positive lymph nodes. *Clin. Cancer Res.*, **11**:8623–8631, 2005.

Coe, B. P. et al. Resolving the resolution of array CGH. *Genomics*, **89**:647–653, 2007.

Collins, F. S. and Barker, A. D. Mapping the cancer genome. Pinpointing the genes involved in cancer will help chart a new course across the complex landscape of human malignancies. *Sci. Am.*, **296**:50–57, 2007.

Collins, F. S. et al. A vision for the future of genomics research. *Nature*, **422**:835–847, 2003.

Conradie, R. et al. Restriction point control of the mammalian cell cycle via the cyclin E/Cdk2:p27 complex. *FEBS J.*, **277**:357–367, 2010.

Consortium, E. N. C. O. D. E. P. The ENCODE (ENCyclopedia Of DNA Elements) Project. *Science*, **306**:636–640, 2004.

Cook, P. J. et al. A mathematical model for the age distribution of cancer in man. *Int. J. Cancer*, **4**:93–112, 1969.

Cook, P. R. A model for all genomes: the role of transcription factories. *J. Mol. Biol.*, **395**:1–10, 2010.

Costanzo, M. et al. Charting the genetic interaction map of a cell. *Curr. Opin. Biotechnol.*, **22**:66–74, 2011.

Courtot, M. et al. Controlled vocabularies and semantics in systems biology. *Mol. Syst. Biol.*, **7**:543, 2011.

Cowell, J. and Hawthorn, L. The application of microarray technology to the analysis of the cancer genome. *Curr. Mol. Med.*, **7**:103–120, 2007.

Crick, F. Central Dogma of Molecular Biology. *Nature*, **227**:561–563, 1970.

Croce, C. M. Oncogenes and cancer. *N. Engl. J. Med.*, **358**:502–511, 2008.

Csete, M. and Doyle, J. Bow ties, metabolism and disease. *Trends Biotechnol.*, **22**:446–450, 2004.

Csikász-Nagy, A. Computational systems biology of the cell cycle. *Brief. Bioinform.*, **10**:424–434, 2009.

Csikász-Nagy, A. et al. Cell cycle regulation by feed-forward loops coupling transcription and phosphorylation. *Mol. Syst. Biol.*, **5**:236, 2009.

Cui, J. et al. Two Independent Positive Feedbacks and Bistability in the Bcl-2 Apoptotic Switch. *PLoS One*, **3**:e1469, 2008.

Cui, Q. et al. A map of human cancer signaling. *Mol. Syst. Biol.*, **3**:152, 2007.

Czauderna, T. et al. Editing, validating and translating of SBGN maps. *Bioinformatics*, **26**:2340–2341, 2010.

Dalma-Weiszhausz, D. D. et al. The affymetrix GeneChip platform: an overview. *Methods Enzymol.*, **410**:3–28, 2006.

von Dassow, G. et al. The segment polarity network is a robust developmental module. *Nature*, **406**:188–192, 2000.

David, A. R. and Zimmerman, M. R. Cancer: an old disease, a new disease or something in between? *Nat. Rev. Cancer*, **10**:728–733, 2010.

Davies, J. J. et al. Array CGH technologies and their applications to cancer genomes.

Chromosome Res., **13**:237–248, 2005.

Day, R. S. Treatment sequencing, asymmetry, and uncertainty: protocol strategies for combination chemotherapy. *Cancer Res.*, **46**:3876–3885, 1986.

De Bie, T. et al. Kernel-based data fusion for gene prioritization. *Bioinformatics*, **23**:i125–i132, 2007.

de Jong, H. Modeling and simulation of genetic regulatory systems: a literature review. *J. Comput. Biol.*, **9**:67–103, 2002.

de Tayrac, M. et al. Simultaneous analysis of distinct Omics data sets with integration of biological knowledge: Multiple Factor Analysis approach. *BMC Genomics*, **10**:32, 2009.

Dean, J. and Ghemawat, S. MapReduce: simplified data processing on large clusters. *Commun. ACM*, **51**:107–113, 2008.

Deisboeck, T. S. et al. Multiscale cancer modeling. *Annu. Rev. Biomed. Eng.*, **13**:127–155, 2011.

Dekker, J. et al. Capturing chromosome conformation. *Science*, **295**:1306–1311, 2002.

Desmedt, C. et al. Biological processes associated with breast cancer clinical outcome depend on the molecular subtypes. *Clin. Cancer Res.*, **14**:5158–5165, 2008.

Deutscher, D. et al. Multiple knockout analysis of genetic robustness in the yeast metabolic network. *Nat. Genet.*, **38**:993–998, 2006.

DeVita, V. et al. *Cancer: principle and practice of oncology*. Lippincott Williams Wilkins, 2008.

Devroye, L. et al. *A Probabilistic Theory of Pattern Recognition*, volume 31 of *Applications of Mathematics*. Springer, 1996.

D'haeseleer, P. et al. Genetic network inference: from co-expression clustering to reverse engineering. *Bioinformatics*, **16**:707–726, 2000.

Dhillon, A. S. et al. MAP kinase signalling pathways in cancer. *Oncogene*, **26**:3279–3290, 2007.

Do, J. H. and Choi, D. K. Normalization of microarray data: single-labeled and dual-labeled arrays. *Mol. Cells*, **22**:254–261, 2006.

Dobbin, K. et al. Statistical design of reverse dye microarrays. *Bioinformatics*, **19**:803–810, 2003.

Domon, B. and Aebersold, R. Options and considerations when selecting a quantitative proteomics strategy. *Nat. Biotechnol.*, **28**:710–721, 2010.

Donoho, D. L. et al. High-Dimensional Data Analysis : The Curses and Blessings of Dimensionality. *Statistics*, pages 1–33, 2000.

Dostie, J. et al. Chromosome Conformation Capture Carbon Copy (5C): a massively parallel solution for mapping interactions between genomic elements. *Genome Res.*, **16**:1299–1309, 2006.

Dostie, J. et al. Chromosome conformation capture carbon copy technology. *Curr. Protoc. Mol. Biol.*, **Chapter 21**:Unit 21.14, 2007.

Downward, J. Targeting RAS signalling pathways in cancer therapy. *Nat. Rev. Cancer*, **3**:11–22, 2003.

Draghi, J. A. et al. Mutational robustness can facilitate adaptation. *Nature*, **463**:353–355, 2010.

Drasdo, D. et al. A quantitative mathematical modeling approach to liver regeneration. In Haussinger, D., editor, *Liver Regeneration*. De Gruyter Berlin/Boston, 2011.

Drossel and Schwabl. Self-organized critical forest-fire model. *Phys. Rev. Lett.*, **69**:1629–1632, 1992.

Duarte, N. C. et al. Global reconstruction of the human metabolic network based on genomic and bibliomic data. *PNAS*, **104**:1777–1782, 2007.

Duda, R. O. et al. *Pattern Classification*. Wiley-Interscience, 2001.

Dudoit, S. et al. Comparison of discrimination methods for classification of tumors using gene expression data. *J. Am. Stat. Assoc.*, **97**:77–87, 2002a.

Dudoit, S. et al. Statistical methods for identifying differentially expressed genes in replicated cDNA microarray experiments. *Statistica Sinica*, **12**:111–139, 2002b.

Dunn, G. P. et al. Cancer immunoediting: from immunosurveillance to tumor escape. *Nat. Immunol.*, **3**:991–998, 2002.

Edelman, G. M. and Gally, J. A. Degeneracy and complexity in biological systems. *PNAS*, **98**:13763–13768, 2001.

Edgar, R. et al. Gene Expression Omnibus: NCBI gene expression and hybridization array data repository. *Nucleic Acids Res.*, **30**:207–210, 2002.

Edwards, S. L. et al. Resistance to therapy caused by intragenic deletion in BRCA2. *Nature*, **451**:1111–1115, 2008.

Efcavitch, J. W. and Thompson, J. F. Single-molecule DNA analysis. *Annu. Rev. Anal. Chem. (Palo Alto Calif.)*, **3**:109–128, 2010.

Eid, J. et al. Real-time DNA sequencing from single polymerase molecules. *Science*, **323**:133–138, 2009.

Ein-Dor, L. et al. Outcome signature genes in breast cancer: is there a unique set? *Bioinformatics*, **21**:171–178, 2005.

Ein-Dor, L. et al. Thousands of samples are needed to generate a robust gene list for predicting outcome in cancer. *PNAS*, **103**:5923–5928, 2006.

Eisen, M. B. et al. Cluster analysis and display of genome-wide expression patterns. *PNAS*, **95**:14863–14868, 1998.

Eissing, T. et al. Bistability Analyses of a Caspase Activation Model for Receptor-induced Apoptosis. *J. Biol. Chem.*, **279**:36892–36897, 2004.

Eissing, T. et al. Response to Bistability in Apoptosis: Roles of Bax, Bcl-2, and Mitochondrial Permeability Transition Pores. *Biophys. J.*, **92**:3332–3334, 2007.

Eldredge, N. and Gould, S. Punctuated equilibria: an alternative to phyletic gradualism. In Schopf, T., editor, *Models in Paleobiology*. San Francisco: Freeman Cooper, 1972.

Elliott, M. et al. Current Trends in Quantitative Proteomics. *J. Mass Spectrom.*, **44**:1637–60, 2009.

Ellis, I. O. et al. Pathological prognostic factors in breast cancer. II. Histological type. Relationship with survival in a large study with long-term follow-up. *Histopathology*, **20**:479–489, 1992.

Enderling, H. et al. Mathematical modelling of radiotherapy strategies for early breast cancer. *J. Theor. Biol.*, **241**:158–171, 2006.

Epstein, S. S. Losing the war against cancer: who's to blame and what to do about it. *Int. J. Health Serv.*, **20**:53–71, 1990.

Epstein, S. S. et al. The crisis in U.S. and international cancer policy. *Int. J. Health Serv.*, **32**:669–707, 2002.

Escofier, B. and Pages, J. Multiple factor analysis (AFMULT package). *Computational Statistics & Data Analysis*, **18**:121–140, 1994.

Esquela-Kerscher, A. and Slack, F. J. Oncomirs - microRNAs with a role in cancer.

Nat. Rev. Cancer, **6**:259–269, 2006.

Esteller, M. Cancer epigenomics: DNA methylomes and histone-modification maps. *Nat. Rev. Genet.*, **8**:286–298, 2007.

Esteller, M. Epigenetics in cancer. *N. Engl. J. Med.*, **358**:1148–1159, 2008.

Esteller, M. et al. Inactivation of the DNA-repair gene MGMT and the clinical response of gliomas to alkylating agents. *N. Engl. J. Med.*, **343**:1350–1354, 2000.

Ewing, R. M. et al. Large-scale mapping of human protein-protein interactions by mass spectrometry. *Mol. Syst. Biol.*, **3**:89, 2007.

Fabbri, M. et al. MicroRNAs. *Cancer J.*, **14**:1–6, 2008.

Fan, J.-B. et al. Illumina universal bead arrays. *Methods Enzymol.*, **410**:57–73, 2006.

Fang, Z. and Cui, X. Design and validation issues in RNA-seq experiments. *Brief. Bioinform.*, **12**:280–287, 2011.

Faratian, D. et al. Systems biology reveals new strategies for personalizing cancer medicine and confirms the role of PTEN in resistance to trastuzumab. *Cancer Res.*, **69**:6713–6720, 2009.

Fare, T. L. et al. Effects of atmospheric ozone on microarray data quality. *Anal. Chem.*, **75**:4672–4675, 2003.

Farnham, P. J. Insights from genomic profiling of transcription factors. *Nat. Rev. Genet.*, **10**:605–616, 2009.

Faure, A. et al. Dynamical analysis of a generic Boolean model for the control of the mammalian cell cycle. *Bioinformatics*, **22**:e124–e131, 2006.

Fearon, E. R. and Vogelstein, B. A genetic model for colorectal tumorigenesis. *Cell*, **61**:759–767, 1990.

Feinberg, M. Chemical reaction network structure and the stability of complex isothermal reactors: I. The deficiency zero and deficiency one theorems. *Chem. Eng. Sci.*, **42(10)**:2229–2268, 1987.

Fell, D. A. *Understanding the Control of Metabolism*. Portland Press, London, 1997.

Felsher, D. W. Oncogene addiction versus oncogene amnesia: perhaps more than just a bad habit? *Cancer Res.*, **68**:3081–6; discussion 3086, 2008.

Ferrell, J. E., Jr et al. Modeling the cell cycle: why do certain circuits oscillate? *Cell*, **144**:874–885, 2011.

Fidler, I. J. and Kripke, M. L. Metastasis results from preexisting variant cells within a malignant tumor. *Science*, **197**:893–895, 1977.

Fiebitz, A. et al. High-throughput mammalian two-hybrid screening for protein-protein interactions using transfected cell arrays. *BMC Genomics*, **9**:68, 2008.

Fields, S. and Song, O. A novel genetic system to detect protein-protein interactions. *Nature*, **340**:245–246, 1989.

Figeys, D. Mapping the human protein interactome. *Cell Res.*, **18**:716–724, 2008.

Filipowicz, W. et al. Mechanisms of post-transcriptional regulation by microRNAs: are the answers in sight? *Nat. Rev. Genet.*, **9**:102–114, 2008.

Flusberg, B. A. et al. Direct detection of DNA methylation during single-molecule, real-time sequencing. *Nat. Methods*, **7**:461–465, 2010.

Foekens, J. A. et al. Multicenter validation of a gene expression-based prognostic signature in lymph node-negative primary breast cancer. *J. Clin. Oncol.*, **24**:1665–1671, 2006.

Forbes, S. A. et al. The Catalogue of Somatic Mutations in Cancer (COSMIC). *Curr. Protoc. Hum. Genet.*, **Chapter 10**:Unit 10.11, 2008.

Forbes, S. A. et al. COSMIC: mining complete cancer genomes in the Catalogue of

Somatic Mutations in Cancer. *Nucleic Acids Res.*, **39**:D945–D950, 2011.

Franke, L. et al. Reconstruction of a functional human gene network, with an application for prioritizing positional candidate genes. *Am. J. Hum. Genet.*, **78**:1011–1025, 2006.

Franks, S. J. et al. Modelling the early growth of ductal carcinoma in situ of the breast. *J. Math. Biol.*, **47**:424–452, 2003.

Fraser, K. et al. *Microarray Image Analysis: An Algorithmic Approach.* Chapman and Hall/CRC, 2010.

Freeman, J. L. et al. Copy number variation: new insights in genome diversity. *Genome Res.*, **16**:949–961, 2006.

Frigola, J. et al. Epigenetic remodeling in colorectal cancer results in coordinate gene suppression across an entire chromosome band. *Nat. Genet.*, **38**:540–549, 2006.

Fullwood, M. J. et al. Next-generation DNA sequencing of paired-end tags (PET) for transcriptome and genome analyses. *Genome Res.*, **19**:521–532, 2009a.

Fullwood, M. J. et al. An oestrogen-receptor-alpha-bound human chromatin interactome. *Nature*, **462**:58–64, 2009b.

Fullwood, M. J. et al. Chromatin interaction analysis using paired-end tag sequencing. *Curr. Protoc. Mol. Biol.*, **Chapter 21**:Unit 21.15.1–Unit 21.1525, 2010.

Furnival, G. M. and Wilson, R. W. Regressions by leaps and bounds. *Technometrics*, **16**:499–511, 1974.

Fussenegger, M. et al. A mathematical model of caspase function in apoptosis. *Nat. Biotechnol.*, **18**:768–774, 2000.

Futreal, P. A. et al. A census of human cancer genes. *Nat Rev Cancer*, **4**:177–183, 2004.

Galluzzi, L. et al. Molecular mechanisms of cisplatin resistance. *Oncogene*, 2011.

Gambacorti-Passerini, C. et al. Multicenter independent assessment of outcomes in chronic myeloid leukemia patients treated with imatinib. *J. Natl. Cancer Inst.*, **103**:553–561, 2011.

Gao, Y. and Church, G. Improving molecular cancer class discovery through sparse non-negative matrix factorization. *Bioinformatics*, **21**:3970–3975, 2005.

Garaj, S. et al. Graphene as a subnanometre trans-electrode membrane. *Nature*, **467**:190–193, 2010.

Garber, M. E. et al. Diversity of gene expression in adenocarcinoma of the lung. *PNAS*, **98**:13784–13789, 2001.

Gatenby, R. A. A change of strategy in the war on cancer. *Nature*, **459**:508–509, 2009.

Gatenby, R. A. and Gawlinski, E. T. A reaction-diffusion model of cancer invasion. *Cancer Res.*, **56**:5745–5753, 1996.

Gatenby, R. A. and Gillies, R. J. Why do cancers have high aerobic glycolysis? *Nat. Rev. Cancer*, **4**:891–899, 2004.

Gatenby, R. A. and Gillies, R. J. A microenvironmental model of carcinogenesis. *Nat. Rev. Cancer*, **8**:56–61, 2008.

Gatenby, R. A. et al. Adaptive therapy. *Cancer Res.*, **69**:4894–4903, 2009.

Gaudet, S. et al. A Compendium of Signals and Responses Triggered by Prodeath and Prosurvival Cytokines. *Mol. Cell. Proteomics*, **4**:1569–1590, 2005.

Gavin, A.-C. et al. Functional organization of the yeast proteome by systematic analysis of protein complexes. *Nature*, **415**:141–7, 2002.

Gavrilets, S. *Fitness Landscapes and the Origin of Species*. Princeton, NJ: Princeton University Press., 2004.

Gedela, S. Integration, warehousing, and analysis strategies of Omics data. *Methods Mol. Biol.*, **719**:399–414, 2011.

Gehlenborg, N. et al. Visualization of omics data for systems biology. *Nat. Methods*, **7**:S56–S68, 2010.

Gentleman, R. C. et al. Bioconductor: Open software development for computational biology and bioinformatics. *Genome Biol.*, **5**:R80, 2004.

Gerlinger, M. and Swanton, C. How Darwinian models inform therapeutic failure initiated by clonal heterogeneity in cancer medicine. *Br. J. Cancer*, **103**:1139–1143, 2010.

Gerlinger, M. et al. Intratumor Heterogeneity and Branched Evolution Revealed by Multiregion Sequencing. *N. Engl. J. Med.*, **36**:883–892, 2012.

Gerstein, M. and Jansen, R. The current excitement in bioinformatics-analysis of whole-genome expression data: how does it relate to protein structure and function? *Curr. Opin. Struct. Biol.*, **10**:574–584, 2000.

Geva-Zatorsky, N. et al. Oscillations and variability in the p53 system. *Mol. Syst. Biol.*, **2**:2006 0033, 2006.

Ghosh, S. et al. Software for systems biology: from tools to integrated platforms. *Nat. Rev. Genet.*, **12**:821–832, 2011.

Giallourakis, C. et al. Disease gene discovery through integrative genomics. *Annu. Rev. Genomics Hum. Genet.*, **6**:381–406, 2005.

Gibney, E. R. and Nolan, C. M. Epigenetics and gene expression. *Heredity*, **105**:4–13, 2010.

Glenn, T. C. Field guide to next-generation DNA sequencers. *Mol. Ecol. Resour.*, 2011.

Goecks, J. et al. Galaxy: a comprehensive approach for supporting accessible, reproducible, and transparent computational research in the life sciences. *Genome Biol.*, **11**:R86, 2010.

Goldbeter, A. and Koshland, D., Jr. Ultrasensitivity in biochemical systems controlled by covalent modification. Interplay between zero-order and multistep effects. *J. Biol. Chem.*, **259**:14441–14447, 1984.

Goldhirsch, A. et al. Strategies for subtypes–dealing with the diversity of breast cancer: highlights of the St. Gallen International Expert Consensus on the Primary Therapy of Early Breast Cancer 2011. *Ann. Oncol.*, **22**:1736–1747, 2011.

Goldie, J. H. and Coldman, A. J. A mathematic model for relating the drug sensitivity of tumors to their spontaneous mutation rate. *Cancer Treat. Rep.*, **63**:1727–1733, 1979.

Golub, T. R. et al. Molecular classification of cancer: class discovery and class prediction by gene expression monitoring. *Science*, **286**:531–537, 1999.

González, I. et al. Highlighting relationships between heterogeneous biological data through graphical displays based on regularized canonical correlation analysis. *J. Biol. Syst.*, **17**:173–199, 2009.

Gonzalez-Angulo, A. M. et al. Future of personalized medicine in oncology: a systems biology approach. *J. Clin. Oncol.*, **28**:2777–2783, 2010.

Gorban, A. Multigrid Integrators on Multiscale Reaction Networks. In *Talk given at Algorithms for Approximation VI, Ambleside, the Lake District, UK*. 2009.

Gorban, A. and Zinovyev, A. Principal Graphs and Manifolds. In Olivas, E. et al., editors, *Handbook of Research on Machine Learning Applications and Trends: Algorithms, Methods and Techniques.* Information Science Reference, 2009.

Gorban, A. et al., editors. *Principal Manifolds for Data Visualisation and Dimension Reduction, LNCSE 58.* Springer, 2008.

Gorban, A. et al. Asymptotology of chemical reaction networks. *Chem. Eng. Sci.,* **65**:2310–2324, 2010.

Gorban, A. N. and Radulescu, O. Dynamical robustness of biological networks with hierarchical distribution of time scales. *IET Syst. Biol.,* **1**:238–246, 2007.

Gorban, A. N. and Zinovyev, A. Principal manifolds and graphs in practice: from molecular biology to dynamical systems. *Int. J. Neural Syst.,* **20**:219–232, 2010.

Gorban, I., A.N. Karlin. *Invariant manifolds for physical and chemical kinetics. In: Lecture Notes in Physics, vol. 660.* Springer, Berlin, Heidelberg, New York., 2005.

Gordon, A. D. *Classification.* Chapman & Hall/CRC, 1999.

Griffin, N. M. et al. Label-free, normalized quantification of complex mass spectrometry data for proteomic analysis. *Nat. Biotechnol.,* **28**:83–89, 2010.

Gromov, M. *Metric Structures for Riemannian and Non-Riemannian Spaces. Progress in Mathematics 152.* Birkhauser Verlag, 1999.

Guedj, M. et al. A refined molecular taxonomy of breast cancer. *Oncogene,* 2011.

Guiot, C. et al. Morphological instability and cancer invasion: a 'splashing water drop' analogy. *Theor. Biol. Med. Model.,* **4**:4, 2007a.

Guiot, C. et al. Physical aspects of cancer invasion. *Phys. Biol.,* **4**:P1–P6, 2007b.

Guldberg, C. and Waage, P. Studies Concerning Affinity. *C. M. Forhandlinger: Videnskabs-Selskabet i Christiana,* **35**, 1864.

Gunderson, K. L. et al. Decoding randomly ordered DNA arrays. *Genome Res.,* **14**:870–877, 2004.

Guo, Z. et al. Towards precise classification of cancers based on robust gene functional expression profiles. *BMC Bioinformatics,* **6**:58, 2005.

Guo, Z. et al. Edge-based scoring and searching method for identifying condition-responsive protein-protein interaction sub-network. *Bioinformatics,* **23**:2121–2128, 2007.

Guyon, I. and Elisseeff, A. An introduction to variable and feature selection. *J. Mach. Learn. Res.,* **3**:1157–1182, 2003.

Guyon, I. et al. Gene selection for cancer classification using support vector machines. *Mach. Learn.,* **46**:389–422, 2002.

Gygi, S. P. et al. Quantitative analysis of complex protein mixtures using isotope-coded affinity tags. *Nat. Biotechnol.,* **17**:994–999, 1999.

Göndör, A. and Ohlsson, R. Chromosome crosstalk in three dimensions. *Nature,* **461**:212–217, 2009.

Göndör, A. et al. High-resolution circular chromosome conformation capture assay. *Nat. Protoc.,* **3**:303–313, 2008.

Haibe-Kains, B. et al. *genefu: Relevant Functions for Gene Expression Analysis, Especially in Breast Cancer.,* 2011. R package version 1.4.0.

Halperin, Y. et al. Allegro: analyzing expression and sequence in concert to discover regulatory programs. *Nucleic Acids Res.,* **37**:1566–1579, 2009.

Han, J.-D. J. et al. Evidence for dynamically organized modularity in the yeast protein-protein interaction network. *Nature,* **430**:88–93, 2004.

Han, L. et al. Mathematical modeling identified c-FLIP as an apoptotic switch in death receptor induced apoptosis. *Apoptosis,* **13**:1198–204, 2008.

Hanahan, D. and Weinberg, R. A. The hallmarks of cancer. *Cell*, **100**:57–70, 2000.

Hanahan, D. and Weinberg, R. A. Hallmarks of cancer: the next generation. *Cell*, **144**:646–674, 2011.

Haraksingh, R. R. et al. Genome-wide mapping of copy number variation in humans: comparative analysis of high resolution array platforms. *PLoS One*, **6**:e27859, 2011.

Harris, T. D. et al. Single-molecule DNA sequencing of a viral genome. *Science*, **320**:106–109, 2008.

Hartigan, J. *Clustering algorithms*. Wiley, New-York, 1975.

Hartwell, L. H. et al. From molecular to modular cell biology. *Nature*, **402**:C47–C52, 1999.

Hastie, T. et al. *The elements of statistical learning: data mining, inference, and prediction*. Springer, 2001.

Haury, A.-C. et al. The influence of feature selection methods on accuracy, stability and interpretability of molecular signatures. *PLoS One*, **6**:e28210, 2011.

Hedges, L. V. and Olkin, I. *Statistical methods for meta-analysis*. Academic Press, 1985.

Heiner, M. et al. Model validation of biological pathways using Petri nets–demonstrated for apoptosis. *Biosystems*, **75**:15–28, 2004.

Heinrich, R. et al. Mathematical models of protein kinase signal transduction. *Mol. Cell*, **9**:957–970, 2002.

Helleday, T. The underlying mechanism for the PARP and BRCA synthetic lethality: clearing up the misunderstandings. *Mol. Oncol.*, **5**:387–393, 2011.

Henley, C. L. Self-organized percolation. a simpler model. *Bull. Am. Phys. Soc.*, **34**:838, 1989.

Hess, K. R. et al. Pharmacogenomic predictor of sensitivity to preoperative chemotherapy with paclitaxel and fluorouracil, doxorubicin, and cyclophosphamide in breast cancer. *J. Clin. Oncol.*, **24**:4236–4244, 2006.

Hillmer, A. M. et al. Comprehensive long-span paired-end-tag mapping reveals characteristic patterns of structural variations in epithelial cancer genomes. *Genome Res.*, **21**:665–675, 2011.

Hirt, B. V. et al. The effects of a telomere destabilizing agent on cancer cell-cycle dynamics-Integrated modelling and experiments. *J. Theor. Biol.*, **295**:9–22, 2012.

Hoehme, S. and Drasdo, D. A cell-based simulation software for multi-cellular systems. *Bioinformatics*, **26**:2641–2642, 2010.

Hoff, D. D. V. et al. Pilot study using molecular profiling of patients' tumors to find potential targets and select treatments for their refractory cancers. *J. Clin. Oncol.*, **28**:4877–4883, 2010.

Hoffmann, A. et al. The IkappaB-NF-kappaB signaling module: temporal control and selective gene activation. *Science*, **298**:1241–1245, 2002.

Hoheisel, J. D. Microarray technology: beyond transcript profiling and genotype analysis. *Nat. Rev. Genet.*, **7**:200–210, 2006.

Hohlbein, J. et al. Surfing on a new wave of single-molecule fluorescence methods. *Phys. Biol.*, **7**:031001, 2010.

Holm, S. A simple sequentially rejective multiple test procedure. *Scandinavian Journal of Statistics*, **6**:65–70, 1979.

Hornberg, J. J. et al. Principles behind the multifarious control of signal transduction. ERK phosphorylation and kinase/phosphatase control. *FEBS J.*, **272**:244–

258, 2005.

Hornberg, J. J. et al. Cancer: a Systems Biology disease. *Biosystems*, **83**:81–90, 2006.

Hornberger, J. et al. Economic analysis of targeting chemotherapy using a 21-gene RT-PCR assay in lymph-node-negative, estrogen-receptor-positive, early-stage breast cancer. *Am. J. Manag. Care*, **11**:313–324, 2005.

Hotelling, H. Relation between two sets of variates. *Biometrika*, **28**:322–377, 1936.

Houchmandzadeh, B. et al. Establishment of developmental precision and proportions in the early Drosophila embryo. *Nature*, **415**:798–802, 2002.

Hoyer, P. O. Non-negative Matrix Factorization with sparseness constraints. *J. Mach. Learn. Res.*, **5**:1457–1469, 2004.

Hu, X. et al. Genetic alterations and oncogenic pathways associated with breast cancer subtypes. *Mol. Cancer Res.*, **7**:511–522, 2009.

Hu, Z. et al. The molecular portraits of breast tumors are conserved across microarray platforms. *BMC Genomics*, **7**:96, 2006.

Hua, F. et al. Effects of Bcl-2 levels on Fas signaling-induced caspase-3 activation: molecular genetic tests of computational model predictions. *J. Immunol.*, **175**:985–95, 2005.

Huang, C. Y. and Ferrell, J. E. Ultrasensitivity in the mitogen-activated protein kinase cascade. *PNAS*, **93**:10078–10083, 1996.

Hudis, C. Trastuzumab–mechanism of action and use in clinical practice. *N. Engl. J. Med.*, **357**:39–51, 2007.

Hull, D. et al. Taverna: a tool for building and running workflows of services. *Nucleic Acids Res.*, **34**:W729–W732, 2006.

Hupé, P. et al. Analysis of array CGH data: from signal ratio to gain and loss of DNA regions. *Bioinformatics*, **20**:3413–3422, 2004.

Hupé, P. *Biostatistical algorithms for omics data in oncology - Application to DNA copy number microarray experiments*. Ph.D. thesis, AgroParisTech, 2008.

Hyman, E. et al. Impact of DNA amplification on gene expression patterns in breast cancer. *Cancer Res.*, **62**:6240–6245, 2002.

Hyvärinen, A. et al. *Independent component analysis*. Wiley Interscience, 2001.

Iafrate, A. J. et al. Detection of large-scale variation in the human genome. *Nat. Genet.*, **36**:949–951, 2004.

Ideker, T. and Sharan, R. Protein networks in disease. *Genome Res.*, **18**:644–652, 2008.

Ideker, T. et al. A new approach to decoding life: systems biology. *Annu. Rev. Genomics Hum. Genet.*, **2**:343–372, 2001.

Ideker, T. et al. Discovering regulatory and signalling circuits in molecular interaction networks. *Bioinformatics*, **18 Suppl 1**:S233–S240, 2002.

Ideker, T. E. et al. Discovery of regulatory interactions through perturbation: inference and experimental design. *Pac. Symp. Biocomput.*, pages 305–316, 2000.

Ihekwaba, A. E. C. et al. Synergistic control of oscillations in the NF-kappaB signalling pathway. *Syst. Biol. (Stevenage)*, **152**:153–160, 2005.

International Cancer Genome Consortium et al. International network of cancer genome projects. *Nature*, **464**:993–998, 2010.

International HapMap Consortium. The International HapMap Project. *Nature*, **426**:789–796, 2003.

International Human Genome Sequencing Consortium. Initial sequencing and analysis of the human genome. *Nature*, **409**:860–921, 2001.

International Human Genome Sequencing Consortium. Finishing the euchromatic sequence of the human genome. *Nature*, **431**, 2004.

Ioannidis, J. P. A. et al. Repeatability of published microarray gene expression analyses. *Nat. Genet.*, **41**:149–155, 2009.

Irizarry, R. A. et al. Exploration, normalization, and summaries of high density oligonucleotide array probe level datas. *Biostatistics*, **4**:249–264, 2003.

Irizarry, R. A. et al. Comparison of Affymetrix GeneChip expression measures. *Bioinformatics*, **22**:789–794, 2006.

Ishkanian, A. S. et al. A tiling resolution DNA microarray with complete coverage of the human genome. *Nat. Genet.*, **36**:299–303, 2004.

Izhikevich, E. M. and Edelman, G. M. Large-scale model of mammalian thalamo-cortical systems. *PNAS*, **105**:3593–3598, 2008.

Jablonka, A. and Lamb, M. J. The changing concept of epigenetics. *Ann. N. Y. Acad. Sci.*, **981**:82–96, 2002.

Jacob, L. et al. Group lasso with overlap and graph lasso. In *ICML '09: Proceedings of the 26th Annual International Conference on Machine Learning*, pages 433–440. ACM, New York, NY, USA, 2009.

Jain, A. K. et al. Data clustering: a review. *ACM Comput. Surv.*, **31**:3, 1999.

Janoueix-Lerosey, I. et al. Preferential occurrence of chromosome breakpoints within early replicating regions in neuroblastoma. *Cell Cycle*, **4**:1842–1846, 2005.

Jansen, R. et al. Relating whole-genome expression data with protein-protein interactions. *Genome Res.*, **12**:37–46, 2002.

Jenatton, R. et al. Structured Variable Selection with Sparsity-Inducing Norms. *J. Mach. Learn. Res.*, **12**:2777–2824, 2011.

Jensen, L. J. and Bateman, A. The rise and fall of supervised machine learning techniques. *Bioinformatics*, **27**:3331–3332, 2011.

Jeong, H. et al. Lethality and centrality in protein networks. *Nature*, **411**:41–42, 2001.

Jin, F. et al. A yeast two-hybrid smart-pool-array system for protein-interaction mapping. *Nat. Methods*, **4**:405–407, 2007.

Johnson, S. A. and Hunter, T. Kinomics: methods for deciphering the kinome. *Nat. Methods*, **2**:17–25, 2005.

Johnson, W. E. et al. Adjusting batch effects in microarray expression data using empirical Bayes methods. *Biostatistics*, **8**:118–127, 2007.

Jolliffe, I. *Principal component analysis*. Springer-Verlag, New-York, 1996.

Jovanovic, J. et al. The epigenetics of breast cancer. *Mol. Oncol.*, **4**:242–254, 2010.

Junker, B. H. et al. VANTED: a system for advanced data analysis and visualization in the context of biological networks. *BMC Bioinformatics*, **7**:109, 2006.

Kallioniemi, A. CGH microarrays and cancer. *Curr. Opin. Biotechnol.*, **19**:36–40, 2008.

Kallioniemi, A. et al. Comparative genomic hybridization for molecular cytogenetic analysis of solid tumors. *Science*, **258**:818–821, 1992.

Kapp, A. V. et al. Discovery and validation of breast cancer subtypes. *BMC Genomics*, **7**:231, 2006.

Karev, G. P. et al. Mathematical modeling of tumor therapy with oncolytic viruses: effects of parametric heterogeneity on cell dynamics. *Biol. Direct*, **1**:30, 2006.

Karlsson, M. O. et al. Pharmacokinetic/pharmacodynamic modelling in oncological drug development. *Basic Clin. Pharmacol. Toxicol.*, **96**:206–211, 2005.

Karni, S. et al. A network-based method for predicting disease-causing genes. *J. Comput. Biol.*, **16**:181–189, 2009.

Kashtan, N. et al. Topological generalizations of network motifs. *Phys. Rev. E. Stat. Nonlin. Soft. Matter Phys.*, **70**:031909, 2004.

Kass, R. E. and Wasserman, L. A Reference Bayesian Test for Nested Hypotheses and its Relationship to the Schwarz Criterion. *J. Am. Stat. Assoc.*, **90**:928–934, 1995.

Katouli, A. A. and Komarova, N. L. The worst drug rule revisited: mathematical modeling of cyclic cancer treatments. *Bull. Math. Biol.*, **73**:549–584, 2011.

Kaufman, L. and Rousseeuw, P. *Finding groups in data: an introduction to cluster analysis.* John Wiley & Sons, New-York, 1990.

Kearns, J. D. et al. IkappaBepsilon provides negative feedback to control NF-kappaB oscillations, signaling dynamics, and inflammatory gene expression. *J. Cell Biol.*, **173**:659–664, 2006.

Kel, A. et al. ExPlain: finding upstream drug targets in disease gene regulatory networks. *SAR QSAR Environ. Res.*, **19**:481–494, 2008.

Kerr, M. K. and Churchill, G. A. Experimental design for gene expression microarrays. *Biostatistics*, **2**:183–201, 2001.

Khalil, I. G. and Hill, C. Systems biology for cancer. *Curr. Opin. Oncol.*, **17**:44–48, 2005.

Khatri, P. and Drăghici, S. Ontological analysis of gene expression data: current tools, limitations, and open problems. *Bioinformatics*, **21**:3587–3595, 2005.

Kholodenko, B. N. Negative feedback and ultrasensitivity can bring about oscillations in the mitogen-activated protein kinase cascades. *Eur. J. Biochem.*, **267**:1583–1588, 2000.

Kholodenko, B. N. MAP kinase cascade signaling and endocytic trafficking: a marriage of convenience? *Trends Cell. Biol.*, **12**:173–177, 2002.

Kim, E. et al. Insights into the connection between cancer and alternative splicing. *Trends Genet.*, **24**:7–10, 2008.

Kim, H. and Park, H. Sparse non-negative matrix factorizations via alternating non-negativity-constrained least squares for microarray data analysis. *Bioinformatics*, **23**:1495–1502, 2007.

Kimura, M. *The Neutral Theory of Molecular Evolution.* Cambridge University Press, ISBN 0-521-23109-4., 1983.

Kitano, H. *Foundations of Systems Biology.* MIT Press, 2001.

Kitano, H. Computational systems biology. *Nature*, **420**:206–210, 2002a.

Kitano, H. Systems biology: a brief overview. *Science*, **295**:1662–1664, 2002b.

Kitano, H. Cancer robustness: tumour tactics. *Nature*, **426**:125, 2003.

Kitano, H. Biological robustness. *Nat. Rev. Genet.*, **5**:826–837, 2004a.

Kitano, H. Cancer as a robust system: implications for anticancer therapy. *Nat. Rev. Cancer*, **4**:227–235, 2004b.

Kitano, H. Towards a theory of biological robustness. *Mol. Syst. Biol.*, **3**:137, 2007.

Kitano, H. Violations of robustness trade-offs. *Mol. Syst. Biol.*, **6**:384, 2010.

Kitano, H. et al. Using process diagrams for the graphical representation of biological networks. *Nat. Biotechnol.*, **8**:961–966, 2005.

Klamt, S. Generalized concept of minimal cut sets in biochemical networks. *Biosystems*, **83**:233–247, 2006.

Klein, C. A. Random mutations, selected mutations: A PIN opens the door to new genetic landscapes. *PNAS*, **103**:18033–18034, 2006.

Klipp, E. and Liebermeister, W. Mathematical modeling of intracellular signaling pathways. *BMC Neurosci.*, **7 Suppl 1**:S10, 2006.

Knudson, A. G. Mutation and Cancer: Statistical Study of Retinoblastoma. *PNAS*, **68**:820–823, 1971.

Koch, A. L. and Schaechter, M. A model for statistics of the cell division process. *J. Gen. Microbiol.*, **29**:435–454, 1962.

Kohn, K. W. Molecular interaction map of the mammalian cell cycle control and DNA repair systems. *Mol. Biol. Cell*, **10**:2703–2734, 1999.

Kohonen, T. The self-organising map. *Proc. IEEE*, **78**:1464–1479, 1990.

Komarova, D. W. N. *Computational Biology of Cancer: Lecture Notes and Mathematical Modeling.* World Scientific, 2005.

Komarova, N. L. Genomic instability in cancer: biological and mathematical approaches. *Cell Cycle*, **3**:1081–1085, 2004.

Komarova, N. L. et al. Selective pressures for and against genetic instability in cancer: a time-dependent problem. *J. R. Soc. Interface*, **5**:105–121, 2008.

Koonin, E. V. Darwinian evolution in the light of genomics. *Nucleic Acids Res.*, **37**:1011–1034, 2009.

Koren, A. et al. Autocorrelation analysis reveals widespread spatial biases in microarray experiments. *BMC Genomics*, **8**:164, 2007.

Kreeger, P. K. and Lauffenburger, D. A. Cancer systems biology: a network modeling perspective. *Carcinogenesis*, **31**:2–8, 2010.

Kreutz, C. and Timmer, J. Systems biology: experimental design. *FEBS J.*, **276**:923–942, 2009.

Krol, J. et al. The widespread regulation of microRNA biogenesis, function and decay. *Nat. Rev. Genet.*, **11**:597–610, 2010.

Krull, M. et al. TRANSPATH: an integrated database on signal transduction and a tool for array analysis. *Nucleic Acids Res.*, **31**:97–100, 2003.

Krull, M. et al. TRANSPATH: an information resource for storing and visualizing signaling pathways and their pathological aberrations. *Nucleic Acids Res.*, **34**:D546–51, 2006.

Krzywinski, M. et al. Circos: an information aesthetic for comparative genomics. *Genome Res.*, **19**:1639–1645, 2009.

Kurata, H. et al. Module-based analysis of robustness tradeoffs in the heat shock response system. *PLoS Comput. Biol.*, **2**:e59, 2006.

Kuznetsov, Y. *Elements of Applied Bifurcation Theory.* Springer-Verlag, New York, 2004.

LaBaer, J. and Ramachandran, N. Protein microarrays as tools for functional proteomics. *Curr. Opin. Chem. Biol.*, **9**:14–19, 2005.

Lahav, G. et al. Dynamics of the p53-Mdm2 feedback loop in individual cells. *Nat. Genet.*, **36**:147–50, 2004.

Laird, P. W. Principles and challenges of genome-wide DNA methylation analysis. *Nat. Rev. Genet.*, **11**:191–203, 2010.

Lanckriet, G. et al. Learning the kernel matrix with semidefinite programming. *J. Mach. Learn. Res.*, **5**:27–72, 2004a.

Lanckriet, G. R. G. et al. A statistical framework for genomic data fusion. *Bioinformatics*, **20**:2626–2635, 2004b.

Lander, E. S. Array of hope. *Nat. Genet.*, **21**:3–4, 1999.

Langmead, B. et al. Ultrafast and memory-efficient alignment of short DNA sequences to the human genome. *Genome Biol.*, **10**:R25, 2009.

Lavrik, I. N. et al. Analysis of CD95 threshold signaling: triggering of CD95 (FAS/APO-1) at low concentrations primarily results in survival signaling. *J. Biol. Chem.*, **282**:13664–71, 2007.

Lazebnik, Y. What are the hallmarks of cancer? *Nat. Rev. Cancer*, **10**:232–233, 2010.

Lê Cao, K.-A. et al. Sparse canonical methods for biological data integration: application to a cross-platform study. *BMC Bioinformatics*, **10**:34, 2009.

Le Novère, N. et al. Minimum information requested in the annotation of biochemical models (MIRIAM). *Nat. Biotechnol.*, **23**:1509–1515, 2005.

Le Novère, N. et al. The Systems Biology Graphical Notation. *Nat. Biotechnol.*, **27**:735–741, 2009.

Lee, D. D. and Seung, H. S. Learning the parts of objects by non-negative matrix factorization. *Nature*, **401**:788–791, 1999.

Lee, E. et al. Inferring pathway activity toward precise disease classification. *PLoS Comput. Biol.*, **4**:e1000217, 2008.

Lee, J. and Verleysen, M. *Nonlinear Dimensionality Reduction*. Springer, 2007.

Lee, J. W. et al. An extensive comparison of recent classification tools applied to microarray datas. *Comput. Stat. Data An.*, **48**:869–885, 2005.

Lee, S.-I. and Batzoglou, S. Application of independent component analysis to microarrays. *Genome Biol.*, **4**:R76, 2003.

van Leeuwen, I. M. et al. Crypt dynamics and colorectal cancer: advances in mathematical modelling. *Cell Prolif.*, **39**:157–181, 2006.

Legewie, S. et al. Mathematical Modeling Identifies Inhibitors of Apoptosis as Mediators of Positive Feedback and Bistability. *PLoS Comput. Biol.*, **2**:e120, 2006.

Leisch, F. Sweave: Dynamic generation of statistical reports using literate data analysis. In Härdle, W. and Rönz, B., editors, *Compstat 2002 - Proceedings in Computational Statistics*, pages 575–580. Physica Verlag, Heidelberg, 2002.

Lengauer, C. et al. Genetic instabilities in human cancers. *Nature*, **396**:643–649, 1998.

Lesne, A. Robustness: confronting lessons from physics and biology. *Biol. Rev. Camb. Philos. Soc.*, **83**:509–532, 2008.

Lev Bar-Or, R. et al. Generation of oscillations by the p53-Mdm2 feedback loop: a theoretical and experimental study. *PNAS*, **97**:11250–5, 2000.

Levchenko, A. et al. Scaffold proteins may biphasically affect the levels of mitogen-activated protein kinase signaling and reduce its threshold properties. *PNAS*, **97**:5818–5823, 2000.

Levi, F. et al. Implications of circadian clocks for the rhythmic delivery of cancer therapeutics. *Philos. Transact. A. Math. Phys. Eng. Sci.*, **366**:3575–3598, 2008.

Levi, F. et al. Circadian rhythms and cancer chronotherapeutics. In Cesario, A. and Marcus, F., editors, *Cancer Systems Biology, Bioinformatcs and Medicine: Research and Clinical Applications*. Springer, 2011.

Levine, H. A. et al. A mathematical model for the roles of pericytes and macrophages in the initiation of angiogenesis. I. The role of protease inhibitors in preventing angiogenesis. *Math. Biosci.*, **168**:77–115, 2000.

Levy, S. et al. The diploid genome sequence of an individual human. *PLoS Biol.*,

5:e254, 2007.

Li, C. et al. Modelling and simulation of signal transductions in an apoptosis pathway by using timed Petri nets. *J. Biosci.*, **32**:113–27, 2007.

Liao, J. C. et al. Network component analysis: Reconstruction of regulatory signals in biological systems. *PNAS*, **100**:15522–15527, 2003.

Lieberman-Aiden, E. et al. Comprehensive mapping of long-range interactions reveals folding principles of the human genome. *Science*, **326**:289–293, 2009.

Liebermeister, W. Linear modes of gene expression determined by independent component analysis. *Bioinformatics*, **18**:51–60, 2002.

Lievens, S. et al. Mammalian two-hybrids come of age. *Trends Biochem. Sci.*, **34**:579–588, 2009.

Lima-Mendez, G. and van Helden, J. The powerful law of the power law and other myths in network biology. *Mol. BioSyst.*, **5**:1482–1493, 2009.

Liotta, L. A. and Kohn, E. C. Cancer's deadly signature. *Nat. Genet.*, **33**:10–11, 2003.

Lipniacki, T. et al. Mathematical model of NF-kappaB regulatory module. *J. Theor. Biol.*, **228**:195–215, 2004.

Liu, M. et al. Network-based analysis of affected biological processes in type 2 diabetes models. *PLoS Genet.*, **3**:e96, 2007.

Loeb, L. A. et al. Errors in DNA replication as a basis of malignant changes. *Cancer Res.*, **34**:2311–2321, 1974.

Loi, S. et al. Definition of clinically distinct molecular subtypes in estrogen receptor-positive breast carcinomas through genomic grade. *J. Clin. Oncol.*, **25**:1239–1246, 2007.

Lønning, P. E. Breast cancer prognostication and prediction: are we making progress? *Ann. Oncol.*, **18 Suppl 8**:viii3–viii7, 2007.

Lord, C. J. and Ashworth, A. The DNA damage response and cancer therapy. *Nature*, **481**:287–294, 2012.

Lu, P. et al. The extracellular matrix: A dynamic niche in cancer progression. *J. Cell Biol.*, **196**:395–406, 2012.

Luan, B. et al. Characterizing and controlling the motion of ssDNA in a solid-state nanopore. *Biophys. J.*, **101**:2214–2222, 2011.

Lund, E. and Dumeaux, V. Systems epidemiology in cancer. *Cancer Epidemiol Biomarkers Prev*, **17**:2954–2957, 2008.

Luo, Y. et al. Mammalian two-hybrid system: a complementary approach to the yeast two-hybrid system. *BioTechniques*, **22**:350–352, 1997.

Lyapunov, A. *Stability of Motion*. Academic Press, New-York & London, 1966.

Ma, H. et al. The Edinburgh human metabolic network reconstruction and its functional analysis. *Mol. Syst. Biol.*, **3**:135, 2007.

Ma, L. et al. A plausible model for the digital response of p53 to DNA damage. *PNAS*, **102**:14266–71, 2005.

Ma, W. et al. Robustness and modular design of the Drosophila segment polarity network. *Mol. Syst. Biol.*, **2**:70, 2006.

Mac Gabhann, F. and Popel, A. S. Targeting neuropilin-1 to inhibit VEGF signaling in cancer: Comparison of therapeutic approaches. *PLoS Comput. Biol.*, **2**:e180, 2006.

MacBeath, G. Protein microarrays and proteomics. *Nat. Genet.*, **32 Suppl**:526–532, 2002.

Machado, D. et al. Modeling formalisms in Systems Biology. *AMB Express*, **1**:45, 2011.

Macklin, P. et al. Multiscale modelling and nonlinear simulation of vascular tumour growth. *J. Math. Biol.*, **58**:765–798, 2009.

MacLachlan. *Discriminant analysis and statistical pattern recognition*. New York: John Wiley & Sons, 1992.

MacQueen, J. Some Methods for classification and Analysis of Multivariate Observations. In *Proceedings of 5th Berkeley Symposium on Mathematical Statistics and Probability*, pages 281–297. University of California Press, 1967.

Madeira, S. C. and Oliveira, A. L. Biclustering algorithms for biological data analysis: a survey. *IEEE/ACM Trans Comput Biol Bioinform*, **1**:24–45, 2004.

Mairal, J. et al. Online Learning for Matrix Factorization and Sparse Coding. *J. Mach. Learn. Res.*, **11**:19–60, 2010.

Mamanova, L. et al. Target-enrichment strategies for next-generation sequencing. *Nat. Methods*, **7**:111–118, 2010.

Mantzaris, N. V. et al. Mathematical modeling of tumor-induced angiogenesis. *J. Math. Biol.*, **49**:111–187, 2004.

Manu et al. Canalization of gene expression and domain shifts in the Drosophila blastoderm by dynamical attractors. *PLoS Comput. Biol.*, **5**:e1000303, 2009.

MAQC Consortium et al. The MicroArray Quality Control (MAQC) project shows inter- and intraplatform reproducibility of gene expression measurements. *Nat. Biotechnol.*, **24**:1151–1161, 2006.

Mardis, E. R. The impact of next-generation sequencing technology on genetics. *Trends Genet.*, **24**:133–141, 2008a.

Mardis, E. R. Next-generation DNA sequencing methods. *Annu. Rev. Genomics Hum. Genet.*, **9**:387–402, 2008b.

Martin, J. A. and Wang, Z. Next-generation transcriptome assembly. *Nat. Rev. Genet.*, **12**:671–682, 2011.

Martin, N. K. et al. Tumour-stromal interactions in acid-mediated invasion: a mathematical model. *J. Theor. Biol.*, **267**:461–470, 2010.

Marusyk, A. and Polyak, K. Tumor heterogeneity: causes and consequences. *Biochim. Biophys. Acta*, **1805**:105–117, 2010.

Masel, J. and Trotter, M. V. Robustness and evolvability. *Trends Genet.*, **26**:406–414, 2010.

Mattick, J. S. Challenging the dogma: the hidden layer of non-protein-coding RNAs in complex organisms. *BioEssays*, **25**:930–939, 2003.

Mattick, J. S. and Makunin, I. V. Non-coding RNA. *Hum. Mol. Genet.*, **15**:R17–R29, 2006.

Mayawala, K. et al. MAPK cascade possesses decoupled controllability of signal amplification and duration. *Biophys. J.*, **87**:L01–L02, 2004.

Mayr, L. M. and Bojanic, D. Novel trends in high-throughput screening. *Curr. Opin. Pharmacol.*, **9**:580–588, 2009.

McCarthy, A. Third generation DNA sequencing: pacific biosciences' single molecule real time technology. *Chem. Biol.*, **17**:675–676, 2010.

McDougall, S. R. et al. Mathematical modelling of dynamic adaptive tumour-induced angiogenesis: clinical implications and therapeutic targeting strategies. *J. Theor. Biol.*, **241**:564–589, 2006.

Melke, P. et al. A rate equation approach to elucidate the kinetics and robustness of the TGF-beta pathway. *Biophys. J.*, **91**:4368–4380, 2006.

Mello, C. C. and Conte Jr, D. Revealing the world of RNA interference. *Nature*, **43**:338–342, 2004.

Meng, G. et al. A computational evaluation of over-representation of regulatory motifs in the promoter regions of differentially expressed genes. *BMC Bioinformatics*, **11**:267, 2010.

Merks, R. M. H. and Glazier, J. A. Dynamic mechanisms of blood vessel growth. *Nonlinearity*, **19**:C1–C10, 2006.

Metzker, M. L. Sequencing technologies - the next generation. *Nat. Rev. Genet.*, **11**:31–46, 2010.

Michiels, S. et al. Prediction of cancer outcome with microarrays: a multiple random validation strategy. *Lancet*, **365**:488–492, 2005.

Michor, F. et al. Dynamics of colorectal cancer. *Semin. Cancer Biol.*, **15**:484–493, 2005.

Milligan, G. W. and Cooper, M. C. An examination of procedures for determining the number of clusters in a data set. *Psychometrika*, **50**:159–179, 1985.

Mishra, K. P. et al. A review of high throughput technology for the screening of natural products. *Biomed. Pharmacother.*, **62**:94–98, 2008.

Misra, J. et al. Interactive exploration of microarray gene expression patterns in a reduced dimensional space. *Genome Res.*, **12**:1112–1120, 2002.

Mitelman, F. et al. The impact of translocations and gene fusions on cancer causation. *Nat. Rev. Cancer*, **7**:233–245, 2007.

Mitra, A. et al. p53 and retinoblastoma pathways in bladder cancer. *World J. Urol.*, **25**:563 571, 2007.

Molinaro, A. et al. Prediction error estimation: a comparison of resampling methods. *Bioinformatics*, **21**:3301–3307, 2005.

Mootha, V. K. et al. PGC-1α-responsive genes involved in oxidative phosphorylation are coordinately downregulated in human diabetes. *Nat. Genet.*, **34**:267–273, 2003.

Mordelet, F. and Vert, J.-P. ProDiGe: Prioritization Of Disease Genes with multitask machine learning from positive and unlabeled examples. *BMC Bioinformatics*, **12**:389, 2011.

Morley, M. et al. Genetic analysis of genome-wide variation in human gene expression. *Nature*, **430**:743–747, 2004.

Mukherjee, S. *The Emperor of All Maladies: A Biography of Cancer*. Scribner Book Company, 2010.

Nacu, S. et al. Gene expression network analysis and applications to immunology. *Bioinformatics*, **23**:850–858, 2007.

Nakabayashi, J. and Sasaki, A. A mathematical model for apoptosome assembly: The optimal cytochrome c/Apaf-1 ratio. *J. Theor. Biol.*, **242**:280–287, 2006.

Nakamura, Y. ATM: the p53 booster. *Nat. Med.*, **4**:1231–1232, 1998.

National Cancer Institute. A to Z list of cancers. http://cancer.gov/cancertopics/types/alphalist, 2012.

Nature Publishing Group. DNA Technologies - Milestones timeline. *Nature Milestones*, 2007. http://www.nature.com/milestones/miledna/timeline.html.

Navin, N. and Hicks, J. Future medical applications of single-cell sequencing in cancer. *Genome Med.*, **3**:31, 2011.

Navin, N. et al. Tumour evolution inferred by single-cell sequencing. *Nature*, **472**:90–

94, 2011.

Navin, N. E. and Hicks, J. Tracing the tumor lineage. *Mol. Oncol.*, **4**:267–283, 2010.

Negrini, S. et al. Genomic instability–an evolving hallmark of cancer. *Nat. Rev. Mol. Cell Biol.*, **11**:220–228, 2010.

Netterwald, J. The $1000 Genome: Coming Soon? *Drug Discovery & Development*, **13**:14–15, 2010.

Neuvial, P. et al. Spatial normalization of array-CGH data. *BMC Bioinformatics*, **7**:264, 2006.

Nguyen, N.-K. and Williams, E. R. Experimental designs for 2-colour cDNA microarray experiments. *Appl. Stoch. Model. Bus.*, **22**:631–638, 2006.

Niedringhaus, T. P. et al. Landscape of Next-Generation Sequencing Technologies. *Anal. Chem.*, 2011.

Nielsen, C. B. et al. Visualizing genomes: techniques and challenges. *Nat. Methods*, **7**:S5–S15, 2010.

Nikitin, A. et al. Pathway studio–the analysis and navigation of molecular networks. *Bioinformatics*, **19**:2155–2157, 2003.

Noble, W. S. A quick guide to organizing computational biology projects. *PLoS Comput. Biol.*, **5**:e1000424, 2009.

Nooter, K. and Herweijer, H. Multidrug resistance (mdr) genes in human cancer. *Br. J. Cancer*, **63**:663–669, 1991.

Nordling, C. O. A new theory on cancer-inducing mechanism. *Br. J. Cancer*, **7**:68–72, 1953.

Novak, B. and Tyson, J. J. A model for restriction point control of the mammalian cell cycle. *J. Theor. Biol.*, **230**:563–579, 2004.

Novak, P. et al. Epigenetic inactivation of the HOXA gene cluster in breast cancer. *Cancer Res.*, **66**:10664–10670, 2006.

Novak, P. et al. Agglomerative epigenetic aberrations are a common event in human breast cancer. *Cancer Res.*, **68**:8616–8625, 2008.

Novikov, E. and Barillot, E. Software package for automatic microarray image analysis (MAIA). *Bioinformatics*, **23**:639–640, 2007.

Nowell, P. C. The clonal evolution of tumor cell populations. *Science*, **194**:23–28, 1976.

Oberhardt, M. A. et al. Applications of genome-scale metabolic reconstructions. *Mol. Syst. Biol.*, **5**:320, 2009.

Obozinski, G. et al. Group Lasso with Overlaps: the Latent Group Lasso approach. Technical Report inria-00628498, HAL, 2011.

O'Connor, K. C. et al. Modeling suppression of cell death by Bcl-2 over-expression in myeloma NS0 6A1 cells. *Biotechnol. Lett.*, **28**:1919–24, 2006.

Oda, K. and Kitano, H. A comprehensive map of the toll-like receptor signaling network. *Mol. Syst. Biol.*, **2**:2006.0015, 2006.

Oda, K. et al. A comprehensive pathway map of epidermal growth factor receptor signaling. *Mol. Syst. Biol.*, **1**:2005.0010, 2005.

Okasha, S. Why Won't the Group Selection Controversy Go Away? *Brit. J. Philos. Sci.*, **52(1)**:25–50, 2001.

Ong, S.-E. et al. Stable isotope labeling by amino acids in cell culture, SILAC, as a simple and accurate approach to expression proteomics. *Mol. Cell. Proteomics*, **1**:376–386, 2002.

Orme, M. E. and Chaplain, M. A. A mathematical model of the first steps of tumour-related angiogenesis: capillary sprout formation and secondary branching. *IMA J. Math. Appl. Med. Biol.*, **13**:73–98, 1996.

Orme, M. E. and Chaplain, M. A. Two-dimensional models of tumour angiogenesis and anti-angiogenesis strategies. *IMA J. Math. Appl. Med. Biol.*, **14**:189–205, 1997.

Ortiz-Estevez, M. et al. ACNE: a summarization method to estimate allele-specific copy numbers for Affymetrix SNP arrays. *Bioinformatics*, **26**:1827–1833, 2010.

Osella, M. et al. The role of incoherent microRNA-mediated feedforward loops in noise buffering. *PLoS Comput. Biol.*, **7**:e1001101, 2011.

Owen, M. R. et al. Angiogenesis and vascular remodelling in normal and cancerous tissues. *J. Math. Biol.*, **58**:689–721, 2009.

Paatero, P. and Tapper, U. Positive matrix factorization: A non-negative factor model with optimal utilization of error estimates of data values. *Environmetrics*, **5**:111–126, 1994.

Paik, S. Development and clinical utility of a 21-gene recurrence score prognostic assay in patients with early breast cancer treated with tamoxifen. *Oncologist*, **12**:631–635, 2007.

Paik, S. et al. Gene expression and benefit of chemotherapy in women with node-negative, estrogen receptor-positive breast cancer. *J. Clin. Oncol.*, **24**:3726–3734, 2006.

Palsson, B. *Systems Biology: Properties of Reconstructed Networks*. Cambridge University Press, 2006.

Pardee, A. B. A restriction point for control of normal animal cell proliferation. *PNAS*, **71**:1286–1290, 1974.

Parker, J. S. et al. Supervised risk predictor of breast cancer based on intrinsic subtypes. *J. Clin. Oncol.*, **27**:1160–1167, 2009.

Parkhomenko, E. et al. Sparse canonical correlation analysis with application to genomic data integration. *Stat. Appl. Genet. Mol. Biol.*, **8**:Article 1, 2009.

Patil, K. R. and Nielsen, J. Uncovering transcriptional regulation of metabolism by using metabolic network topology. *PNAS*, **102**:2685–2689, 2005.

Pattee, H. The Evolution of Self-Simplifying Systems. In E.Laszlo, editor, *The Relevance of General Systems Theory*. New York: Braziller, 1972.

Patterson, S. D. and Aebersold, R. H. Proteomics: the first decade and beyond. *Nat. Genet.*, **33 Suppl**:311–323, 2003.

Pavesi, G. et al. Weeder Web: discovery of transcription factor binding sites in a set of sequences from co-regulated genes. *Nucleic Acids Res.*, **32**:W199–W203, 2004.

Pavlidis, P. et al. Exploring gene expression data with class scores. *Pac. Symp. Biocomput.*, pages 474–485, 2002a.

Pavlidis, P. et al. Learning Gene Functional Classifications from Multiple Data Types. *J. Comput. Biol.*, **9**:401–411, 2002b.

Pearson, K. On lines and planes of closest fit to systems of points in space. *Philos. Mag.*, **2**:559–572, 1901.

Peirce, S. M. Computational and mathematical modeling of angiogenesis. *Microcirculation*, **15**:739–751, 2008.

Pelleg, D. and Moore, A. X-means: Extending K-means with efficient estimation of the number of clusters. In *Proceedings of the Seventeenth International Conference on Machine Learning*, pages 727–734. Morgan Kaufmann, San Francisco, 2000.

Peng, J. et al. Regularized Multivariate Regression for Identifying Master Predictors with Application to Integrative Genomics Study of Breast Cancer. *Ann. Appl. Stat.*, **4**:53–77, 2010.

Perez-Iratxeta, C. et al. Association of genes to genetically inherited diseases using data mining. *Nat. Genet.*, **31**:316–319, 2002.

Perfahl, H. et al. Multiscale modelling of vascular tumour growth in 3D: the roles of domain size and boundary conditions. *PLoS One*, **6**:e14790, 2011.

Perou, C. M. et al. Molecular portraits of human breast tumours. *Nature*, **406**:747–752, 2000.

Perry, G. H. et al. Diet and the evolution of human amylase gene copy number variation. *Nat. Genet.*, **39**:1256–1260, 2007.

Philippi, N. et al. Modeling system states in liver cells: survival, apoptosis and their modifications in response to viral infection. *BMC Syst. Biol.*, **3**:97, 2009.

de Pillis, L. G. et al. A validated mathematical model of cell-mediated immune response to tumor growth. *Cancer Res.*, **65**:7950–7958, 2005.

Pinkel, D. and Albertson, D. G. Array comparative genomic hybridization and its applications in cancer. *Nat. Genet.*, **37 Suppl**:S11–S17, 2005.

Pinkel, D. et al. High resolution analysis of DNA copy number variation using comparative genomic hybridization to microarrays. *Nat. Genet.*, **20**:207–211, 1998.

Podsypanina, K. et al. Seeding and propagation of untransformed mouse mammary cells in the lung. *Science*, **321**:1841–1844, 2008.

Pollack, J. R. et al. Genome-wide analysis of DNA copy-number changes using cDNA microarrays. *Nat. Genet.*, **23**:41–46, 1999.

Pollack, J. R. et al. Microarray analysis reveals a major direct role of DNA copy number alteration in the transcriptional program of human breast tumors. *PNAS*, **99**:12963–12968, 2002.

Polouliakh, N. et al. G-protein coupled receptor signaling architecture of mammalian immune cells. *PLoS One*, **4**:e4189, 2009.

Popova, T. et al. Genome Alteration Print (GAP): a tool to visualize and mine complex cancer genomic profiles obtained by SNP arrays. *Genome Biol.*, **10**:R128, 2009.

Portales-Casamar, E. et al. JASPAR 2010: the greatly expanded open-access database of transcription factor binding profiles. *Nucleic Acids Res.*, **38**:D105–D110, 2010.

Portela, A. and Esteller, M. Epigenetic modifications and human disease. *Nat. Biotechnol.*, **28**:1057–1068, 2010.

Prahallad, A. et al. Unresponsiveness of colon cancer to BRAF(V600E) inhibition through feedback activation of EGFR. *Nature*, **483**:100–103, 2012.

Praz, V. et al. CleanEx: a database of heterogeneous gene expression data based on a consistent gene nomenclature. *Nucleic Acids Res.*, **32**:D542–D547, 2004.

Prill, R. J. et al. Dynamic properties of network motifs contribute to biological network organization. *PLoS Biol.*, **3**:e343, 2005.

Priness, I. et al. Evaluation of gene-expression clustering via mutual information distance measure. *BMC Bioinformatics*, **8**:111, 2007.

Pruessner, G. and Jensen, H. J. Broken scaling in the forest-fire model. *Phys. Rev. E. Stat. Nonlin. Soft. Matter Phys.*, **65**:056707, 2002.

Puig, O. et al. The tandem affinity purification (TAP) method: a general procedure of protein complex purification. *Methods*, **24**:218–229, 2001.

Pujana, M. A. et al. Network modeling links breast cancer susceptibility and centrosome dysfunction. *Nat. Genet.*, **39**:1338–1349, 2007.

Pushkarev, D. et al. Single-molecule sequencing of an individual human genome. *Nat. Biotechnol.*, **27**:847–852, 2009.

Qu, Z. et al. Dynamics of the cell cycle: checkpoints, sizers, and timers. *Biophys. J.*, **85**:3600–3611, 2003a.

Qu, Z. et al. Regulation of the mammalian cell cycle: a model of the G1-to-S transition. *Am. J. Physiol. Cell. Physiol.*, **284**:C349–C364, 2003b.

Quackenbush, J. Microarray data normalization and transformation. *Nat. Genet.*, **32 Suppl**:496–501, 2002.

R Development Core Team. *R: A Language and Environment for Statistical Computing*. R Foundation for Statistical Computing, Vienna, Austria, 2011. ISBN 3-900051-07-0.

Radulescu, O. et al. Hierarchies and modules in complex biological systems. In *Proceedings of European Conference on Complex Systems, Oxford, UK*. 2006.

Radulescu, O. et al. Model reduction and model comparison for NFkB signalling. In *Proceedings of Foundations of Systems Biology in Engineering (FOSBE-2007), Stuttgart, Germany*. 2007.

Radulescu, O. et al. Robust simplifications of multiscale biochemical networks. *BMC Syst. Biol.*, **2**:86, 2008.

Rahnenführer, J. et al. Calculating the statistical significance of changes in pathway activity from gene expression data. *Stat. Appl. Genet. Mol. Biol.*, **3**:Article16, 2004.

Rajagopalan, D. and Agarwal, P. Inferring pathways from gene lists using a literature-derived network of biological relationships. *Bioinformatics*, **21**:788–793, 2005.

Ramaswamy, S. et al. A molecular signature of metastasis in primary solid tumors. *Nat. Genet.*, **33**:49–54, 2003.

Ramis-Conde, I. et al. Multi-scale modelling of cancer cell intravasation: the role of cadherins in metastasis. *Phys. Biol.*, **6**:016008, 2009.

Rapaport, F. et al. Classification of microarray data using gene networks. *BMC Bioinformatics*, **8**:35, 2007.

Ravdin, P. M. et al. Computer program to assist in making decisions about adjuvant therapy for women with early breast cancer. *J. Clin. Oncol.*, **19**:980–991, 2001.

Raychaudhuri, S. et al. Principal components analysis to summarize microarray experiments: application to sporulation time series. *Pac. Symp. Biocomput.*, pages 455–466, 2000.

Redon, R. et al. Global variation in copy number in the human genome. *Nature*, **444**:444–454, 2006.

Rehm, M. et al. Dynamics of outer mitochondrial membrane permeabilization during apoptosis. *Cell Death Differ.*, **16**:613–623, 2009.

Reidys, C. et al. Generic properties of combinatory maps: neutral networks of RNA secondary structures. *Bull. Math. Biol.*, **59**:339–397, 1997.

Reinders, J. et al. Genome-wide, high-resolution DNA methylation profiling using bisulfite-mediated cytosine conversion. *Genome Res.*, **18**:469–76, 2008.

Resendis-Antonio, O. et al. Modeling core metabolism in cancer cells: surveying the topology underlying the Warburg effect. *PLoS One*, **5**:e12383, 2010.

Rhodes, D. R. et al. Large-scale meta-analysis of cancer microarray data identifies common transcriptional profiles of neoplastic transformation and progression.

PNAS, **101**:9309–9314, 2004a.

Rhodes, D. R. et al. ONCOMINE: a cancer microarray database and integrated data-mining platform. *Neoplasia*, **6**:1–6, 2004b.

Rhodes, D. R. et al. Mining for regulatory programs in the cancer transcriptome. *Nat. Genet.*, **37**:579–583, 2005a.

Rhodes, D. R. et al. Probabilistic model of the human protein-protein interaction network. *Nat. Biotechnol.*, **23**:951–959, 2005b.

Rhodes, D. R. et al. Oncomine 3.0: genes, pathways, and networks in a collection of 18,000 cancer gene expression profiles. *Neoplasia*, **9**:166–180, 2007.

Ribba, B. et al. A multiscale mathematical model of cancer, and its use in analyzing irradiation therapies. *Theor. Biol. Med. Model.*, **3**:7, 2006.

Richter, B. G. and Sexton, D. P. Managing and analyzing next-generation sequence data. *PLoS Comput. Biol.*, **5**:e1000369, 2009.

Rigaill, G. et al. ITALICS: an algorithm for normalization and DNA copy number calling for Affymetrix SNP arrays. *Bioinformatics*, **24**:768–774, 2008.

Rigaut, G. et al. A generic protein purification method for protein complex characterization and proteome exploration. *Nat. Biotechnol.*, **17**:1030–1032, 1999.

Risso, D. et al. GC-Content Normalization for RNA-Seq Data. *BMC Bioinformatics*, **12**:480, 2011.

Rivals, I. et al. Enrichment or depletion of a GO category within a class of genes: which test? *Bioinformatics*, **23**:401–407, 2007.

Rodriguez-Brenes, I. A. and Peskin, C. S. Quantitative theory of telomere length regulation and cellular senescence. *PNAS*, **107**:5387–5392, 2010.

Rodríguez-Paredes, M. and Esteller, M. Cancer epigenetics reaches mainstream oncology. *Nat. Med.*, **17**:330–339, 2011.

Roepstorff, P. and Fohlman, J. Proposal for a common nomenclature for sequence ions in mass spectra of peptides. *Biomed. Mass Spectrom.*, **11**:601, 1984.

Roschke, A. V. et al. Karyotypic complexity of the NCI-60 drug-screening panel. *Cancer Res.*, **63**:8634–8647, 2003.

Ross, P. L. et al. Multiplexed protein quantitation in Saccharomyces cerevisiae using amine-reactive isobaric tagging reagents. *Mol. Cell. Proteomics*, **3**:1154–1169, 2004.

Roth, V. The generalized LASSO. *IEEE Trans. Neural Netw.*, **15**:16–28, 2004.

Rothberg, J. M. et al. An integrated semiconductor device enabling non-optical genome sequencing. *Nature*, **475**:348–352, 2011.

Roweis, S. T. and Saul, L. K. Nonlinear dimensionality reduction by locally linear embedding. *Science*, **290**:2323–6, 2000.

Rual, J.-F. et al. Towards a proteome-scale map of the human protein-protein interaction network. *Nature*, **437**:1173–1178, 2005.

Rudin, L. I. et al. Nonlinear total variation based noise removal algorithms. *Physica D*, **60**:259–268, 1992.

Rusk, N. and Kiermer, V. Primer: Sequencing - the next generation. *Nat. Methods*, **5**:15, 2008.

Sabbah, C. et al. SMETHILLIUM: spatial normalization method for Illumina infinium HumanMethylation BeadChip. *Bioinformatics*, **27**:1693–1695, 2011.

Sachidanandam, R. et al. A map of human genome sequence variation containing 1.42 million single nucleotide polymorphisms. *Nature*, **409**:928–933, 2001.

Sahin, O. et al. Modeling ERBB receptor-regulated G1/S transition to find novel targets for de novo trastuzumab resistance. *BMC Syst. Biol.*, **3**:1, 2009.

Sakai, W. et al. Secondary mutations as a mechanism of cisplatin resistance in BRCA2-mutated cancers. *Nature*, **451**:1116–1120, 2008.

Samaga, R. et al. Computing combinatorial intervention strategies and failure modes in signaling networks. *J. Comput. Biol.*, **17**:39–53, 2010.

Satzinger, H. Theodor and Marcella Boveri: chromosomes and cytoplasm in heredity and development. *Nat. Rev. Genet.*, **9**:231–238, 2008.

Sawyers, C. L. The cancer biomarker problem. *Nature*, **452**:548–552, 2008.

Scherer, A., editor. *Batch Effects and Noise in Microarray Experiments: Sources and Solutions.* John Wiley and Sons, 2009.

Schlatter, R. et al. ON/OFF and beyond–a boolean model of apoptosis. *PLoS Comput. Biol.*, **5**:e1000595, 2009.

Schoeberl, B. et al. Computational modeling of the dynamics of the MAP kinase cascade activated by surface and internalized EGF receptors. *Nat. Biotechnol.*, **20**:370–375, 2002.

Schölkopf, B. and Smola, A. J. *Learning with Kernels: Support Vector Machines, Regularization, Optimization, and Beyond.* MIT Press, Cambridge, MA, 2002.

Schölkopf, B. et al. Kernel principal component analysis. In Schölkopf, B. et al., editors, *Advances in Kernel Methods - Support Vector Learning*, pages 327–352. MIT Press, 1999.

Schölkopf, B. et al. *Kernel Methods in Computational Biology.* MIT Press, The MIT Press, Cambridge, Massachussetts, 2004.

Schones, D. E. and Zhao, K. Genome-wide approaches to studying chromatin modifications. *Nat. Rev. Genet.*, **9**:179–191, 2008.

Schumacher, A. et al. Microarray-based DNA methylation profiling: technology and applications. *Nucleic Acids Res.*, **34**:528–542, 2006.

Schuster, S. C. Next-generation sequencing transforms today's biology. *Nat. Methods*, **5**:16–18, 2007.

Schwartz, M. A. and Baron, V. Interactions between mitogenic stimuli, or, a thousand and one connections. *Curr. Opin. Cell Biol.*, **11**:197–202, 1999.

Scott, M. S. et al. Identifying regulatory subnetworks for a set of genes. *Mol. Cell. Proteomics*, **4**:683–692, 2005.

Shadforth, I. et al. Protein and peptide identification algorithms using MS for use in high-throughput, automated pipelines. *Proteomics*, **5**:4082–4095, 2005.

Shann, Y. et al. Genome-Wide Mapping and Characterization of Hypomethylated Sites in Human Tissues and Breast Cancer Cell Lines. *Genome Res.*, **18**:791–801, 2008.

Shannon, P. et al. Cytoscape: a software environment for integrated models of biomolecular interaction networks. *Genome Res.*, **13**:2498–2504, 2003.

Shawe-Taylor, J. and Cristianini, N. *Kernel Methods for Pattern Analysis.* Cambridge University Press, New York, NY, USA, 2004.

Sherr, C. J. Cancer cell cycles. *Science*, **274**:1672–1677, 1996.

Sherr, C. J. Principles of Tumor Suppression. *Cell*, **116**:235–246, 2004.

Sherry, S. T. et al. dbSNP: the NCBI database of genetic variation. *Nucleic Acids Res.*, **29**:308–311, 2001.

Shevade, S. K. and Keerthi, S. S. A simple and efficient algorithm for gene selection using sparse logistic regression. *Bioinformatics*, **19**:2246–2253, 2003.

Shi, J. et al. A survey of optimization models on cancer chemotherapy treatment planning. *Ann. Oper. Res.*, 2011.

Shi, L. et al. The MicroArray Quality Control (MAQC)-II study of common practices for the development and validation of microarray-based predictive models. *Nat. Biotechnol.*, **28**:827–838, 2010.

Shibata, T. and Fujimoto, K. Noisy signal amplification in ultrasensitive signal transduction. *PNAS*, **102**:331–336, 2005.

Shieh, A. C. and Swartz, M. A. Regulation of tumor invasion by interstitial fluid flow. *Phys. Biol.*, **8**:015012, 2011.

Shieh, A. C. et al. Tumor cell invasion is promoted by interstitial flow-induced matrix priming by stromal fibroblasts. *Cancer Res.*, **71**:790–800, 2011.

Shih, T. and Lindley, C. Bevacizumab: an angiogenesis inhibitor for the treatment of solid malignancies. *Clin. Ther.*, **28**:1779–1802, 2006.

Shinar, G. and Feinberg, M. Structural sources of robustness in biochemical reaction networks. *Science*, **327**:1389–1391, 2010.

Shoval, O. and Alon, U. SnapShot: network motifs. *Cell*, **143**:326–3e1, 2010.

Simonis, M. et al. An evaluation of 3C-based methods to capture DNA interactions. *Nat. Methods*, **4**:895–901, 2007.

Slonim, D. K. From patterns to pathways: gene expression data analysis comes of age. *Nat. Genet.*, **32 Suppl**:502–508, 2002.

Smyth, G. K. Linear models and empirical Bayes methods for assessing differential expression in microarray experiments. *Stat. Appl. Genet. Mol. Biol.*, **3**:Article3, 2004.

Sohler, F. et al. New methods for joint analysis of biological networks and expression data. *Bioinformatics*, **20**:1517–1521, 2004.

Solinas-Toldo, S. et al. Matrix-based comparative genomic hybridization: Biochips to screen for genomic imbalances. *Genes Chromosomes Cancer*, **20**:399–407, 1997.

Soneson, C. et al. Integrative analysis of gene expression and copy number alterations using canonical correlation analysis. *BMC Bioinformatics*, **11**:191, 2010.

Sonnenschein, C. and Soto, A. M. Why systems biology and cancer? *Semin. Cancer Biol.*, **21**:147–149, 2011.

Sørlie, T. et al. Gene expression patterns of breast carcinomas distinguish tumor subclasses with clinical implications. *PNAS*, **98**:10869–10874, 2001.

Sørlie, T. et al. Repeated observation of breast tumor subtypes in independent gene expression data sets. *PNAS*, **100**:8418–8423, 2003.

Sotiriou, C. et al. Breast cancer classification and prognosis based on gene expression profiles from a population-based study. *PNAS*, **100**:10393–10398, 2003.

Southern, E. et al. Molecular interactions on microarrays. *Nat. Genet.*, **21**:5–9, 1999.

Speybroeck, L. V. From Epigenesis to Epigenetics: The Case of C. H. Waddington. *Ann. N. Y. Acad. Sci.*, **981**:61–81, 2002.

Spurrier, B. et al. Reverse-phase protein lysate microarrays for cell signaling analysis. *Nat. Protoc.*, **3**:1796–1808, 2008.

Srebrow, A. and Kornblihtt, A. R. The connection between splicing and cancer. *J. Cell Sci.*, **119**:2635–2641, 2006.

Stafford, P. *Methods in Microarray Normalization.* CRC Press, 2007.

Stein, L. D. The case for cloud computing in genome informatics. *Genome Biol.*, **11**:207, 2010.

Stelzl, U. et al. A human protein-protein interaction network: a resource for annotating the proteome. *Cell*, **122**:957–968, 2005.

Stephens, P. J. et al. Massive genomic rearrangement acquired in a single catastrophic event during cancer development. *Cell*, **144**:27–40, 2011.

Steuer, R. et al. The mutual information: detecting and evaluating dependencies between variables. *Bioinformatics*, **18 Suppl 2**:S231–S240, 2002.

Stoddart, D. et al. Nucleobase recognition in ssDNA at the central constriction of the alpha-hemolysin pore. *Nano. Lett.*, **10**:3633–3637, 2010.

Stokes, C. L. and Lauffenburger, D. A. Analysis of the roles of microvessel endothelial cell random motility and chemotaxis in angiogenesis. *J. Theor. Biol.*, **152**:377–403, 1991.

Stommel, J. M. et al. Coactivation of receptor tyrosine kinases affects the response of tumor cells to targeted therapies. *Science*, **318**:287–290, 2007.

Stransky, N. et al. Regional copy number-independent deregulation of transcription in cancer. *Nat. Genet.*, **38**:1386–1396, 2006.

Stratton, M. and Rahman, N. The emerging landscape of breast cancer susceptibility. *Nat. Genet.*, **40**:17–22, 2008.

Stratton, M. R. et al. The cancer genome. *Nature*, **458**:719–724, 2009.

Strebhardt, K. and Ullrich, A. Paul Ehrlich's magic bullet concept: 100 years of progress. *Nat. Rev. Cancer*, **8**:473–480, 2008.

Stucki, J. W. and Simon, H.-U. Mathematical modeling of the regulation of caspase-3 activation and degradation. *J. Theor. Biol.*, **234**:123–131, 2005.

Subramanian, A. et al. Gene set enrichment analysis: a knowledge-based approach for interpreting genome-wide expression profiles. *PNAS*, **102**:15545–15550, 2005.

Suter, B. et al. Two-hybrid technologies in proteomics research. *Curr. Opin. Biotechnol.*, **19**:316–323, 2008.

Sutherland, H. and Bickmore, W. A. Transcription factories: gene expression in unions? *Nat. Rev. Genet.*, **10**:457–466, 2009.

Swan, G. W. and Vincent, T. L. Optimal control analysis in the chemotherapy of IgG multiple myeloma. *Bull. Math. Biol.*, **39**:317–337, 1977.

Swanson, K. R. et al. Virtual and real brain tumors: using mathematical modeling to quantify glioma growth and invasion. *J. Neurol. Sci.*, **216**:1–10, 2003.

Taby, R. and Issa, J.-P. J. Cancer epigenetics. *CA Cancer J. Clin.*, **60**:376–392, 2010.

Talagrand, M. New concentration inequalities for product spaces. *Inventionnes Math.*, **126**:505–563, 1996.

Talmadge, J. E. Clonal selection of metastasis within the life history of a tumor. *Cancer Res.*, **67**:11471–11475, 2007.

Tamayo, P. et al. Interpreting patterns of gene expression with self-organizing maps: methods and application to hematopoietic differentiation. *PNAS*, **96**:2907–2912, 1999.

Tanaka, G. et al. Mathematical modelling of prostate cancer growth and its application to hormone therapy. *Philos. Transact. A. Math. Phys. Eng. Sci.*, **368**:5029–5044, 2010.

Tanizawa, H. and Noma, K.-I. Unravelling global genome organization by 3C-seq. *Semin. Cell. Dev. Biol.*, 2011.

Taylor, C. F. et al. The minimum information about a proteomics experiment (MIAPE). *Nat. Biotechnol.*, **25**:887–893, 2007.

Taylor, C. F. et al. Guidelines for reporting the use of mass spectrometry in proteomics. *Nat. Biotechnol.*, **26**:860–861, 2008.

Taylor, J. S. and Raes, J. Duplication and divergence: the evolution of new genes and old ideas. *Annu. Rev. Genet.*, **38**:615–643, 2004.

Teer, J. K. and Mullikin, J. C. Exome sequencing: the sweet spot before whole genomes. *Hum. Mol. Genet.*, 2010.

Tenenbaum, J. B. et al. A global geometric framework for nonlinear dimensionality reduction. *Science*, **290**:2319–23, 2000.

Terentiev, A. A. et al. Dynamic proteomics in modeling of the living cell. Protein-protein interactions. *Biochemistry (Mosc)*, **74**:1586–1607, 2009.

Thiery, J. P. et al. Epithelial-mesenchymal transitions in development and disease. *Cell*, **139**:871–890, 2009.

Thom, R. *Structural Stability and Morphogenesis (English translation of the 1972 French edition)*. W.A. Benjamin, Inc., Reading, MA, 1975.

Thomas, D. C. et al. Use of pathway information in molecular epidemiology. *Hum. Genomics*, **4**:21–42, 2009.

Thomas, R. On the relation between the logical structure of systems and their ability to generate multiple steady states and sustained oscillations. In *Series in Synergetics*. Springer, 1981.

Thompson, E. W. and Haviv, I. The social aspects of EMT-MET plasticity. *Nat. Med.*, **17**:1048–1049, 2011.

Thompson, J. F. and Milos, P. M. The properties and applications of single-molecule DNA sequencing. *Genome Biol.*, **12**:217, 2011.

Tian, L. et al. Discovering statistically significant pathways in expression profiling studies. *PNAS*, **102**:13544–13549, 2005.

Tian, T. et al. The origins of cancer robustness and evolvability. *Integr. Biol. (Camb)*, **3**:17–30, 2011.

Tibshirani, R. Regression shrinkage and selection via the lasso. *J. R. Stat. Soc. Ser. B*, **58**:267–288, 1996.

Tibshirani, R. The lasso method for variable selection in the Cox model. *Stat. Med.*, **16**:385–395, 1997.

Tibshirani, R. et al. Estimating the number of clusters in a data set via the gap statistics. *J. R. Stat. Soc. Ser. B*, **63**:411–423, 2001.

Tibshirani, R. et al. Sparsity and smoothness via the fused lasso. *J. R. Stat. Soc. Ser. B Stat. Methodol.*, **67**:91–108, 2005.

Tiffin, N. et al. Integration of text- and data-mining using ontologies successfully selects disease gene candidates. *Nucleic Acids Res.*, **33**:1544–1552, 2005.

Todorovic-Rakovic, N. TGF-beta 1 could be a missing link in the interplay between ER and HER-2 in breast cancer. *Med. Hypotheses*, **65**:546–551, 2005.

Tomfohr, J. et al. Pathway level analysis of gene expression using singular value decomposition. *BMC Bioinformatics*, **6**:225, 2005.

Tomita, M. et al. E-CELL: software environment for whole-cell simulation. *Bioinformatics*, **15**:72–84, 1999.

Tomizaki, K. et al. Protein-protein interactions and selection: array-based techniques for screening disease-associated biomarkers in predictive/early diagnosis. *FEBS J.*, **277**:1996–2005, 2010.

Tompa, M. et al. Assessing computational tools for the discovery of transcription factor binding sites. *Nat. Biotechnol.*, **23**:137–144, 2005.

Torkkola, K. Feature Extraction by Non-Parametric Mutual Information Maximization. *J. Mach. Learn. Res.*, **3**:1415–1438, 2003.

Tournier, L. and Chaves, M. Uncovering operational interactions in genetic networks using asynchronous Boolean dynamics. *J. Theor. Biol.*, **260**:196–209, 2009.

Tranchevent, L.-C. et al. A guide to web tools to prioritize candidate genes. *Brief. Bioinform.*, **11**, 2010.

Trumpp, A. and Wiestler, O. D. Mechanisms of Disease: cancer stem cells–targeting the evil twin. *Nat. Clin. Pract. Oncol.*, **5**:337–347, 2008.

Trédan, O. et al. Drug resistance and the solid tumor microenvironment. *J. Natl. Canc. Inst.*, **99**:1441–1454, 2007.

Tsui, I. F. L. et al. Public databases and software for the pathway analysis of cancer genomes. *Cancer Inform.*, **3**:379–397, 2007.

Tubio, J. M. C. and Estivill, X. Cancer: When catastrophe strikes a cell. *Nature*, **470**:476–477, 2011.

Turner, B. M. Cellular Memory and the Histone Code. *Cell*, **111**:285–291, 2002.

Turner, F. S. et al. POCUS: mining genomic sequence annotation to predict disease genes. *Genome Biol.*, **4**:R75, 2003.

Tusher, V. G. et al. Significance analysis of microarrays applied to the ionizing radiation response. *PNAS*, **98**:5116–5121, 2001.

Tut, V. et al. Cyclin D1 expression in transitional cell carcinoma of the bladder: correlation with p53, waf1, pRb and Ki67. *Br. J. Cancer*, **84**:270–275, 2001.

Tyson, J. J. et al. Sniffers, buzzers, toggles and blinkers: dynamics of regulatory and signaling pathways in the cell. *Curr. Opin. Cell Biol.*, **15**:221–231, 2003.

Ulitsky, I. et al. DEGAS: de novo discovery of dysregulated pathways in human diseases. *PLoS One*, **5**.e13367, 2010.

Vaidya, J. S. Breast cancer: an artistic view. *Lancet Oncol.*, **8**:583–585, 2007.

Van't Veer, L. J. and Bernards, R. Enabling personalized cancer medicine through analysis of gene-expression patterns. *Nature*, **452**:564–570, 2008.

Vanunu, O. et al. Associating genes and protein complexes with disease via network propagation. *PLoS Comput. Biol.*, **6**:e1000641, 2010.

Vapnik, V. N. *Statistical Learning Theory*. Wiley, New York, 1998.

van't Veer, L. J. et al. Gene expression profiling predicts clinical outcome of breast cancers. *Nature*, **415**:530–536, 2002.

Venables, J. P. Aberrant and Alternative Splicing in Cancer. *Cancer Res.*, **64**:7647–7654, 2004.

Venter, J. C. et al. The sequence of the human genome. *Science*, **291**:1304–1351, 2001.

Vera-Licona, P. et al. OCSANA: Optimal Cut Sets Algorithm for Network Analysis. *In preparation*, 2012.

Vermeulen, L. et al. Cancer stem cells - old concepts, new insights. *Cell Death Differ.*, **15**:947–58, 2008.

van de Vijver, M. J. et al. A gene-expression signature as a predictor of survival in breast cancer. *N. Engl. J. Med.*, **347**:1999–2009, 2002.

Vilar, J. M. G. et al. Signal processing in the TGF-beta superfamily ligand-receptor network. *PLoS Comput. Biol.*, **2**:e3, 2006.

van Vlodrop, I. J. H. et al. Analysis of promoter CpG island hypermethylation in cancer: location, location, location! *Clin Cancer Res*, **17**:4225–4231, 2011.

Voduc, K. D. et al. Breast cancer subtypes and the risk of local and regional relapse. *J. Clin. Oncol.*, **28**:1684–1691, 2010.

Vogelstein, B. and Kinzler, K. W. The multistep nature of cancer. *Trends Genet.*, **9**:138–141, 1993.

Vogelstein, B. and Kinzler, K. W. Cancer genes and the pathways they control. *Nat. Med.*, **10**:789–799, 2004.

Waaijenborg, S. et al. Quantifying the association between gene expressions and DNA-markers by penalized canonical correlation analysis. *Stat. Appl. Genet. Mol. Biol.*, **7**:Article3, 2008.

Waddington, C. H. *The Strategy of the Genes.* London : George Allen & Unwin., 1957.

Wadman, M. James Watson's genome sequenced at high speed. *Nature*, **452**:788, 2008.

Wagner, A. *Robustness and Evolvability in Living Systems.* Princeton Univ. Press, 2005.

Wallace, E. V. B. et al. Identification of epigenetic DNA modifications with a protein nanopore. *Chem. Commun. (Camb.)*, **46**:8195–8197, 2010.

Waltemath, D. et al. Minimum Information About a Simulation Experiment (MI-ASE). *PLoS Comput. Biol.*, **7**:e1001122, 2011a.

Waltemath, D. et al. Reproducible computational biology experiments with SED-ML - The Simulation Experiment Description Markup Language. *BMC Syst. Biol.*, **5**:198, 2011b.

Wang, E. *Cancer Systems Biology.* Chapman & Hall/CRC, 2010.

Wang, K. et al. Estimation of tumor heterogeneity using CGH array data. *BMC Bioinformatics*, **10**:12, 2009.

Wang, S. E. et al. A mathematical model quantifies proliferation and motility effects of TGFbeta on cancer cells. *Computational and Mathematical Methods in Medicine*, **10**, 2007.

Wang, Y. et al. Gene-expression profiles to predict distant metastasis of lymph-node-negative primary breast cancers. *Lancet*, **365**:671–679, 2005.

Watanabe, T. et al. Endogenous siRNAs from naturally formed dsRNAs regulate transcripts in mouse oocytes. *Nature*, **453**:539–543, 2008.

Weigelt, B. et al. Breast cancer molecular profiling with single sample predictors: a retrospective analysis. *Lancet Oncol.*, **11**:339–349, 2010.

Weinberg, R. A. *The biology of cancer.* Garland Science, Taylor & Francis Group, LLC, 2007.

Weinberger, K. Q. et al. Distance metric learning for large margin nearest neighbor classification. In Weiss, Y. et al., editors, *Adv. Neural. Inform. Process Syst.*, volume 18. MIT Press, Cambridge, MA, 2006.

Weinstein, I. B. Cancer. Addiction to oncogenes–the Achilles heal of cancer. *Science*, **297**:63–64, 2002.

Werner, S. L. et al. Stimulus specificity of gene expression programs determined by temporal control of IKK activity. *Science*, **309**:1857–1861, 2005.

Westerhoff, H. V. et al. Systems biochemistry in practice: experimenting with modelling and understanding, with regulation and control. *Biochem. Soc. Trans.*, **38**:1189–1196, 2010.

Westfall, P. H. and Young, S. S. *Resampling-based multiple testing: Examples and methods for p-value adjustment.* John Wiley and Sons, 1993.

Wheeler, D. A. et al. The complete genome of an individual by massively parallel DNA sequencing. *Nature*, **452**:872–876, 2008.

Whitacre, J. M. and Bender, A. Networked buffering: a basic mechanism for distributed robustness in complex adaptive systems. *Theor. Biol. Med. Model.*, **7**:20, 2010.

Whitesell, L. and Lindquist, S. L. HSP90 and the chaperoning of cancer. *Nat. Rev. Cancer*, **5**:761–772, 2005.

Wiechec, E. and Hansen, L. L. The effect of genetic variability on drug response in conventional breast cancer treatment. *Eur. J. Pharmacol.*, **625**:122–130, 2009.

Wirapati, P. et al. Meta-analysis of gene expression profiles in breast cancer: toward a unified understanding of breast cancer subtyping and prognosis signatures. *Breast Cancer Res.*, **10**:R65, 2008.

Wirtz, D. et al. The physics of cancer: the role of physical interactions and mechanical forces in metastasis. *Nat. Rev. Cancer*, **11**:512–522, 2011.

Witten, D. M. et al. A penalized matrix decomposition, with applications to sparse principal components and canonical correlation analysis. *Biostatistics*, **10**:515–534, 2009.

Wittmann, D. M. et al. Transforming Boolean models to continuous models: methodology and application to T-cell receptor signaling. *BMC Syst. Biol.*, **3**:98, 2009.

Workman, C. et al. A new non-linear normalization method for reducing variability in DNA microarray experiments. *Genome Biol.*, **3**:research0048, 2002.

Wright, G. W. and Simon, R. M. A random variance model for detection of differential gene expression in small microarray experiments. *Bioinformatics*, **19**:2448–2455, 2003.

Wu, F. T. H. et al. Modeling of growth factor-receptor systems from molecular-level protein interaction networks to whole-body compartment models. *Methods Enzymol.*, **467**:461–497, 2009.

Wu, Z. et al. A Model Based Background Adjustment for Oligonucleotide Expression Arrays. Technical report, John Hopkins University, Department of Biostatistics Working Papers, Baltimore, MD, 2003.

Xing, E. et al. Distance Metric Learning with Application to Clustering with Side-Information. In S. Becker, S. T. and Obermayer, K., editors, *Adv. Neural. Inform. Process Syst.*, volume 15, pages 505–512. MIT Press, Cambridge, MA, 2003.

Xu, X. et al. The tandem affinity purification method: an efficient system for protein complex purification and protein interaction identification. *Protein Expr. Purif.*, **72**:149–156, 2010.

Yang, Y. H. and Speed, T. Design issues for cDNA microarray experiments. *Nat. Rev. Genet.*, **3**:579–588, 2002.

Yang, Y. H. et al. Normalization for cDNA microarray data: a robust composite method addressing single and multiple slide systematic variation. *Nucleic Acids Res.*, **30**:e15, 2002.

Yau, C. et al. A statistical approach for detecting genomic aberrations in heterogeneous tumor samples from single nucleotide polymorphism genotyping data. *Genome Biol.*, **11**:R92, 2010.

Ylstra, B. et al. BAC to the future! or oligonucleotides: a perspective for micro array comparative genomic hybridization (array CGH). *Nucleic Acids Res.*, **34**:445–450, 2006.

Yuan, M. and Lin, Y. Model selection and estimation in regression with grouped variables. *J. R. Stat. Soc. Ser. B*, **68**:49–67, 2006.

Zambelli, F. et al. Pscan: finding over-represented transcription factor binding site motifs in sequences from co-regulated or co-expressed genes. *Nucleic Acids Res.*, **37**:W247–W252, 2009.

Zanella, F. et al. High content screening: seeing is believing. *Trends Biotechnol.*, **28**:237–245, 2010.

Zeitouni, B. et al. SVDetect: a tool to identify genomic structural variations from paired-end and mate-pair sequencing data. *Bioinformatics*, **26**:1895–1896, 2010.

Zetterberg, A. et al. What is the restriction point? *Curr. Opin. Cell Biol.*, **7**:835–842, 1995.

Zhang, L. et al. Multiscale agent-based cancer modeling. *J. Math. Biol.*, **58**:545–559, 2009.

Zhou, T. et al. Evolutionary dynamics and highly optimized tolerance. *J. Theor. Biol.*, **236**:438–447, 2005.

Zi, Z. and Klipp, E. Constraint-based modeling and kinetic analysis of the Smad dependent TGF-beta signaling pathway. *PLoS One*, **2**:e936, 2007.

Zinovyev, A. et al. BiNoM: a Cytoscape plugin for manipulating and analyzing biological networks. *Bioinformatics*, **24**:876–877, 2008.

Zou, H. and Hastie, T. Regularization and variable selection via the Elastic Net. *J. R. Stat. Soc. Ser. B*, **67**:301–320, 2005.

Zou, H. et al. Sparse Principal Component Analysis. *J. Comput. Graph. Stat.*, **15**:265–286, 2006.

Index

ABL1, 6
acidification, 259
adaptive landscape, 291
adaptive response, 326
adaptive treatment, 297
agent-based modelling, 211
alternative splicing, 38, 362
altruism, 292
amplification, 26, 61
analysis of variance, 107
ANOVA, *see* analysis of variance
APC, 263
apoptosis, *see* cell death
array CGH, 58, 60, 61, 112, 113, 162
ATM, 31, 38, 251, 264
attractor, 223, 326, 327
    strange attractor, 327

BAF profile, 63, 68
balance parameters, 321
basin of attraction, 327
BAX, 249
Bayes' rule, 169
BCL2, 29, 249
BCR, 6
BCR-ABL1, 5
between-group sum of squares, 138
betweenness, 341
bifurcation, 228, 230
bioinformatics tools, 105
BioPAX, 118
bisulfite conversion, 72
Boolean model, 222, 223
bowtie structure, 282
    bowtieness, 281
BRAF, 239
BRAF(V600E), 353

BRCA1, BRCA2, 7, 11, 27, 31, 251, 264, 301, 348, 353
breast cancer subtypes, 128, 130

canalisation, 297, 299
cancer
    bladder, 29, 31, 37, 129, 156, 159, 160, 285
    breast, 2–5, 7, 9, 11, 12, 25, 27, 31, 37, 38, 47, 51, 66, 68, 127, 128, 130–132, 134, 135, 140, 142, 143, 145, 146, 153, 155, 162, 165–167, 175, 176, 183, 185, 194, 195, 198, 203, 239, 242, 263, 264, 269, 340, 343, 344, 353
    carcinoma, 25, 31, 43, 127, 159, 239, 242, 245, 247, 264, 351
    colon, 6, 25, 29, 242, 247, 253, 254, 263, 353
    Ewing's sarcoma, 29, 103, 264, 302
    familial, 27, 253, 348
    gastrointestinal, 247
    glioblastoma, 25, 247
    Kaposi's sarcoma, 29
    kidney, 247
    leukaemia, 2, 4, 25, 28, 142, 153, 163, 171, 269, 293, 296, 349
    lung, 4, 25, 31, 142, 347
    lymphoma, 25, 29, 142, 153, 352
    medulloblastoma, 25
    molecular classification of, 129
    neuroblastoma, 25, 28, 29, 46, 61
    neuroendocrine, 247
    paediatric, 27–29, 61, 302
    prevention, 347
    renal, 351

T - #0008 - 160425 - C8 - 234/156/25 [27] - CB - 9781439831441 - Gloss Lamination